Springer Biographies

The books published in the Springer Biographies tell of the life and work of scholars, innovators, and pioneers in all fields of learning and throughout the ages. Prominent scientists and philosophers will feature, but so too will lesser known personalities whose significant contributions deserve greater recognition and whose remarkable life stories will stir and motivate readers. Authored by historians and other academic writers, the volumes describe and analyse the main achievements of their subjects in manner accessible to nonspecialists, interweaving these with salient aspects of the protagonists' personal lives. Autobiographies and memoirs also fall into the scope of the series.

Thomas Ward

People, Places, and Mathematics

A Memoir

Thomas Ward 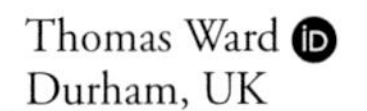
Durham, UK

ISSN 2365-0613 ISSN 2365-0621 (electronic)
Springer Biographies
ISBN 978-3-031-39073-9 ISBN 978-3-031-39074-6 (eBook)
https://doi.org/10.1007/978-3-031-39074-6

Cover illustration: Front cover: the author and his brother James near Port Nolloth, December 1981 (© Thomas Ward)

This Springer imprint is published by the registered company Springer Nature Switzerland AG
The registered company address is: Gewerbestrasse 11, 6330 Cham, Switzerland

To the memory of Alan and Honor and
the future of Tania, Adele, and Raphael

Preface

I was prompted to write some memories of the unpredictable incidences and coincidences that shaped my professional life most particularly by an exchange on social media that resulted in a joint publication. This pleasant but absurd juxtaposition of the new and the old—the question was one that Euler certainly could have asked in 1736, and possibly could have answered—and the extraordinary role played by technology in our lives also pushed me to reflect on the objects and places, as well as the people, that have shaped my professional life. So I have done my best to look back and try to understand various threads and incidents of my working life, and the people, ideas, places, and objects that accompanied them.

One of those threads was woven by advances in our ability to compute and communicate. From slide rules and log tables in childhood to programmable calculators, from the push and poke commands of machine code to effortless experimentation using free computer algebra packages on generic laptops, from a hand-written undergraduate project on the fundamental group to complex books with camera-ready copy produced by the authors, from unrecorded lectures on blackboards on the campus of Warwick University in the early 1980s to routine use of video conferencing, technology has dramatically changed the life of anyone who works in education.

The more consequential threads involve people. The one constant through all the changes was the central role of personal connections and collaborations. It is the people not the propositions, the love not the lemmas, that I am really seeking to record and celebrate. It has been my great privilege to enjoy some long collaborations, and it is unarguably the case that my mathematical life would have ground to a halt entirely once I took on senior university roles without the encouragement and energy of some of those collaborators.

> Among scientists, mathematicians have probably been the most interested, historically, in personal connections—witness the Erdös Number and fascination with mathematical ancestry through the Mathematics Genealogy Project.
> (Norman Richert, from [255])

The Erdös number alluded to records the distance in the collaboration graph from the prolific Hungarian mathematician Paul Erdös. His own number is zero by definition, a co-author has Erdös number 1, any of their co-authors who has not published with Erdös has Erdös number 2, and so on [357]. My Erdös number is 2 because I have published with Florian Luca. What became the Mathematics Genealogy Project started when American mathematician Harry Coonce was interested in finding out the name of his PhD supervisor's supervisor (in North American parlance, his adviser's adviser). It is now hosted at North Dakota State University, and comprises a database of around a quarter of a million mathematicians, showing the supervisory relationship. It takes an inclusive view of what constitutes 'Mathematics' and illuminates some of the great ebbs and flows of mathematical history—and the almost parental way in which the supervisory relationship is seen in Mathematics [156].

Concerning what is and is not contained here, I could do no better than to use another's words once again to explain the extent to which this is not a sequential autobiography.

> The narrative will not always be chronological but often discursive when advantageous. Nor will it tell much about my private life, which is of sparse general interest, though I will pay tribute to my family and the support they have given me.
> (Chapman Pincher, in the introduction to his autobiography [243])

Sparse interest to others perhaps, but I cannot begin an account of my professional life without saying that every single move from one city or country to another since our marriage in 1990 has been driven by my professional roles and has often been to the huge disruption of Tania's. This is not the place to dwell on it, but the huge inequity with which many couples navigate the professional world is certainly manifest in our lives.

I have enjoyed the diaries of some political figures but have never understood how anyone has the time to record replete notes day by day in the rush of life. Nor does the linear sequential passing of the days make much sense in hindsight—the density of incident in life is far from uniform, and the threads of meaning loop backwards and forwards in time.

These notes also inevitably involve some thoughts and memories concerning higher education and the universities in which I have worked. I entered the profession at the end of the 1980s, having spent most of that decade as a student. Despite relentless downward pressure on the funding of university education in the United Kingdom throughout that period and the decade to follow, the understanding of both the educational and the research contribution of higher education as a primarily public good was barely contested. Margaret Thatcher made significant changes in the direction of greater accountability, removing the notion of academic tenure, replacing the more paternal University Grants Committee with the Universities Funding Council, introducing full cost fees for international students in 1981, and dramatically cutting the funding for home students. Perhaps most significantly of all, under her government Sir Peter Swinnerton-Dyer was tasked with creating a national Research Assessment Exercise to guide the allocation of core research funding. This bracing wind had undoubted impact, which some welcomed and some viewed as destructive. It certainly began the process that continues to this day, in which many UK universities are driven to become fierce competitors on a global stage. Some comments from an overview of her legacy published in the Times Higher Education Supplement in 2013 give a small sense of the contested heritage [140].

The former Vice-Chancellor of the private University of Buckingham, awarded its Royal Charter in 1983, was effusive.

> Before Mrs Thatcher, universities were very similar to public utilities—run for the benefit of staff with government money. Now they are stellar. (Professor Terence Kealey)

Someone who worked in Margaret Thatcher's policy unit before going on to become a Member of Parliament and Higher Education Minister saw her changes as essential enabling steps to building a world-class higher education system.

> [Her] extraordinary achievements [set the scene] for the world-class higher education sector we have today. (David Willetts)

The then chairman of the Universities Funding Council and architect of the research funding changes welcomed the creation of more accountability and competition for research resources.

> She was a great prime minister, and she did much to change the atmosphere of higher education. Universities were spending money wastefully, so the Research Assessment Exercise was essentially invented by me and was instituted so money could be divided up in a fair way. (Professor Peter Swinnerton-Dyer)

The former head of the Higher Education Quality Council who went on to become Vice-Chancellor of Southampton Solent University and then a Professor of Higher Education Policy at Liverpool Hope University was among those who saw her impact as something like Henry VIII's dissolution of the monasteries.

> The cuts in 1981 were a disaster for British higher education—some of [the] worst things that have ever happened to higher education. (Professor Roger Brown)

At the time the focus was understandably on the dramatic cuts in funding, but the bigger picture was that this era marked the beginning of a journey from a shared understanding of higher education as a public good to higher education as a private advantage. The consequences are still unfolding forty years later and, if anything, are more strongly disputed now than then.

The impetus to record some reflections on a career, the first half of which was mathematical, was also certainly strengthened by the death of both my parents. In writing an obituary for our father and archiving some of the documents they both left behind, I realised how easy it is for even the sequence of events in a working life to slip away, let alone the motivations behind those events, or the emotions with which they were freighted.

> What am I now that I was then?
> May memory restore again and again
> The smallest color of the smallest day:
> Time is the school in which we learn,
> Time is the fire in which we burn.
> (From 'Calmly We Walk Through This April's Day' by Delmore Schwartz)

I also found that in starting to think about the people who have been part of my own particular small story, it became painfully apparent that all too many of them are no longer with us. Part of the surprise in each case is the idea of a life suddenly transformed, from something so obviously unbounded and capable of branching out in any of a multitude of directions into a record left of a finite collection of memories, words, photographs, and documents. The gap between the two qualitatively different things, between the infinite and the finite, seems at times unbridgeable. It finds an echo in the difference between instances of a phenomenon in Mathematics—examples—and a richly portentous proof that the phenomenon is universally true.

I am more than conscious of the problematic legacy, even impudence, of recording these thoughts in the 'most awkward literary genre' of something like a scientific autobiography.[1] The fact that many mathematicians, often distinguished ones, have written various sort of memoirs and apologiæ created another push. I hope never to be guilty of false pride nor of false modesty: My honest calibration is that while it is perfectly reasonable that I have enjoyed a career as a professional mathematician, I make no pretence that it has been one of significant impact. The careers of the ordinary are both more relatable and, possibly, more interesting than the careers of the exceptional. I was never a person fought over by elite universities as a mathematician. Just as the politician Michael Heseltine was once described with patrician disdain as someone who 'had to buy all his furniture',[2] as a mathematician I had to apply for my own jobs. Thus this is a sketchy memoir of an ordinary mathematician who nonetheless derived great enjoyment from Mathematics and the people who practice it. It is, in particular, somewhat random about what is included and what is omitted, and the many wonderful colleagues and students not mentioned are absent purely for reasons of brevity.

The period of time I will write about covers huge changes in the world beyond the parochial concerns of higher education in the UK. Since my birth, 36 countries have gained their independence from the United Kingdom, and some of the events that seemed relevant to my life predate my birth significantly and reach back to the period before Prime Minister Harold Macmillan's 'Winds of change' speech to the Parliament of South Africa in 1960. Inevitably, this means that some of the words quoted and used here, or references cited here, will reflect attitudes that have not been acceptable for many years. I have made no attempt to sanitise anything, but do apologise for any inadvertent offence caused.

---

[1] The biochemist Edwin Chargaff in his review of 'The Double Helix' by James D. Watson [47] pointed out that 'scientific autobiography [...] belongs to a most awkward literary genre [...] compounded in the case of scientists, of whom many lead monotonous and uneventful lives and who, besides, often do not know how to write. Though I have no profound knowledge of this field, most scientific autobiographies that I have seen give me the impression of having been written for the remainder tables of the bookstores, reaching them almost before they are published.'

[2] This sweeping put-down, puzzling outside the upper classes of the United Kingdom, is often attributed to the politician and military historian Alan Clark; in fact in Clark's diaries it is attributed to the peer Michael Jopling [54, p. 162].

More significantly, I suffer from the myopia induced by the legacy of colonialism and empire that shaped my own experiences. The countries and societies I grew up in were largely defined by colonial conquest and imposition. The deep history of pre-colonial Africa was barely mentioned in my schooling. The Ashante Kingdom and Fante Confederacy in Ghana; the Lunda kingdom of Kazembe and the influence of matriarchal leaders like Mwenya Mukulu in Zambia; the ancient Kingdom of Zimbabwe and the city of Great Zimbabwe; the ancient Kings—and the extraordinary modern activist Regina Twala's life in eSwatini—all are things I only ever began to really learn anything about in adulthood.

> All these words from the seller, but not one word from the sold. The Kings and Captains whose words moved ships. But not one word from the cargo. The thoughts of the "black ivory", the "coin of Africa", had no market value. Africa's ambassadors to the New World have come and worked and died, and left their spoor, but no recorded thought. (From 'Barracoon' by Zora Neale Hurston [155])

The education I experienced—in Government schools in independent Zambia and in an explicitly anti-colonial and politically engaged school in Swaziland—was dominated by borders and realities defined by colonialism. I did not take up the option to study some of the history of Mathematics while at Warwick University, not for any lack of interest but because of the abundance of other possibilities that interested me more. In the modules I did take—on the rare occasions history was mentioned—this was firmly located in a canonical story as it was widely understood at the time. Superficially this was something along the lines of Pythagoras, Eudoxus, Euclid, a mysterious lengthy gap, a brief tip of the hat in the direction of the Arab etymology of the words 'algebra' and 'algorithm', followed by Newton, Fermat, and then the great explosion of European Mathematics. The same educational journey now would have a far more global and nuanced point of view, but there is much still to do.

In the pages to follow, accounts are made of some of the history of higher education in England, and views are expressed on issues faced by the sector. Needless to say these do not come from an expert historical perspective, and I can only apologise for misunderstandings or outright errors. Above all, the opinions expressed here are entirely mine and do not reflect any of the institutions for which I have worked or organisations to which I have belonged.

More importantly, in writing this I have come across stories from my own childhood that have the texture of actual memory but are revealed by diaries to be reliant on later accounts. I have done my best to record the small number of family incidents mentioned accurately, but am more than conscious that in writing them down they take on a possibly spurious authenticity for others.

It is more than difficult to make an invidious selection of people to thank for assistance in putting this modest collection of memories together, but a few who have been great sources of encouragement stand out to such an extent that I must: My three siblings Kristina Baker, Sheena Brook, and James Ward for their help with early family history; Rémi Lodh and Francesca Ferrari at Springer for sage advice at an early stage and help in the production process respectively; and Tania Barnett for reading and commenting on an entire draft. It is yet more so to single out anyone to thank for the substance of my purpose here—which was to record the threads of my mathematical life—beyond the people mentioned in the text, but I would like to particularly thank Manfred Einsiedler for an enjoyable collaboration and friendship that is now approaching its thirtieth year, Alan Wicks for his inspirational mathematics teaching many years ago, Klaus Schmidt for his patient supervision of my doctoral efforts, Doug Lind for so many enjoyable moments both mathematical and personal since we met on the 16th of July 1986, Richard Miles for a long and productive collaboration, and Shaun Stevens for the mathematics we did together and the role he played in the development of the Mathematics department at UEA.

Finally, a sensible discipline for any writer is to have some concept of the intended reader, and on this I am entirely clear.

> I carry within a kind of inner statue, a statue sculpted since childhood, that gives my life a continuity and is the most intimate part of me, the hardest kernel of my character. (From the autobiography of François Jacob [157])

This is written for me and the statue within, to remember parts of the mathematical and professional life I have been privileged to live, with affection and gratitude—and before memory fades.

Durham, UK
July 10, 2023

Thomas Ward

Photograph permissions are recorded in each caption; where this is not mentioned the photograph is used with permission from the Ward family or was taken by the author.

Contents

1

Dorset and Ghana

"Mrs. Ward, you have a fine son, but he's a whopper!" I imagine my mother was already more than aware of the second observation, made by a doctor in the Yeatman Hospital in Sherborne shortly after my arrival in the world on the 3rd of October, 1963. Work obligations meant my father was still in Ghana at the time, and there was some confusion about the agreed name. So it was that the birth of Matthew Boulton Ward was announced in Ghana, and the birth of Thomas Boulton Ward was announced in Dorset. The event was recorded in my father's diary, and the name was later amended in the same firm handwriting that he maintained through almost his entire adult life (Fig. 1.1).

My father's assumption was related to a distant and possibly apocryphal family connection to Matthew Boulton (1728–1809), a partner of James Watt, a prominent founding member of the Lunar society [302], and a significant industrialist [12]. My mother's assumption was perhaps more apostolic, given an extant older brother named James.

One of the stories I later heard was that our mother managed to deliver sweets to the three older siblings using paper aeroplanes from the hospital window. We returned to Coasters Cottage, the small house in rural Dorset that played a role in all our lives and imaginations that far exceeded the amount of time we spent there (Fig. 1.2). Our parents had purchased it in the early 1950s and named it after the distinctive Atlantic waves on Takoradi beach in Ghana. Throughout our childhood this was the place we returned to for periods of leave, usually every two years, and was the house our parents retired to in 1986. Here we were exposed on various short visits to a different and unfamiliar life of hot water bottles, electric blankets, a small Parkray coal fire that heated

T. Ward, *People, Places, and Mathematics*, Springer Biographies,
https://doi.org/10.1007/978-3-031-39074-6_1

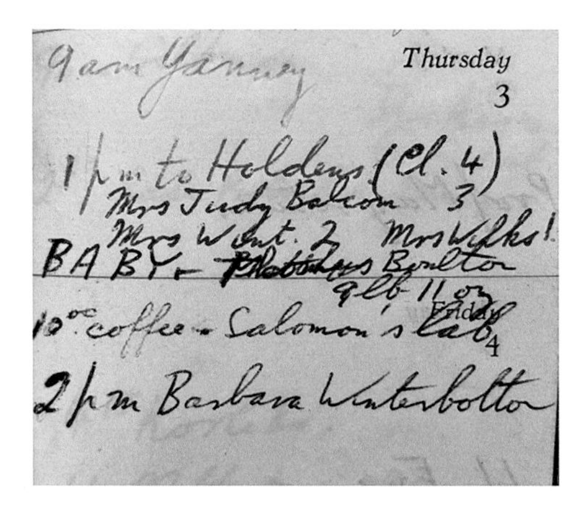
Thursday
3
1 p.m. to Holdens (Cl. 4)
Mrs Judy Balcom 3
Mrs Wilks 1
BABY
9 lb 11 oz
10 o'c coffee – Salomon's lab
Friday
4
2 p.m. Barbara Winterbottom

Fig. 1.1 My father's diary entry for the 3rd of October 1963

Fig. 1.2 Sheena, James, and Kristina welcoming the new arrival in the garden of Coasters Cottage

the water, snow, television, wellington boots on muddy walks, and a little van that did a weekly delivery of exotic things like chocolate biscuits. Part of the liminal nature of our lives was that for the four children Coasters Cottage had a permanence that no other house did, but we never really lived there. This was slightly different for our parents, as they always intended to retire there and ended up living in the cottage longer than in any other house.

Fig. 1.3 The last time the whole family gathered: In the garden of Coasters Cottage on the 24th of May 2008

It was the only house they ever owned, and over the years various changes were made to make it more habitable. In 1954 one of our father's cousins, the architect Duncan Thomson, helped with the changes needed to remove a primitive 'Elsan' chemical toilet and replace it with more modern plumbing. An extension with cedar wood walls was added, and we all helped build a large garage from a kit with large pre-fabricated concrete wall pieces. After retirement they had a more substantial extension built in 1987, increasing the size of the kitchen just enough to open a dishwasher door without this making it impossible to get through the room.

For our parents Coasters Cottage spanned a period that encompassed visits of relatives before the birth of my older sister in 1954 and regular gatherings including great-grandchildren seven decades later. We last all gathered there in 2008 (Fig. 1.3). Our father lived there until a few months before his death in 2021, and its final sale in 2022 marked the end of its powerful symbolic place as a—largely notional—permanent home in a life of temporary contracts and rented houses.

The plan to reunite the family following my arrival was for our mother, the four children, and a young woman called Doreen Hunter to travel to Ghana by Elder Dempster ship from Liverpool. At the last minute the shipping line wrote to say I could not travel as I had no smallpox vaccination. The local doctor in Cerne Abbas said I was far too young to tolerate the vaccination, so there was an impasse. In the end the letter our doctor wrote sufficed, but it was a sufficiently stressful incident that our mother remembered it years later. President J. F. Kennedy was assassinated while we were at sea, and my older sister Kristi remembers 'the hush that spread through the dining room' when the frightening news came through. We were on the last voyage of the Apapa and, despite bad weather, reached Tema Harbour in Ghana after ten days, where I met my father for the first time. The vessel was the third incarnation of the name Apapa, the name of a local government area in Lagos. The first Apapa was built in 1914. She was torpedoed by German submarine U-96 off the Welsh coast on the 28th of November 1917 while sailing back to Liverpool from West Africa. Of the 249 people on board, 40 passengers and 37 crewmen lost their lives. The second Apapa was built in 1926; as part of Convoy SL 53 she was bombed and sunk in the Atlantic Ocean to the West of County Mayo by a Focke-Wulf Fw 200 Condor aircraft with the loss of 23 people. She was travelling from Lagos to Liverpool. The third Apapa was built in 1948 by Vickers–Armstrong in Barrow, and was later sold to a shipping line in Hong Kong in 1968, where she was renamed Taipooshan.

The Elder Dempster shipping line had origins in the middle of the nineteenth Century but existed under this name from 1932 to 2000. It mainly operated between Liverpool and ports in West Africa, and branched out in several ways. For the period 1935 to 1940 it was a partner in Elders Colonial Airways in Nigeria. The precursor company Dempster played a role in one of the darkest episodes in colonial history. Towards the end of the nineteenth Century Dempster was contracted by King Leopold II of Belgium to transport supplies to and from the Congo Free State. A junior shipping clerk called Edmund Morel working in Antwerp for the Dempster company noticed huge discrepancies between the private records of what was being transported and the public ones, which were designed to create the fiction of ordinary trade. There was a steady export of weapons, ammunitions, and manacles in one direction, and an entirely disproportionate import of ivory, rubber, and other valuable commodities in the other. He deduced correctly that no trade was taking place at all, but an organised system of mass exploitation on an almost unimaginable scale. Morel joined forces with the diplomat and Irish Nationalist Roger Casement to campaign against slavery in the Congo Free State and started the Congo Reform Association. The campaign, and the influential 'Casement Report' detailing the extent of the abuses in the Congo Free State, did eventually force King Leopold II to sell the Congo Free State to the government of Belgium.[1] Morel went on to be jailed for his pacifist activism during the First World War, and became a Labour Member of Parliament in 1922.

The name of the shipping line, Elder Dempster, was often heard at home because our mother was fond of a version of the old music hall song 'My old man's a dustman'. This became widely popular when the Scottish skiffle singer Lonnie Donegan's version became the first record to go straight in at the top of the UK music charts in the early 1960s. For our mother it was an earlier version that apparently became popular at Birmingham University in the 1940s as a result of Mining students who would travel to and from West Africa on Elder Dempster ships that she would quote.

> My old man's a fireman now what do you think of that
> He wears cor blimey trousers and a little cor blimey hat
> He wears a flipping muffler around his flipping throat
> Because my old man's a fireman on an Elder Dempster Boat.

Lonnie Donegan is quoted as saying 'My Old Man's A Fireman (on the Elder Dempster Line) became a student's union song in Birmingham, and then it was picked up by the troops in World War 1 as: My old man's a dustman, he fought in the Battle of Mons. He killed ten thousand Germans with only a hundred bombs.' It is likely that the latter version is earlier however, as the Elder Dempster line only came into existence in the 1930s. The name still has both a musical and a West African connection as there is a current Nigerian gospel singer called Elder Dempster.

The University of Ghana at the time was quite a small community, and one of the congratulation letters came from the Vice-Chancellor, Conor Cruise O'Brien (Fig. 1.4). His own story is remarkable, and this was relatively soon after he had been dismissed from his secondment from Ireland's United Nations (UN) Delegation to be a special representative to Secretary General Dag Hammarskjöld in the contested Katanga region of the newly independent Congo. This has its own tragic history, and O'Brien had been involved in controversial decisions to change the role of UN troops from peacekeeping to peace enforcement. He had in particular played a role in Operation Morthor in 1961, where Indian troops under the *Opération des Nations Unies au Congo* (ONUC) were directly engaged in conflict with Katangan secessionist troops. By the time O'Brien was withdrawn, Secretary General Dag Hammarskjöld

[1] https://archive.org/details/CasementReport.

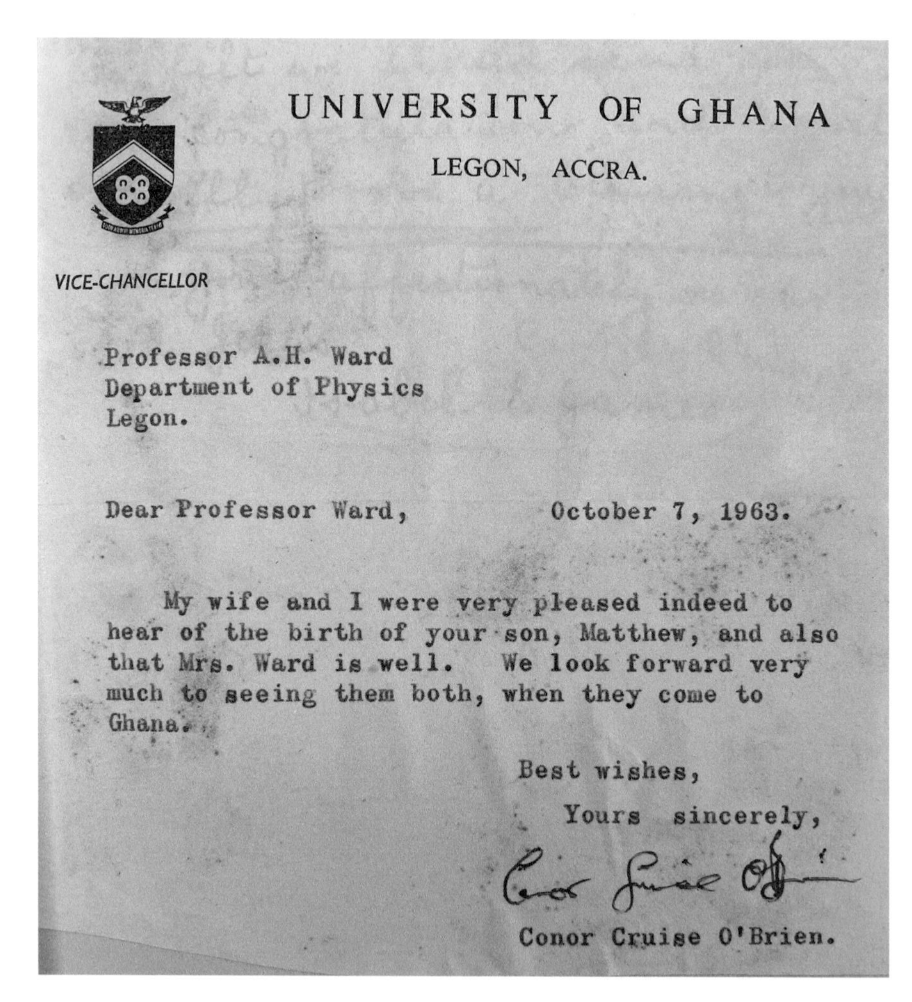

UNIVERSITY OF GHANA

LEGON, ACCRA.

VICE-CHANCELLOR

Professor A.H. Ward
Department of Physics
Legon.

Dear Professor Ward, October 7, 1963.

My wife and I were very pleased indeed to hear of the birth of your son, Matthew, and also that Mrs. Ward is well. We look forward very much to seeing them both, when they come to Ghana.

Best wishes,

Yours sincerely,

Conor Cruise O'Brien.

Fig. 1.4 A note of congratulation from Vice-Chancellor Conor Cruise O'Brien

had been killed in a plane crash near Ndola while flying to meet one of the secessionist leaders to discuss a ceasefire following the failure of Operation Morthor. The crash of the Douglas DC-6 carrying Hammarskjöld was never definitively explained as far as I know, and various theories implicating outside forces ranging from the CIA to Belgian mining companies have been put forward. O'Brien went on to write his own account of his time in Katanga [230], but later archival research has suggested that this account was somewhat disingenuous [163].

Birmingham

Our parents met at the University of Birmingham in the 1940s, where they were both studying Physics. Our father completed a shortened wartime degree which included instruction in electronics and radio alongside Physics, and trained as a Flight Mechanic with the Air Training Corps during his time as a student. After basic training he spent a vacation learning to drive tank transporters. These were probably the M19 Diamond T truck and trailer combination, built in the United States and supplied under the lend-lease arrangements. He particularly remembered the physical strength involved in using the diaphragm sprung dry disk clutch and the two unsynchronised manual gearboxes requiring double declutching. In another vacation he trained with the Royal Electrical and Mechanical Engineers, working on fault-finding the standard 1056 army radio receiver. For one winter he was sent to Malvern College, which had been taken over by the army to house the Telecommunications Research Establishment where Bernard Lovell led the team that developed the HS2 radar system. Our father went on to do a PhD in nuclear scattering, funded in part by the Department of Scientific and Industrial Research (DSIR), and served with the Home Guard during this time. His meticulous diary, maintained from the start of his university studies, is full of allusions to the extraordinary period of history they both lived through (see Figs. 1.5 and 1.6).

Our father's diary gives an impression of how Birmingham University had at the time attracted some influential figures in nuclear physics (Fig. 1.7). Tony Hilton Royle Skyrme (1922–1987) was a British physicist who had worked on diffusion equipment for isotope separation as part of the Manhattan project. He turned down a Fellowship at Oxford to work with Peierls at the University of Birmingham, where he became a research fellow. His approach to modelling the effective interaction between nucleons in nuclei by a zero-range potential is still widely used in understanding nuclear structures, and in the equations of

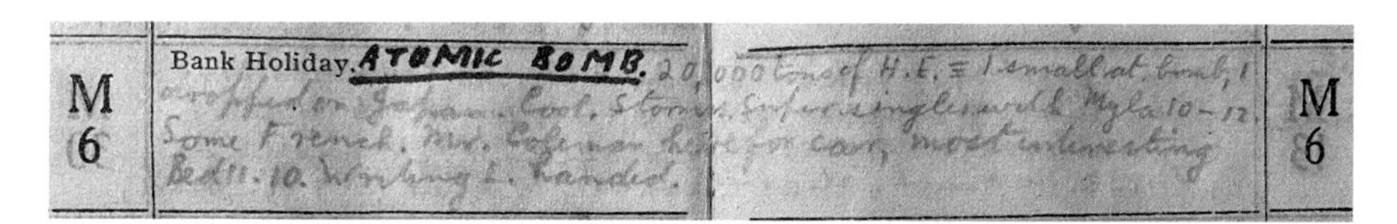

Fig. 1.5 The first atomic bomb with its estimated power of '20,000 tons of H.E.' (high explosive, meaning the quantity of trinitrotoluene which, if detonated, would provide the same discharge of energy) as recorded on the 6th of August 1945, in ink rather than his usual pencil

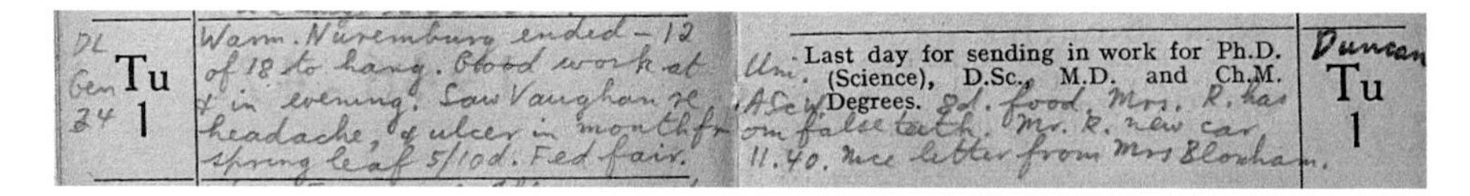
Tu 1

Last day for sending in work for Ph.D. (Science), D.Sc., M.D. and Ch.M. Degrees.

Tu 1

Fig. 1.6 A diary entry for the 1st of October 1946, recording the death sentence ('12 of 18 to hang') passed at the Nuremberg trials on Martin Bormann, Hans Frank, Wilhelm Frick, Hermann Göring, Alfred Jodl, Ernst Kaltenbrunner, Wilhelm Keitel, Joachim von Ribbentrop, Alfred Rosenberg, Fritz Sauckel, Arthur Seyss-Inquart, and Julius Streicher

TIME TABLE, 1946—1947

Time	Monday	Tuesday	Wednesday	Thursday	Friday	Saturday
9–10	9.30 Skyrme	Peierls	Dr. Kynch	Peierls	Prof Peierls	
10–1				Adv. Lab.	10.15. Cyclotron meeting	
1.15–2		Dr. Horne				
2–3		Dr Bleuler				
4.15			Colloquia			

184

Fig. 1.7 Schedule during postgraduate study at the University of Birmingham, showing sessions with Tony Skyrme, Rudolf Peierls, and Konrad Bleuler

state for neutron stars. He is best known for the first topological soliton model of a particle, now called the skyrmion. Sir Rudolf Ernst Peierls, CBE FRS (1907–1995) was a German-born British physicist. He played a significant role in Tube Alloys (Britain's nuclear weapon programme) and in the Manhattan Project. His short paper in 1940 with Otto Robert Frisch (1904–1979), an Austrian-born British physicist, was the first account of how an atomic bomb could be constructed from a small quantity of Uranium 235 rather than an impossibly large amount. Their conclusion, at the time a surprise, was that fast neutron bombardment of Uranium-235 led to a critical mass of about a kilogram (more or less the size of a golf ball). Long after the war Peierls became concerned about the implications of nuclear weapons and was involved in the Pugwash Conferences on Science and World Affairs. Konrad Bleuler (1912–1992) was a Swiss theoretical physicist who completed a PhD in 1942 at ETH Zürich under the supervision of Michel Plancherel. Through the 1940s and 1950s he visited Rome, Birmingham, Stockholm, Helsinki, and Genoa, working on theoretical particle physics and quantum field theory.

The Physics department at Birmingham, led by Mark Oliphant, had produced two of the most significant scientific contributions to the war effort. The first was the invention of the cavity magnetron, which made it possible to generate high intensity short wavelength radio beams. This gave the British radar technology a significant advantage over German radar, and was described by C. P. Snow many years later as 'the most valuable English scientific innovation in the Hitler war' [280, p. 58]. The second was the work leading to the short memorandum by Rudolf Peierls and Otto Frisch on the practical possibility of a nuclear weapon.

Birmingham was a centre of intense activity in nuclear physics at the time, and the experience of being part of this story influenced our father greatly. He acquired an extraordinary range of skills, as he learned to blow his own glass equipment, build Geiger counters, coincidence counters, and cloud chambers, learned to touch type, and so on. The typing grew to become one of the sounds of our childhood, as our father spent so many evenings in the study, typing on a large Hermes typewriter late into the evening. He had in fact learned the basics of typing earlier, when the oldest of his sisters, Kathleen, gave him a typewriter to take into the bomb shelter as the tapping sound helped keep him calm. He also took a workshop course at Birmingham, learning the rudiments of lathes, welding, riveting, and screwcutting, acquiring skills he would use throughout his life. He made several visits to the Atomic Energy Research Establishment (AERE) at Harwell to collect fresh samples of radioactive Hafnium for the experiments connected to his thesis. The contacts with the AERE led to a temporary job in their instrumentation division, and then to his first post-doctoral research position in Copenhagen. The significance of nuclear technology was clear to all, but even then our father was interested in the peaceful use of nuclear physics, not particularly as a source of energy but in medical physics.

Our mother, Elizabeth Honor Shedden, graduated in Physics from the University of Birmingham in 1948 and went on to spend a year in teacher training. For the academic years 1948–1950 she was teaching Physics and Junior Science at Sutton Coldfield High School for Girls and for the year 1949–1950 our father was working at the Finsen Institute in Copenhagen. They were married on the 23rd of September 1950 at the Quinton Evangelical Free Church in Birmingham and drove off in our mother's Austin 7 afterwards. They both talked later about the sense of relief after they left the wedding, because our mother's family were members of an Exclusive Brethren group that demanded minimal contact with anyone who was not a member. Our mother's involvement in the early days of Birmingham University Evangelical Christian Union while studying there had already been a source of great difficulty for her. When she decided to be baptised within that group she explained her conviction to the Brothers in leadership in the Brethren, who tried to bully her out of the decision in a meeting she later simply called 'stormy'. Her own mother Phoebe, widowed when our mother was three years old, was at times open to small acts of subversion against the stifling control of the Brothers in leadership. One of many things forbidden by the Brethren at the time were wireless sets (which were described as 'things of the Devil', 'open sewers in your home', and so on). Phoebe asked her physicist daughter to make her a disguised one in a wooden box, with an embroidered cloth cover to hide it when needed.

Our parent's short honeymoon in the Cotswolds was interrupted by lengthy visits to the AERE at Harwell to collect and test radioactive measuring equipment before they both returned to Denmark. It was certainly not the last time that our mother showed a more than average degree of forbearance in the course of a long and happy marriage.

As the rules tightened in the Exclusive Brethren our mother was more or less entirely cut off from her family and that community, leading to periods of letters returned unopened and a withdrawal of contact, particularly painful in the case of her sister Barbara. These experiences might incline one to bitterness and efforts to find some other kind of exclusive society in which to find an identity, but our mother's vision expanded rather than contracted. She ran a household open and welcoming to all in each of the countries we lived, and worked hard to build bridges between churches and between cultures. She also found concrete ways to support a respectful and anti-colonial approach to science teaching in Africa, producing a series of secondary school Physics textbooks using images, illustrations, and examples from Ghana [308, 309]. Much later both our parents updated these for SI units and revised curricula [310–314]. In fact our mother was a prolific writer. In addition to the Physics books and several books for the Africa Christian Press, she wrote an adventure story 'Touchdown to Adventure' [177] under a pseudonym and a book aimed at younger readers on the big bang and evolution from the perspective of a scientist of faith called 'The Story of Creation' [317]. Many years later she wrote 'AIDS, sex and family planning—a Christian view' jointly with my older sister Kristi [318] in response to the emergent public health crisis in the early days of the HIV pandemic.

It may or may not be entirely coincidental that my mother-in-law Ruth Barnett (née Michaelis) also made productive use of yet more ghastly and tragic early life circumstances. Ruth was a *Kindertransport*, rescued with her older brother from Berlin in 1939 when she was four years old. The two of them were sponsored by the Religious Society of Friends (the Quakers), and eventually were sent to the Friends' School in Saffron Walden. Her own remarkable story is recorded in a brief profile [353] and a memoir [14]. These terrible early experiences could readily translate into a closed and defensive personality, but Ruth used her experiences to teach schoolchildren about the perils of bigotry of all sorts, and was active in campaigning for the rights of refugees and traveller communities, and in raising the profile of other genocides from Armenia to Rwanda. Ruth's years of service with several organisations including the Holocaust Educational Trust led to her being awarded a 'Member of the Most Excellent Order of the British Empire' (MBE) in the 2020 New Year's Honours list.

University College of the Gold Coast

The University College of the Gold Coast was founded on August the 11th of 1948 as an affiliate college of the University of London. This flowed from the West Africa Commission of the Asquith Commission on Higher Education in the Colonies chaired by the Rt. Hon. Walter Elliot; the initial proposal of a single large University in Nigeria led to strong protests from the Gold Coast [7, 97]. It became a full university with its own degree-awarding powers in 1961.

The family had already been in Ghana for more than a decade when we landed at Tema harbour in 1963, as our father was at this time working in the Physics department of the University of Ghana. He had been offered a lectureship at the University of Birmingham at the end of his post-doctoral years in Denmark, but they both felt a calling to support the development of both science education and the church in the newly independent countries of Africa. The initial appointment in 1951 was for two years, and they were to spend the next 35 years on a succession of short-term contracts in three different countries. One of the stories passed on to me is that Mary Hartley, author of a high school text on plane geometry [144] and at the time a lecturer in Mathematics at the University, looked into the cot of the new arrival when we rejoined our father there and announced that I would be a mathematician.

Mary was a remarkable person and great friend to the family. She was born in 1923 and died in 1998; my student Gary Morris and I dedicated a paper to her memory [221] the same year. After Pendleton High School for Girls she studied at Newnham College in Cambridge University, gaining a first class degree in Part I of the Mathematical Tripos, winning the status of

'Senior Optime' in Part II, and Honours in Part III. She won scholarships and studentships in each year of her studies, and went on to complete a PhD in 1951 [145]. She spent the years 1950–57 as Assistant Lecturer in Mathematics at the University of London, and moved to the University College of Ghana at Legon as Lecturer in Mathematics in 1958 after her mother's death. She was elected to the position of Warden of Volta Hall, and by 1967 four of her former students had completed PhDs and returned to be staff members. She was influential in the development of school Mathematics across the continent, and brought together a group of writers to create modern mathematical textbooks using Ghanaian and West African images. She held several offices including President of the Mathematical Association in Ghana. In 1973 her concerns about the teaching of Mathematics in rural schools in Ghana led her to resign from a comfortable life at the University and begin ten years working at St Francis Secondary School in Jirapah, a school for girls in the North-West of the country. After 25 years in Ghana she returned to England, where she continued to mark A-level examination papers and ran classes in Family History for the University of the Third Age in Cambridge.

Our father's expertise in nuclear Physics proved to be timely. His early work on the health consequences arising from the use of Thorotrast as a contrast medium in cerebral angiography in Copenhagen [304, 306] led to a programme of work using monkeys to assess the health impacts of Strontium-90 and Yttrium-90 for human health [70–72, 305]. When the University moved from borrowed buildings at Achimota to its permanent site in Legon, he was able to design a large 'Health Physics and Radioisotope Unit' with several research fellows. Funding from the University of Ghana itself was difficult, but he was supported by the International Atomic Energy Agency (IAEA). This allowed him to maintain extensive contacts in nuclear research internationally. The IAEA agreed to fund a Chief Technician for the Health Physics and Radioisotope Unit, and the appointment of a Canadian called Jack Marr to this position began a long collaboration. Together they used generous IAEA funding to build up a substantial laboratory, with sophisticated scintillation counters allowing them to measure Carbon-14 and Tritium, opening up a wide range of applications of radio tracer work in agriculture, geology, and health research.

On the 11th of April 1958 the French Prime Minister Félix Gaillard ordered an atomic bomb test, which would make France the fourth nuclear power after the United States of America, the Soviet Union, and the United Kingdom. As soon as it became clear that this would take place in French Algeria, our father began preparations to test the resulting radioactive fallout in Ghana. This became the so-called *Gerboise Bleue* test, a Plutonium device with

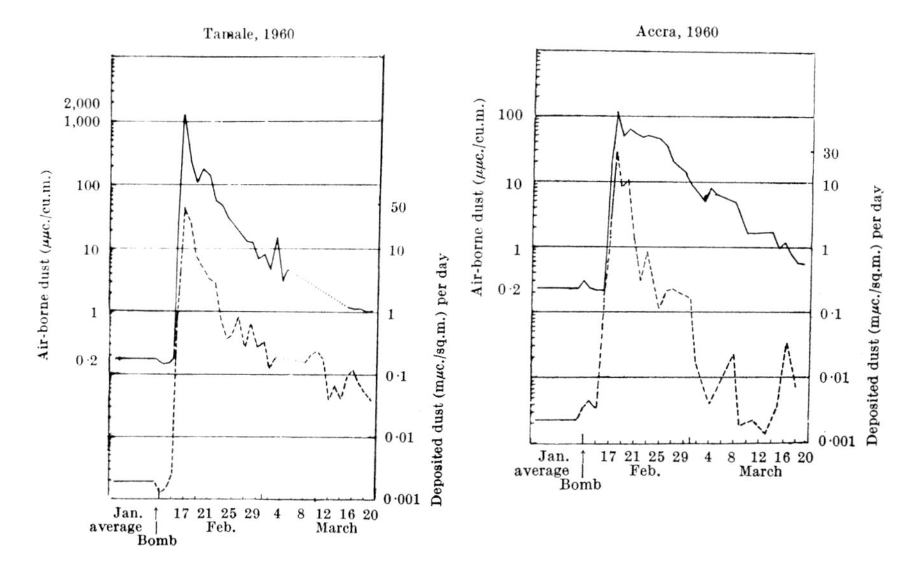

Fig. 1.8 Graphs showing radioactive fallout at Tamale and Accra in 1960 from the French *Gerboise Bleue* nuclear test in Reggane, French Algeria. (From [307] with permission under STM Permission Guidelines, ©Springer Nature)

a 70 kilotonne yield. Apart from the geopolitical considerations at play, this involved a particularly striking instance of the myopia induced by the colonial experience. Many maps and accounts of the time—and doubtless many still—depict huge areas of the Algerian Sahara as empty desert devoid of humans. The colonial view of Africa as a *tabula rasa* needing to become populated by the imperialist's ideas became a literal perception that there was nobody there.

There was a strong official reaction to our father's publication of the radioactive fallout data from the French government, including at one point a military delegation demanding to see the equipment used and verify its calibration. Once this rather tense confrontation came down to one scientist talking to another agreement was quickly reached that the published fallout figures were indeed accurate (Fig. 1.8). The resulting report to the Cabinet in Ghana estimated that the fallout in Accra was around 2% of a level that would then be regarded as dangerous to human life, though the full extent of various biological processes capable of concentrating radioactive elements in the food chain was only partially understood. The figure of 2% also has to be seen in light of the fact that Accra is almost three thousand kilometres South of Reggane, giving some sense of the scale of impact across Algeria, Mali, Libya, Niger, and many other countries.

In total France went on to detonate a further 16 atomic bombs in the Algerian Sahara, with both long and short term health impacts on huge numbers of people in Algeria and Libya particularly. A study conducted from Oran University in Algeria suggested tens of thousand of Algerians experienced sufficient radiation to have consequences for their health and for the health of their descendants, and a detailed study in the Libyan village of Fezzan also showed a high level of health consequences.

The Soviet Union series of 57 nuclear tests between September and November of 1961 also left their footprint on Ghana 8000 km away, and the radioisotope unit was well placed to record this (Fig. 1.9).

Within a few years of my birth the complex politics of the nuclear ambitions of President Kwame Nkrumah collided with our father's sound scientific mind and clear ethical outlook, resulting in an involuntary departure from the country when it came to light that our father was on a list of people to be 'PI-ed' (declared a Prohibited Immigrant, and thereby deported). There is a brief account of the political context in a study by Abena Dove Osseo-Asare [236], and a description of our father's work in revealing the damage done by French atomic weapons tests in the Sahara desert in his profile [355]. The Vice-Chancellor at the time, Conor Cruise O'Brien, also left the country in 1965 having fallen out with the increasingly authoritarian President Kwame Nkrumah over the conflict between the ideology of Nkrumahism and the principles of academic freedom.

Our parents did explore the possibility of returning to England, but two powerful forces worked against this. One was that our father had shifted his research activities to serve Ghana as much as possible. Using radioactive tracers to measure the movement and persistence of systemic insecticides in cacao trees, for example, was of importance for the economy of Ghana but was less than ideal for academic appointments in UK Physics departments. The second was much weightier. The original impetus that took our mother from a childhood in an Exclusive Brethren household in Birmingham, our father from a post-doctoral role in Denmark studying the health effects of Strontium as an X-ray contrast medium, and both of them from the Birmingham University Evangelical Christian Union to teach and research in Ghana was undimmed.

> A few months before the end [of his contract at the Finsen Institute in Denmark], he wrote an application for an advertised position as lecturer in Physics at The University College of the Gold Coast. It is hard now to remember what his motivation was. I think it arose partly from a wish to teach Physics in a new and exciting milieu—partly from a Christian commitment to the Gospel in Africa—

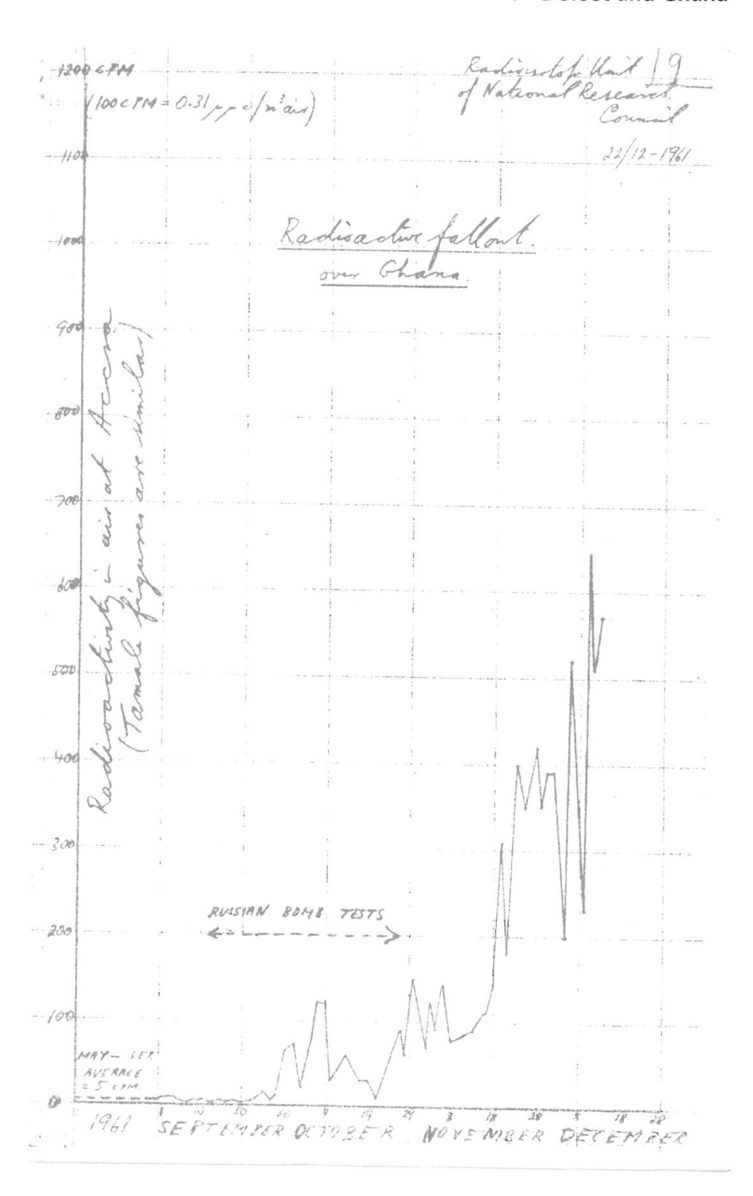

Fig. 1.9 Radioactive fallout over Ghana from the Soviet Union tests in Semipalatinsk (Kazakhstan) and in the Novaya Zemlya archipelago

and partly from a wish to help the emerging independent democracies of the Commonwealth.
(From our mother's unpublished autobiographical notes)

They consciously sought to outwork their Christian faith in practical ways, and the wish to contribute to the development of science education and research in Africa took them next to Zambia, to be part of the founding of the new University of Zambia (UNZA). The three referees named in our father's application for the chair of Physics in UNZA give a flavour of the context both geographically and politically: Prof. Alan Nunn May, Head of the Department of Physics at the University of Ghana, who had moved to Ghana after spending more than six years in prison in England; Prof. R. W. H. Wright, an expert in the relatively new discipline of equatorial aeronomy (the word was coined by the geophysicist and mathematician Sydney Chapman for the study of the upper atmosphere, and Prof. Wright used back-scattering from lasers to study the ionosphere and winds in the upper atmosphere) who had moved to the University of the West Indies; and Mr E. L. Quartey, the Chief Electrical Engineer for the city of Accra.

Alan Nunn May was recruited by James Chadwick to a team at Cambridge University working on a possible heavy water reactor; this team was merged into the Manhattan Project in the United States. He was transferred along with the rest of the Cambridge team to the Montreal Laboratory developing a reactor at Chalk River. He returned to his lecturing post in London in September 1945. The documents obtained due to the defection of the GRU cypher clerk Igor Gouzenko the same month implicated Nunn May, leading to his arrest in March 1946. The information he had passed on concerned reactors but not weapons. He was sentenced to ten years of hard labour, and was released in 1952. He was unable to work at any University in Britain as a result, and spent some years working for a scientific instruments company. He joined the University of Ghana in 1961. While the secrets he shared had far less significance than those of Klaus Fuchs or Julius and Ethel Rosenberg, his arrest was the first public demonstration that the Soviet Union had obtained nuclear secrets using espionage. It also contributed to American mistrust of British security, and the fact that Gouzenko's defection and the subsequent arrests came to light during debate on the Atomic Energy Act of 1946 in the United States led to the section entitled 'Dissemination of Information' becoming 'Control of Information', and the doctrine that some information related to nuclear energy and weapons should be 'Restricted Data' or 'born secret'. Nunn May did not regret his actions, saying later that he was "wholeheartedly concerned with securing victory over Nazi Germany and Japan, and the furtherance of the development of the peaceful uses of atomic energy" in an interview for the *New York Times* in 1952.

Our father had been sufficiently concerned about working in a department headed by Nunn May to contact the British High Commission in Accra and ask about what consequences this might have for his own future career. They were reassuring, reflecting both the passage of time and possibly a degree of sympathy with Nunn May's decisions from some parts of the diplomatic world.

In family letters our parents mention several leaving events that others organised in the last few days before the family left Ghana, and these say much about their life there. They also illustrate the extent to which the University College of the Gold Coast was modelled on a College tradition transplanted from Cambridge.

Our last week in the country began with a party at the Ambassador Hotel in Accra with the Ghana Society of Professional Engineers, and our mother enjoyed meeting for the first time some of the people our father had worked with as part of the Electrical Engineering side of his interests. He maintained links to both Physics and Electrical Engineering, and remained a member of the Institute of Electrical and Electronic Engineers (IEEE) throughout his adult life.

The first of the University events was arranged by Alan Nunn May as head of department, and consisted of a party at Legon Hall (the first residence built at the University College of the Gold Coast) to which the department and our father's former students had been invited: A 'This is your life' event that reflected fifteen years of dedicated work teaching Physics. Our mother later wrote 'I cannot imagine anything which could have given Alan greater pleasure.' They also presented a copy of the book 'The Growth of Ghana', a travelling clock, and an 'oware' board carved out of a single piece of wood, which is still in the family.[2] Our father had to slip away from the party for a time in order to take evening prayers at Akuafo Hall, the second residential hall built at the University.

After the party at Legon Hall the Christian Union students at Trinity Hall, which at the time had no electric power, arranged an event to thank them for their contribution to students in Ghana, and gave them an inscribed Bible and traditional African sandals. Reverend Andoh, who had christened me two years earlier, also attended. The ceremony ended with the students singing Jeremiah Rankin's nineteenth Century hymn 'God be with you 'til we meet again' as they filed out, each shaking hands with our parents and stepping out into the darkness. It is a hymn that moved our mother to tears many years later.

[2]Oware is a strategy game played in a circular sequence of pits using seeds or pebbles. It has multiple names and variants, and is sufficiently ancient that there is no certainty as to its origins. Among many origin stories is the traditional belief that an oware game between a man and woman took so long that they decided to marry in order to finish the game.

The second leaving event was personal rather than professional, organised by their dear friend Mary Hartley who claimed to have invited the 'non-physical pagan' friends as there had already been the event organised by Alan Nunn May, and there was to be another for Christian friends. On this occasion it was our mother who had to slip away for a little, also to lead evening prayers at Akuafo Hall. It was not really 'pagan' at all, and the guests included Father Koster, Master of Mensah Sarbah Hall (the fifth of the five traditional residential halls at the University) and the Vice-Chancellor Prof. Abrahams. Dr Hartley was Warden of Volta Hall, still the only all-female hall at the University of Ghana Legon campus, with motto 'Akokor bere nso nim adekyee' (the hen also knows the approach of dawn).

The final event was at the home of Colonel Philemon Quaye, who went on to become Chief of Naval Staff of the Ghana Navy and Scripture Union International Council chairman. This was for Scripture Union friends, who prayed particularly for safety on the journey as it involved transit through the Congo during the *Crise congolaise*, and it had not been possible to obtain visas for the required transit. This period of conflict between 1960 and 1965 in the (then) Republic of the Congo began almost immediately after independence from Belgium and more or less ended with the whole country ruled by Joseph-Désiré Mobutu. It included several civil wars, and was one of many proxy African Cold War proxy conflicts, with the Soviet Union and the United States backing opposing factions [366].

There were other groups and gifts, including at the last minute two students from the Ghana Medical Students' Association who came to the house with a carved lampstand of a tortoise and a snail which I remember well. The family left the country laden with gifts and good wishes (Fig. 1.10).

Flying to Zambia

The journey from Legon to Lusaka is also revealing about the practical logistics of travel in Africa in the 1960s, and our mother left detailed notes about it.

Ghana had a national airline founded in 1958, and the route to London used a BOAC Boeing 377 Stratocruiser with adapted livery. Flying to Zambia was quite involved and considerably less luxurious. The journey started on the 1st of February 1966 in a Ghana Airways Ilyushin Il-18 turboprop for the short flight to Lagos (Fig. 1.11). My older sister obtained the autograph of the Russian pilot. The next leg was on a Fokker Friendship to the tiny airport at Douala in the Republic of Cameroon. Then on to Léopoldville (now Kinshasa), where our parents approached immigration with considerable anxiety—only to find

Fig. 1.10 With my parents in 1965, our father wearing a traditional 'Kente' cloth leaving gift

Ghana's Ambassador to the Congo was there to greet them. He had received a telex from Flagstaff House (the presidential palace in Accra, which was both the residence and the office of the president) saying that the family would be travelling through. Clearly President Nkrumah's displeasure with our father did not extend to his entire staff. The Ambassador arranged for the whole family—with the addition of a young Egyptian doctor and a young English 'Voluntary Service Overseas' (VSO)[3] student our parents had taken under their wing on the journey—to be placed in the VIP lounge, and an aide dealt with

[3]Voluntary Service Overseas is a charity placing volunteers in educational establishments to live and work alongside local educators to expand opportunities for children globally.

Fig. 1.11 The Ghana Airways Ilyushin Il-18V. (With permission from Ron Dupas at https://1000aircraftphotos.com)

all the passports, health cards, and luggage. The Ambassador then piled the family into his *Corps Diplomatique* car. He had even managed to arrange hotel rooms, a near impossibility at the time because of the *Crise congolaise*. The aide who collected us all in the morning was a graduate of Akuafo Hall, who had fond memories of our parents from the University. All this kindness from the Ambassador went far beyond what might have been expected.

Next there was a comparatively enormous Sabena Boeing 707 for the relatively short flight to Élisabethville (now Lubumbashi). Once again a guardian angel appeared at Customs, this time the Zambian *Chargé d'affaires* in Élisabethville, who knew the family was travelling and wanted to be sure we made the connection to the next part of the journey. The next flight was a Vickers Viscount to Ndola in Zambia, where we were met by our mother's first cousin Robin, who worked on the Copperbelt in Northern Zambia. We all drove in his enormous Dodge through Kitwe, where Robin queued for some hours to obtain more petrol as rationing was already strict. After a few weeks in Chingola with our cousins the journey continued—a bus to Ndola because there was no chance of obtaining petrol for the trip, then a flight to Lusaka on a plane that was 'small and old' in our mother's account, and the beginning of the next phase of our lives. A telling detail in our mother's notes from this journey is that the bus to Ndola was really a South African Railways one, which was on its way ultimately to Johannesburg—so we were seated at the front, with the African passengers at the back. Our mother waited for the stewardess to sit down at the front, and then crept back down the nearly empty bus to talk to what turned out to be the relief driver and a domestic science teacher. The odious tentacles of apartheid reached even as far as a northern part of independent Zambia.

Our parents' own story is far more interesting than mine, and some of our father's part in it is captured in a brief obituary [346] and a profile [355].

2
Lusaka

Lusaka became the capital city of Northern Rhodesia in 1935, replacing Livingstone. It was deliberately planned by the colonial administration to be a garden city, with planned tree planting, open spaces, and wide streets.

> [Lusaka was] so spacious and so rich in trees and lawns and brilliant gardens that it looks much more like a garden suburb than a city. (District Officer Kenneth Bradley [35][1])

Early memories of my life in Lusaka are not authentically mine—they come entirely from older siblings. One that seems to be corroborated is that I had a liking for warm milk with sugar, and an insistence that this be served in the middle of the night with a specific spoon. I have no direct memory of this, but do remember the spoon: A small and distinctively shaped teaspoon with a black plastic handle. Apparently my noisy demands would sometimes wake up my older sister Kristi, who would demand "what are you doing to poor little Thomas?" of my no doubt exhausted mother if she had produced a mug of warm milk with the wrong spoon, causing outrage.

I was enrolled in Woodlands Primary School in Lusaka at the start of 1969. Several friends from my nursery school also started at the same time. The

[1]Kenneth Bradley was also the author of a plan for Lusaka written for its official opening as part of the Jubilee celebrations for King George V in 1935, and printed in a limited edition for private circulation. It has been reprinted together with an introduction by Robert Home [36]. The sequence of chapter headings in the plan give a flavour of the era: Unveiling; Origins; The Plan; The Buildings; The Native Side; The Airport; The Trees; Appendix—List of Trees and Shrubs.

T. Ward, *People, Places, and Mathematics*, Springer Biographies,
https://doi.org/10.1007/978-3-031-39074-6_2

two friends I remember most clearly from the early years were Patulani Nyalugwe and Chantal Krishnadasan. Patulani was a great friend, though apparently I had hospitalised him briefly while at nursery school as a result of an incautiously thrown stone. In the first year of primary school one of us was violently sick on the other's desk—but I cannot remember which of us was the recipient and which the deliverer of this unwelcome gift. Fifty years later I learned from a Woodlands alumni group that Patulani had passed away.

Mathematics at Woodlands in the early 1970s consisted largely of arithmetic of a rather traditional sort. Initially this was not a problem, but within a few years it seems I was not doing well at this. As a result my father built a wooden multiplication table to help me practice for daily mental arithmetic tests. Many years later I made something similar for our own children, an addition and a multiplication table with pieces made of stained and varnished MDF and a simple pine frame.

One consistent theme from my primary school reports was of impatience leading to poor outcomes (Fig. 2.1). "Often careless in arithmetic", "Never takes his time to work", "often careless especially in arithmetic", "good but very careless", "carelessness and poor handwriting", "lost marks through carelessness", "inclined to rush and makes silly mistakes" and similar phrases peppered my primary school reports until Grade 6, when a class teacher called Mrs Holman somehow made things click into place and I became "a very conscientious and responsible pupil". Few other teachers stand out in memory. Mrs Kachalia at some point hit my outstretched hands with a bamboo stick—a humiliating incident memorable in itself, but with no surrounding detail. In particular, I have no memory of what transgression I was being punished for. Others experienced much worse things, and one of the family memories my siblings remember much more than I do is blood being drawn from the knuckles of the child of a cabinet minister by a ruler with a metallic strip down one edge, which did not go down well.

The music teacher Mrs Pamela Rybicki left a more positive and less painful impression. Like the layers of Carboniferous rock deep under our current home, her music lessons left a layer of songs in me that populated a fertile and romantic imagination. Songs and hymns like 'When a knight won his spurs', 'Guide me oh thou great Jehovah', 'Sancta Maria', 'Morning has broken', and 'It's a Long Way to Tipperary' were all part of the culture of our childhood. Many years later my brother and I, along with other former pupils in multiple countries, were able to send cards to Pamela Rybicki before she passed away in May 2020, thanking her for all she had given us. There were other dedicated and kind teachers at Woodlands Primary, but few left as strong an impression.

Progress:

Reading and comprehension. Satisfactory progress made. Thomas reads with understanding.

Number. Thomas has a good understanding of all number work covered this year, but he still sometimes writes his figures back to front.

Writing. Room for much improvement. He often rushes his work.

General. When Thomas has taken the time and trouble, he has produced good work. He must learn to curb his impatience.

D. Pinnell
Class Teacher

HEAD'S REMARKS:

More haste, less speed!

B.C. Hobbs
Head of School

Fig. 2.1 A woodlands school report from 1969

At some point in the early 1970s—it must have been after 1972 when he came into power—the complex part of the cold war being played out in that part of Africa ended up with primary school children, me included, being given round badges bearing the smiling face of President Kim Il-Sung of North Korea. Our mother remembered the slogan around the perimeter as 'Throw yourself on the bosom of Kim Il-Sung', though my badge is long since lost and I cannot recall this myself. We were also given a small red book—not I think the full version of Chairman Mao's famous 'Little Red Book' but a shorter pamphlet with some key sayings. Chinese consumer goods became available in the shops in Lusaka. 'Flying Pigeon' bicycles and milky 'White Rabbit' sweets became a common sight. For a brief period leading up to the celebration of the tenth anniversary of Zambia's independence in 1974 we were all taught to march by North Korean soldiers, but in the end I was not one of the groups of primary school children chosen to join a parade of marches in military uniforms for the event.

It is clear from family correspondence if not from my own memories that by Grade 6 I was not finding life easy at Woodlands Primary School socially, so I was moved for the short period September to December of 1974 to a small private school called Nkhwazi Primary. Here (I was told later) I recovered some confidence, though no memory of it remains. Our father's diaries record this period of 1974 and 1975 as being a time of considerable financial strain for the household, and I imagine even a term at a private school was a considerable additional burden. For Grade 7, the final year of primary school, I was moved back into a government school called Northmead Primary School at the start of 1975. Northmead was a well-organised and highly disciplined environment, with a strong focus on behaviour and academic performance which suited me well. The one personal transgression I remember was being found with a paperback 'Mad magazine' book, which for some reason caused real outrage in the rather strait-laced but basically kindly teacher Mr Lakey.

For the first few years in Zambia our father continued joint research projects with Jack Marr in Ghana. Some of this was carried out over short-wave radio, and the expression 'Dad is talking to Mars' became the reason for some evening absences. These were infrequent—apart from events related to Scripture Union or Lusaka Baptist Church, many of them hosted at our house Leopard's Head, most evenings found both our parents hard at work in a large shared study. They never became part of the typical expatriate 'club' world, partly because of their lifelong teetotalism.

At home, in addition to a great many books of quite extraordinary diversity, there were interesting objects reflecting the scientific interests of both parents. These included a hand cranked Brunsviga Nova calculator and several slide rules. Most exciting of all, someone from The University of Zambia (UNZA) Physics department took some of us stargazing using a portable Questar reflecting telescope (see Fig. 2.2). This added a further layer of excitement to the already dazzling night skies. Lusaka, like much of Zambia, is on the Central African Plateau, and is 1280 metres above sea level. At the time light pollution was negligible, and we had a large garden. This, combined with clean dry air for much of the year, made the night sky vivid and dramatic. When I peer up through a soup of polluted moist air saturated with scattered light from my balcony in Durham, 90 m above sea level, it is all too easy to forget the full import of the Psalmist's imagery, written about desert skies long before any artificial illumination.

Fig. 2.2 A Brunsviga nova (Photo by Maksym Kozlenko—Own work, CC BY-SA 4.0, https://commons.wikimedia.org/w/index.php?curid=98511987) and a Questar telescope. (By Hmaag—Own work, CC BY-SA 3.0, https://commons.wikimedia.org/w/index.php?curid=24406926)

> He determines the number of the stars and calls them each by name. Great is our Lord and mighty in power; his understanding has no limit. (Psalm 147:4–5; New International Version)

The constellations visible in the Southern Hemisphere are one of the few things that make me nostalgic for Zambia and sub-Saharan Africa—something I only realised many years later on a Mathematics trip to Australia. Our cosmological neighbours the Magellanic Clouds are probably the closest thing I have to Proust's madeleines. The other candidates for this position are the sounds of crickets at night and distinctive sub-tropical petrichor.

The household also featured a large second-hand Gestetner duplicating machine. The smell, noise, and feel of the master page, the hand-cranked machine, and the copy pages emerging are a vivid memory distinct to that period of time and location. Collating and stapling was done on a large table, and the variety of documents involved reflected our parents' wide interests. Scripture Union minutes, Science Teacher Association papers, family Christmas letters, draft sections of books, and much else passed through this remarkable machine.

Photography was a cumbersome and costly business of course. Our father had a manual single lens reflex Asahi Pentax 'Spotmatic' camera which came out for significant events. He also used an Eumig C3 clockwork 8mm ciné camera, and the four of us were sometimes asked to do things like walking up a hill again in order for this to be captured on film. Along with this clockwork camera there was an Eumig P8 projector, which I remember most often being used in Coaster Cottage. In order to have a reasonable image size, our father cut a hole in the wall between the tiny kitchen and the sitting room for the projector to shine through. These jerky silent films, the accompanying clattering sound of the projector mechanism, a white-painted softboard screen wedged into the window frame, and the six of us squeezed into the tiny sitting room are part of the fabric of my childhood. Occasionally there would be additions to the family ciné films in the form of Charlie Chaplin silent movies available on 8mm film.

The event with the Questar telescope—together with a great deal of both knowledge and enthusiasm from both parents—triggered my early interest in astronomy. One of the family stories that my mother retold many times involved me bursting into their room early one morning to regale two bleary-eyed parents with an excited (and coherent, in her optimistic memory) account of the red shift and Hubble's law. This seems to have been something of a habit—I do not remember the incident, but one of my sisters found the inscription 'Thomas, aged 6 (?) woke us early one morning to say "Did you know it took more than 50 years to translate linear B?" (see p. 208)' in our mother's handwriting on the inside cover of an old copy of a book about the archaeological discoveries of Heinrich Schliemann and Sir Arthur Evans in Crete [59]. While that incident is lost to memory, the extraordinary accounts and speculations of the Minoan civilization always interested me, and I was delighted when a Mathematics conference I attended in Heraklion in June of 2013 included a tour of Knossos.

I certainly enjoyed reading in a large astronomy book, which was probably a copy of the Larousse 'Encyclopædia of Astronomy'. It was definitely more up to date than the other great source of written knowledge in the house, the ten red volumes comprising 'The Children's Encyclopædia' edited by Arthur Mee [198] (Fig. 2.3).[2] The edition we had probably dated from the 1950s or so, and much of the world it depicted from the 1920s. It did not include the discovery of Pluto in 1930, and it was more or less silent on nuclear Physics. It did however have cogent accounts of aspects of the theory of natural selection and the evolutionary origins of the diversity of life on earth. The British Empire was largely depicted as benign and noble in its intentions, and some of the articles had more than a whiff of eugenic ideas built on the superiority of certain peoples. The politics and views on society were equally of their time, but there was a great deal to entertain in the distinctive red volumes.

[2]Nicholas Tredell's literary biography of C. P. Snow points out that in Snow's childhood home there were 'only about 50 books; perhaps the most important was the *Children's Encyclopædia* (1910) produced by Arthur Mee' [298, p. 4]. That edition of the encyclopedia was an eight volume set of bound editions of a fortnightly magazine [197].

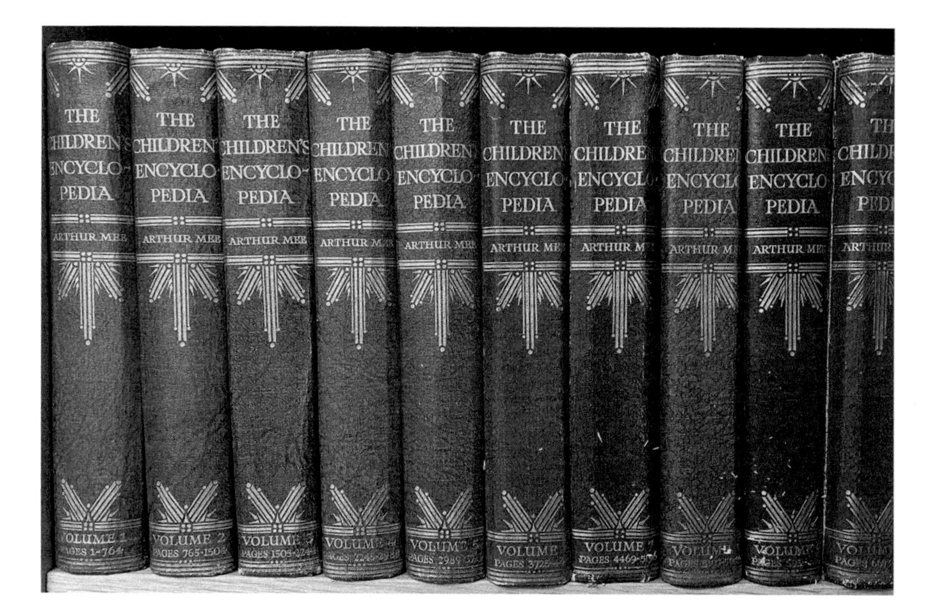

Fig. 2.3 A copy of Arthur Mee's Children's Encyclopædia purchased many years later

The house had a gramophone, and we all enjoyed long-playing records. It was a rare treat for our parents to stop working for a time and listen to a record, which might be classical music or the ingenious comic records of Tom Lehrer or Flanders and Swann. By the early 1970s our sisters began bringing home other records, the most memorable for me being those by Seals and Crofts, Simon and Garfunkel, the Equals, and the Beatles.

When my primary school class were asked to illustrate what we wanted to be when we grew up, my choice was clear (Fig. 2.4). The way I depicted stars may have reflected the fact that a few years later my short-sightedness meant I had to wear spectacles even to read from a blackboard.

By chance I somehow ended up in possession of the same piece of work for the entire class, and it is an insight into several aspects of life in Zambia at the end of the 1960s. The entrenched attitudes that lay behind the boys opting to be pilots, astronauts, skin divers and the girls air hostesses, teachers, and nurses were obvious. The fact that several of my classmates in land-locked Zambia hoped to become skin-divers is testament to the powerful hold that 'The Undersea World of Jacques Cousteau' held on all our imaginations. This television series ran from 1966 and chronicled the voyages of the French naval officer Jacques-Yves Cousteau and the crew of the Calypso, a former mine-sweeper that had been adapted and fitted with an underwater observation

Fig. 2.4 Aspirations from Woodlands Primary School

chamber built around the prow for oceanographic use. Television itself was in its infancy in the country, domestic video recorders unheard of, and there was a single channel. This meant everyone with access to a television would watch the same programme at the same time, amplifying the impact (Fig. 2.5). We never possessed a television while we lived in Zambia, but our friends the Shamwana family did—and we could watch Cousteau there, along with the original series of Star Trek and Doctor Who.

In 1971 Wendy Bond, wife of the District Commissioner Mick Bond at the time of independence in 1964, founded a magazine called *Orbit* [28] (Fig. 2.6). This featured high quality graphic art and a solid scientific basis adapted to a

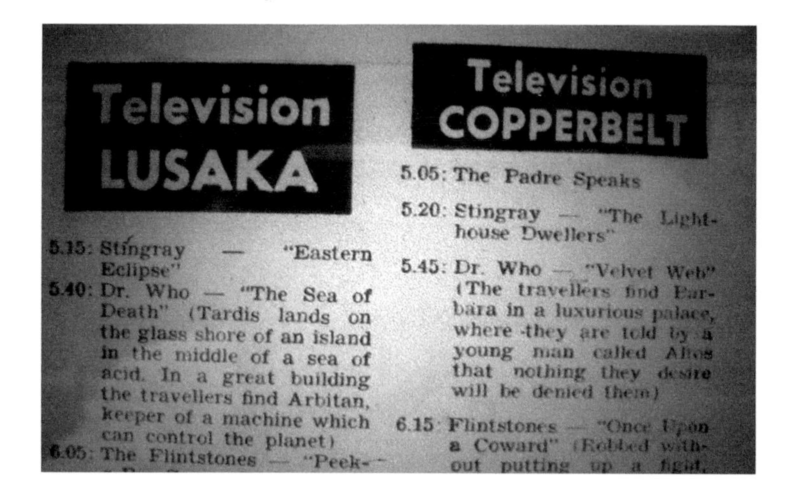

Television
LUSAKA

5.15: Stingray — "Eastern Eclipse"
5.40: Dr. Who — "The Sea of Death" (Tardis lands on the glass shore of an island in the middle of a sea of acid. In a great building the travellers find Arbitan, keeper of a machine which can control the planet)
6.05: The Flintstones — "Peek-

Television
COPPERBELT

5.05: The Padre Speaks
5.20: Stingray — "The Lighthouse Dwellers"
5.45: Dr. Who — "Velvet Web" (The travellers find Barbara in a luxurious palace, where they are told by a young man called Altos that nothing they desire will be denied them)
6.15: Flintstones — "Once Upon a Coward" (Robbed without putting up a fight,

Fig. 2.5 Zambia broadcasting services's first night schedule 13th March 1966. (Wikipedia: Jon Preddle-Attribution-NonCommercial-NoDerivs 3.0 Unported)

Zambian audience. She had done so after the Minister of State for Technical Education, Valentine Shula Musakanya, asked for suggestions of a suitable science magazine.

> When science and technology replace witchcraft here, as they must, everyone should have the knowledge to understand them and not just to believe blindly, replacing one superstition by another. (Valentine Musakanya in 1970, as recalled by Mick Bond[3])

These magazines included some science fiction stories with a Zambian angle alongside basic science articles, a page of responses to science queries, and a 'young farmers' page of advice from the Ministry of Agriculture.

There were other children's magazines in the household. My sisters had a subscription to *Bunty* magazine, and my brother to the *New Hotspur*. These arrived tightly rolled in the post from the UK. As the youngest of four I read them all, and story lines like the 'Four Marys' from *Bunty* became part of my reading.

[3]https://www.britishempire.co.uk/article/fromnorthernrhodesiatozambia.htm.

Fig. 2.6 The cover of the first *Orbit* magazine in 1971 (Photo reproduced with permission from John Freeman)

Leopard's Head

The family had been met at Lusaka airport by the University Registrar and the Buildings Officer, who drove us to some 'Transit Quarters' in the Oppenheimer Building. This had been the home of a College of Social Service and was being acquired by the new University. The Dean of Students, Reverend MacPherson also lived on the site with his large family.

Reverend Fergus MacPherson (9th August 1921–25th June 2002) was a Church of Scotland minister who trained at Edinburgh University. He began his missionary work at Mufulira in Northern Rhodesia and moved on to a mission station at Lubwa in 1951, where the future President Kenneth Kaunda was born. Reverend MacPherson opposed the creation in 1953 of the Central African Federation which brought together Nyasaland (later Malawi), Northern Rhodesia (Zambia) and Southern Rhodesia (Zimbabwe) under a colonial government led by Sir Roy Welensky. MacPherson moved to Nyasaland in 1954 to become principal of the Livingstonia Institution. As resistance to the Federation grew, a state of emergency was declared and Europeans were urged to leave Northern Nyasaland. MacPherson remained, and an arrangement was made for Rhodesian aircraft to fly over the mission station. He was asked to lay out stones reading 'I' if all was well and 'IV' if the station had been attacked. Instead he laid out stones reading 'Ephesians 2.14', referring to the New Testament verse 'For he is our peace, who hath made both one, and hath broken down the middle wall of partition between us' in the King James Version. After a few years as a minister at St Ninian's in Greenock he moved to newly independent Zambia, initially as a teacher and then as Dean of Students at UNZA. He returned to Scotland in 1978 and went on to senior roles in the Church of Scotland and the Council of Churches in Britain and Ireland.

The logistics of life in Zambia at that time were not easy, as petrol rationing meant the family were only allowed four gallons a month. For the first few years our father cycled to UNZA, and we used bicycles to get to and from school—with our mother carrying me in a little seat until I could ride myself. The housing that the University proposed was in Chelston, nine miles from the campus. We quickly moved for a short period to a house on Lake Road, and it came to light that an old house in a district called Woodlands had become available, described as 'a very bad house but in a lovely garden'. The University had purchased this house from the Mitchell-Hegg family. Our father decided to commit to this house, called Leopard's Head, without seeing it, because it was within easy cycling distance of both Woodlands Primary and Kabulonga

Fig. 2.7 Leopard's Head in 1966

Secondary schools. The four children rode in the back of a large open truck which moved the entire household to Leopard's Head on Leopards Hill Road, with a huge garden and a primitive swimming pool. The house itself was in the 'Cape Dutch' style of architecture, with distinctive white painted gables and a huge green corrugated iron roof (Fig. 2.7). It was a sprawling household, with a separate pump room and borehole.

The garden was several acres in extent, and the more formal part included a large oval lawn with a huge mature brachystegia tree. The garden also included guava, mango, frangipani, and avocado trees—and a lychee tree large enough for the children to sit in and feast on fresh lychees. There were both African and Indian mango trees, and the guavas were so prolific that boxes of them were given away each year. The less formal part of the garden was quite wild, and edged with a line of tall eucalyptus globulus (blue gum) trees. We would often find pellets of mouse bones regurgitated by the owls who perched on their high branches at the foot of these trees.

Internally the house was indeed 'bad' in some ways: The kitchen had no work surfaces or shelves, and a family of bush babies lived in the roof. Their droppings had made the main storage cupboard unusable.

> But for our kind of living, very much outside and very much child-dominated, it is ideal. The furniture is old and decrepit even by our standards, with (I very much fear) mice in the upholstery on the sitting room chairs! (From our mother's correspondence)

It was an outdoor life, and we all enjoyed the space available in the garden as well as the extraordinary diversity of plant and animal life in it (Fig. 2.10 shows the four of us with a friend in one of the trees). Tentative chameleons moving slowly along a branch of one of the frangipani trees that lined the drive, long black 'chongololo' millipedes, and large African land snails were pleasant things to encounter.

We knew enough to be cautious about some of the less friendly creatures—venomous snakes including black mambas, gaboon vipers, puff adders, and Mozambique spitting cobras were rare but not unheard of. We each had a cat, and they occasionally would bring alarming gifts into the house by operating as a team to bat something like a live scorpion back and forth between them across wax polished cement floors. These were generally intended for our mother, as they had the wisdom to understand which member of the household was the source of their food. Outside their own hunting, the main source of protein for the cats was usually kapenta (small Tanganyika sardine) or the heads of tilapia fish that our mother prepared using an aluminium pressure cooker.

Antlions

Among the most interesting animals for me were the antlions (Myrmeleontidae), the larval stage of the antlion lacewing. These would dig small conical pits in the sand by throwing grains of sand clear of the pit, and then bury themselves at the bottom. Any small insect that started to fall over the edge of the pit would dislodge grains of sand, causing it to slide further down in its own tiny avalanche. Once at the bottom the fierce tiny jaws of the antlion grabbed hold, the victim was sucked dry, and the empty skin thrown clear of the pit. I would sit in the sunshine and nudge tiny grains of sand in order to watch the tiny predator emerge at the bottom, jaws snapping, hoping for prey.

Fig. 2.8 A tapestry of Leopard's Head in Lusaka made by our mother

Life Outside

The extent to which the garden and trees were important to all of us, but to our mother particularly, is reflected in a tapestry she made of Leopard's Head many years later (Fig. 2.8). The view shown is from the large lawn, which for us was behind the house. The back of the house as we lived in it must once have been the rather grand front, with the high gable showing and a dirt road with a small roundabout for turning cars or horse-drawn carts. There were two small lawns directly in front of the entrance, with steps leading down to the dirt road and then more steps leading onto a larger lawn. For us this was all at the back, and our normal entrance to the house was through a small garage or the kitchen door. Leopards Hill Road must have been built later, overlaying a more modern city of paved roads onto the situation of the house. This meant there was a long dirt driveway which passed what would have been the kitchen garden and orchard.

Our diet was something that would be more or less familiar to British people from that era, adapted to work around missing ingredients. From 1965 the Unilateral Declaration of Independence (UDI) in Rhodesia made the import and export of goods extremely difficult for Zambia, leading to frequent shortages in the shops. Items like Marmite were highly treasured and carried back in suitcases from infrequent trips abroad. There were however some more interesting things to eat all around us. The fruit in the garden certainly, and friends at school from families with parents who had moved from India would sometimes bring in delicious *Gulab Jamun* (fried balls of a milky dough soaked in rosewater syrup). Once a year the flying ants emerged from the ground in vast numbers, and for days afterwards street sellers would sell portions of 'inswa', flying ants stripped of their wings and fried in oil.

The swimming pool was a round cement structure up a slope from the house, and was extremely simple. It may even have originally been used as a water reservoir, and the slope up to it was probably simply the spoils from when it was dug out. The pattern of use seemed to be to fill it and use it until it went green with algae, then drain it and start again. Once the house was connected to potable mains water rather than the supply of rather murky water from the borehole, this was unaffordable as well as irresponsible. Money was tight, and we certainly could not run to a proper cleaning pump and filter, so our father devised one himself. I was too young to be aware of this process, but my older brother James remembers it clearly and shared his memories. Our father bought an old Bedford truck water pump from a junk yard, and found a way to mount this on its side in a frame that would not rust. A vertical drive shaft to the pump was fabricated from wooden poles with some flexible tube joints all held together with 'Jubilee' clips, and this was run from a small electric motor connected to the mains supply and housed under a small zinc galvanised roof. He located a damaged water tank half of which was still usable, and partially buried this in the ground nearby. This was filled with sieved fine sand, then a layer of river sand, fine gravel, and ending with a layer of ordinary gravel. This drained from the bottom back into the pool, maintaining a clean pool for the years we lived there. When it was running the spinning drive shaft under the water was a source of fascination, and was one of many enjoyments of a childhood largely unconstrained by health and safety concerns.

Our father was endlessly inventive, so over the years the household was enhanced in several ways. Swings, treehouses, and climbing poles were added to the garden, making good use of the many mature trees. Several versions of electric cars that we could drive in were made from junk and old car batteries, a practice he returned to in retirement. Grandchildren and great-grandchildren enjoyed driving round the garden of Coasters Cottage in electric cars he built for them decades later (Fig. 2.9). His approach to engineering had some distinctive features. He certainly used more Jubilee clips than I would, and had greater faith in the strength of 'Araldite' (a brand of epoxy resin) repairs than most. Above all he liked to reuse and repair, and gave far more consideration to function than to form.

Fig. 2.9 With my father making running repairs to the cable and pinion steering of an electric car for the great-grandchildren at Coasters Cottage on the 28th of July 2019

A tall slide was constructed for the swimming pool, with a wooden frame and aluminium sheeting to slide down. The aluminium sheet often became too hot to touch in the sun, so we all learned to splash water up from the pool to cool it enough to be used. A small boat with an electric outboard motor was built, stabilised by a car's lead-acid battery as both power source and keel inset into the flat base, sealed in with hot tar. The sides of this boat were made of hardboard, made waterproof with thick oil paint. Boldest of all, a 'foefie slide'—at the time I doubt we were aware this was an Afrikaans word—was erected that ran from a tree and across the pool. When the pool was full this was great fun, ending in a soft but dramatic landing in water. It was far more exciting when the pool was empty in the Winter, allowing a sliding landing in bare feet on the dusty cement floor of the pool. Even by the standards of the time this was a particularly relaxed approach to health and safety, and there were some accidents. A house guest broke his arm and, potentially far more seriously, my older brother James ended up hitting the back of his head hard enough to induce a serious concussion. Nowadays this might have led to

Fig. 2.10 The four of us with a friend in the treehouse in the kitchen garden at Leopard's Head in the late 1960s

computer aided tomography scans to check for bleeds, but in 1974 in Lusaka our mother simply monitored a high temperature and recurrent nausea for several anxious days. I remember the hushed voices and darkened room, and in hindsight this could have been far worse (Fig. 2.10).

A less enjoyable aspect of the swimming pool was dealing with what we called 'water scorpions'—Nepidæ, with the appearance of a fearsome stinging tail and the reality of a painful bite. This involved one or more brave swimmers getting in and walking round and round the circular pool to create a circular current, which moved them into the centre. Then a reluctant volunteer walked across the diameter of the pool with a net made of mosquito netting between two frames of hardboard to pick up the creatures and dump them over the side. This means of dealing with the problem was not something I remember being discussed, but the fact that it works is far from obvious and is not as one might imagine a centripetal phenomenon due to the relative buoyancy of the creatures in water. It is apparently related to the so-called 'tea leaf paradox'—why do tea leaves migrate to the bottom and centre of a cup when it is stirred, rather than collecting along the outer surface? The first real insight came from the physicist James Thomson in 1857, when he identified the role played by 'a diminution of velocity of rotation in the lowest stratum by friction on the bottom' in the source of the trade winds [295]. The full explanation involving the weaker centrifugal (pseudo-)forces nearer the bottom 'wo die Strömungsgeschwindigkeit des Wassers durch Reibung reduziert ist' (where the speed of flow of the water is reduced by friction) resulting in a secondary helical flow that moves outwards on the surface, down on the outer edge, and inwards towards the bottom surface, was finally given by Albert Einstein in 1926 [96].

Leopard's Head was always busy, and often included long term guests. This might start simply with our mother meeting a young VSO teacher from rural Zambia who had come to Lusaka Baptist Church on the first Sunday of the school holidays, and end with a person who spent the school holidays with us for years to come. This practical spirit of hospitality left a deep legacy in all of us in multiple ways, and one of those teachers became my brother-in-law.

Making and Repairing

We all learned to repair and to build things, starting with the maintenance of our own bicycles and the making of wooden guns, and building up to helping with more ambitious car and household repairs. The sight or feel of specific tools bring back memories of handing them to our father lying under one or other of the cars, and we learned the basics of things that have vanished from ordinary life for most of us—bleeding the brakes, using feeler gauges for spark plug gaps, or wielding a grease gun. Of the four of us only my brother James carries out car repairs at a serious level now, and I still feel a small pang of guilt when I ask a garage to do something that our father would do himself. The most evocative tool sets were a black folded tool wrap that had come with the Peugeot 404, and an extensive and good quality 'King Dick' socket set in a metal case, which included objects that seemed massive in a child's hands, like a long torque wrench.

The urge to try and repair things rather than replace them remains in all of the four children, manifested in different ways. Some of this may have its origins in the fact that, though wealthy by local standards, the finances of the household required careful management. To place that in some context, the household at Leopard's Head included two gardeners. Extreme income inequality remains a stubborn problem for the country—a study in 2017 using the 2010 Gini coefficient placed Zambia as the seventh most unequal country in the world [25].

Fig. 2.11 Our father with a Meccano aeroplane in the early 1930s

There was however something more profound than economy at work, innate in our parents. Our father built complex projects in Meccano over almost the entire span of the 96 years he lived, and always took pleasure in anything we made or repaired (Fig. 2.11). He never really stopped making things, and the last time I visited him in Coasters Cottage he asked me to help complete the assembly of pieces of wooden doll's house furniture he was making for his great-grandchildren. These were carefully made from thin sheets of wood, but his eyesight was not up to the task of drilling the holes and screwing in the tiny brass door hinges. We still use some of his phrases relating to his enormous collection of hand tools—'Use the proper tool' speaks for itself; 'Feeding wood to the termites' was used if someone left a wooden-handled tool outside.

Our mother grew up in real hardship in Birmingham, as her father had died in 1929 when she was a small child, after a long period of ill health due to being gassed in the trenches. Her mother Phoebe managed to bring up two daughters on her own by working as a cleaner and letting a room to lodgers. Our mother talked at times about her mother's painful hands with skin cracked from scrubbing other people's floors.

These multiple threads came together to produce a household in which the first instinct was to repair not replace, and few activities were measured using the metric of material gain. Spiritual, ethical, and environmental values dominated the household long before the fragility of the planet during the anthropocene was in our minds. Whether it was food left uneaten, a nut rounded by a spanner of the wrong size, a nail bent by a wrongly directed hammer blow, or a screw head damaged by incautious use of a screwdriver, waste was something seen as innately and substantially wrong, and the starting assumption was that almost anything could be repaired. Our mother in particular was fond of repeating Gandhi's phrase 'Live more simply that all may simply live'.

James had an early—and life-long—interest in cars and the engineering inside them. He undertook an ambitious and successful attempt to resurrect a completely derelict Daihatsu Trimobile delivery van when he was aged about twelve (Fig. 2.12). This was a small three-wheeled delivery van with a single-cylinder 10 horsepower two-stroke engine. The level of work required to get this running included a complete engine rebuild. Acquiring replacement piston rings from Japan was no small task in the early 1970s, involving lengthy

Fig. 2.12 The whole family and my brother's Trimobile in 1971; the slide into the swimming pool is in the background and the pump-house to the right

exchanges of correspondence and complex money transfers. It was a real achievement when it was up and running, and the garden was large enough to allow it to be driven extensively, to the huge enjoyment of James, our friends Michael Jearey, Tom Phillips, the four Shamwana boys, and me.

Many years later James restored a Mark 2 Jaguar saloon, which was used as the wedding car when he married Janine Bill in 1984 in Swaziland. It is still in use at their home in Howick in KwaZulu-Natal.

UNZA

The early beginnings of what became colonial Northern Rhodesia lay with Lewanika, King of the Lozi people in Barotseland. Fearful of attack from the Portuguese to his West and the Ndebele to his East, he sought alliance and friendship of Britain. As a result he was persuaded to sign the Lochner Concession on the 27th of June 1890, which ceded mineral and trading rights across Barotseland not to Britain but to the British South Africa Company (BSAC).

The British South Africa Company was created in 1889 as a result of the amalgamation of the Central Search Association, one of the entities used by Cecil Rhodes to extract as much profit as possible from his ventures in Africa, and the Exploring Company Limited. It continued to exist until 1965, and enormous profits were extracted from Southern Africa during its operations. The main focus of activities was South of the Zambezi River and some of its operations in Northern Rhodesia were constrained by lack of resources. The British Government in effect abdicated responsibility to the BSAC partly in order to avoid being called on to fund any developments, and the historian of Empire John Semple Galbraith argues that BSAC held much of the land that would become Zambia simply as a longer term investment.

> Heavy expenditures in Mashonaland made it impossible for the company to spend much money north of the Zambezi. The company's promoters regarded the trans-Zambezi lands as territory to be held for future speculation rather than immediate development. (From Galbraith [132, p. 204])

Supposedly King Lewanika was tricked into believing this agreement was with Britain by Frank Elliot Lochner, who claimed to be an Ambassador sent by the Queen. The Lochner concession promised King Lewanika a payment of £2,000 annually, weapons, and a royalty of 3% on mineral extraction. It also promised British military protection. There had been earlier agreements of a similar type, and there would be other significant agreements later, generally granting more and more to the BSAC. In just one of the later agreements is there a brief allusion to educational development.

> British South Africa Company further agrees that it will aid and assist in the education and civilization of the native subjects of the King (Lewanika) by the establishment and maintenance and endowment of schools and industrial establishments. (Agreement signed on 17th of October 1900, from [186])

Northern Rhodesia was established as a British Protectorate in 1911, to be administered by BSAC. This arrangement continued until the British Colonial Office took control in 1924. The only educational establishment built during this period was the Barotse National School, founded in 1907. It remained the only Government controlled school in the country up to 1929, when the Jeanes and Agricultural School was opened at Mazabuka [226]. Copper was discovered in 1928, leading to an influx of Europeans, whose education did seem to be a matter of greater interest for the colonial authorities. A revealing insight into how education evolved in Northern Rhodesia over the next few years may be found in an account from Hansard of an exchange in the British Parliament on the 22nd of July 1931.

> ... asked the Under-Secretary of State for the Colonies the amounts spent by the Government of Northern Rhodesia on European and African education, respectively; and the number of European children being educated in that country? (J. F. Horrabin, Labour Member of Parliament for Peterborough)

> At the end of 1930 there were 1,424 European children of school age in Northern Rhodesia, 879 of whom were receiving education in Government controlled schools. Owing to the influx of European families in connection with mining development, the number of European children for whom educational facilities are urgently required is rapidly increasing. The latest estimates show that the Government of Northern Rhodesia provided for the current year £38,497 (less than £5,000 received in fees) for European education. The provision for native education has increased from £348 in the year 1924–25, when the Department of Native Education was first established, to £22,140 in the current estimates largely by way of subvention to mission schools. The total expenditure upon these schools is not known. To this must be added a sum of approximately £7,000 to be spent on the Barotse National School out of the Barotse Trust Fund. Of 265,000 native children of school age, 124,000 are receiving education in mission or Government schools. I would add that the Director of Native Education has recently prepared a scheme which meets with the approval of

> the Governor, for a large extension of educational facilities for natives, by means of Government schools; and increased grants to missions. This scheme, which involves the expenditure of an additional sum of £16,432 in the current year, and an increase of the total expenditure on native education to £71,346 in the year 1935–36 is at present under the consideration of my Noble Friend's expert advisers. (Dr Thomas Shiels, Parliamentary Under-Secretary of State for the Colonies)

The history of the development of educational capacity in Zambia gives some insight into debates about the legacy of empire more generally. Experts have written on this extensively, from the anti-capitalist Guyanese historian Walter Rodney [258] to the 'fully paid-up member of the neo-imperialist gang' Niall Ferguson [126]. Among the more interesting and personal accounts of some of the long shadows cast by empire in Zambia specifically is an auto-ethnographic study by Juliette Bridgette Milner-Thornton of the descendants of white colonialist men and Zambian women, and the way in which they were treated both by the colonial administration of Northern Rhodesia and in independent Zambia [216].

In trying to understand some of the history from my own narrow experiences, a simple analogy comes to mind when I hear apologists for Empire talk about things like the railway networks that were built during the colonial era. Imagine your house has been violently burgled, one of your children killed by the intruders, and many of your belongings spirited away in the middle of the night. When the police investigate the scene it comes to light that during the home invasion one of the burglars did the washing up. Is it reasonable to expect you to place the washing up in some sort of moral balancing scale, and see the good side of the incident as a result?

The birth of the University of Zambia (UNZA) lies at the end of this lengthy history of deliberate under-investment in education by both the British South Africa Company and the Colonial administration. The University was established by the Act of Parliament No. 66 of 1965 a year after Zambia gained independence. For comparison, Fourah Bay College[4] in Sierra Leone was founded in 1827, Makerere University in Uganda as a technical school

[4]Fourah Bay College was affiliated to the University of Durham on 16th May 1876, and an article in the *Durham University Journal* of 16th December 1876 records the aspiration surrounding this: 'The instruction will now be in exact conformity with that pursued at Durham, and the negroes of Western Africa will no longer be unable to avail themselves of a sound university education [...] In conclusion, we can only admire the energy of those who brought to such a successful issue in the midst of such great embarrassments a project which was at the first thought to be impracticable. And it will be no little credit to the University of Durham for history to have to record that she was the first to throw open to Africa the full privileges of a liberal education' (From [1]).

Fig. 2.13 The four of us at the entrance to UNZA around 1970

in 1922, Ibadan College in Nigeria in 1948, and the University College of the Gold Coast in Ghana the same year.

The Canadian political scientist Douglas A. Anglin was made the first Vice-Chancellor of UNZA in July 1965, and President Kenneth Kaunda gave assent to the act in October 1965 (Fig. 2.13). The President was appointed the first Chancellor on 12th July 1966, and the inauguration of the University took place the next day. The main site, the Great East Road Campus, was designed by Julian Elliot as a complex of linked modern buildings designed to be capable of supporting future expansion. The Medical School was built on the Ridgeway Campus, adjacent to the University Teaching Hospital.

When our father joined UNZA much of the campus was still being built, and he enjoyed being able to influence the beginnings of the Physics Department. The University, along with the country of Zambia, played a significant role in the liberation struggles of Southern Africa in several ways, and some of the early students were exiles from South Africa. Nelson Mandela, on his first journey to another country after his release from prison, chose to visit UNZA and make his first University speech there after his release from prison.

3

Swaziland

At the end of 1975 some of the family moved to Swaziland for our father to take up a position at the University of Botswana and Swaziland. The role was once again about building capacity, with the aim of eventually being able to offer a full Physics degree within Swaziland. This also marked the beginning of the always inevitable scattering of the four children in earnest. The eldest, Kristi, stayed on to complete medical training at UNZA in Lusaka; the next child Sheena had moved at the end of the first year studying science at UNZA to embark on a degree at the University of Keele in England. I was fortunate in that the ending of our time in Zambia coincided with the end of primary school, meaning I would in any case be moving to a new school. For my older brother James the timing had not been at all convenient, as he was two years away from O-level examinations at Kabulonga Boys High School in Lusaka. The solution our parents found was for him to go to a boarding school in England, which he did not enjoy. Fortunately he moved to the same school in Swaziland that I was at a year or two later, and like me had a happy time there.

Threading a Needle

The move from Zambia to Swaziland included a final drive in our big white Peugeot 404 estate. These vehicles, with their distinctive Carrozzeria Pininfarina design, were popular in Southern Africa for many years as they seemed to cope well with the large distances and poor road surfaces. The other vehicle in the household at the time was a grey Austin A60 Countryman estate. The last scheduled leave from Ghana was in the Summer of 1965, and

T. Ward, *People, Places, and Mathematics*, Springer Biographies,
https://doi.org/10.1007/978-3-031-39074-6_3

our parents had arranged months in advance to buy the Austin Countryman in England, and to have it fitted with an extra petrol tank. They used it in England for that leave, and then arranged for it to be shipped to Zambia, confident that this would take so long that they would be there first. While it was in transit the predominantly white government of Ian Smith made the Unilateral Declaration of Independence (UDI) for Rhodesia after much dispute with the British government, who insisted that the post-colonial arrangement would involve black majority rule. This reflected the slow evolution in British political views crystalised in the 'Winds of change' speech, itself a mix of principle and *Realpolitik*.

> ...the most striking of all the impressions I have formed since I left London a month ago is of the strength of this African national consciousness. In different places it takes different forms, but it is happening everywhere. The wind of change is blowing through this continent, and, whether we like it or not, this growth of national consciousness is a political fact. We must all accept it as a fact, and our national policies must take account of it. As I have said, the growth of national consciousness in Africa is a political fact, and we must accept it as such. That means, I would judge, that we must come to terms with it. I sincerely believe that if we cannot do so we may imperil the precarious balance between the East and West on which the peace of the world depends. (From the address by British Prime Minister Harold Macmillan to the Parliament of South Africa on the 3rd of February 1960)

Independent Zambia already had very few access routes to the sea, and the closing of borders to Rhodesia created further problems. The shipping route that the Austin Countryman followed from England was by sea to Dar es Salaam in Tanzania and then by road to Lusaka. Remarkably this complex journey worked perfectly, and the car reached the Austin agent in Lusaka in January 1966. When the family left Zambia a decade later this car was given to Margaret Whitehead, the daughter of an UNZA colleague. For the first few years in Zambia petrol rationing was severe, so the Austin Countryman was used very little and a second-hand Austin Mini van was bought; this little vehicle and bicycles became the main method we used to move around.

The Peugeot 404 had been picked up new in Paris in 1968, and transported with the whole family to Cape Town on board the SA Oranje[1] operated by

[1]The vessel was built in 1948 at the Harland & Wolff shipyard in Belfast and named Pretoria Castle. She was sold on the 1st of January 1966 to the South African Marine Corporation (UK) Ltd, and entered service as the SA Oranje in Safmarine livery but retaining Union Castle crew. The renaming ceremony took place at Cape Town on the 2nd February 1966, and the unveiling was carried out by the wife of

Fig. 3.1 James and me on board the SA Oranje in 1968

Union Castle on behalf of Safmarine. We sailed from Southampton on the 21st of June 1968 (Fig. 3.1). Our parents were shocked to discover that on board what they saw as a British flagged vessel owned by a British company the purser insisted on determining all the seating arrangement in order to ensure that there was strict racial segregation at mealtimes, with the white passengers served first. By the second day of the long voyage our parents had obtained permission from the Captain to use the children's playroom each evening at 9 p.m. for a small multi-racial Christian fellowship.

Once landed at Cape Town, and after a lengthy process of getting six people, a new car with French licence plates, and a great deal of luggage through customs and immigration, assembling a roof rack, and loading up, we embarked on a long driving holiday, slowly making our way North to Lusaka. This included the so-called 'Garden Route' part of South Africa, about 300 kilometres of beautiful coastal road along the Southern tip of Africa going East from Cape Town. The Peugeot served the family well and made many lengthy journeys—and as children we all enjoyed lying under blankets on its large roof-rack at the drive-in cinema in Lusaka. When our parents finally left Swaziland in 1986 the same car made the final trip from Manzini to Cape Town, taking

Hendrik Verwoerd, Prime Minister of the Republic of South Africa and often regarded as the supreme architect of apartheid. After 187 sailings she landed at Kaohsuing on the 2nd November 1975 where she was broken up by Chin Tai Steel Enterprises.

our parents to the RMS St Helena for their final trip home to England via the island of St. Helena. The island (which had no airport until 2015) was serviced by the Union Castle line until the frequency of sailings reduced in the 1970s and stopped completely in 1977. In order to maintain supply, the British Government purchased from Vancouver the mixed passenger and cargo ship Northland Prince which had been used between Vancouver and Alaska. The Northland Prince had launched in 1963, and underwent a substantial refurbishment to create rooms for 76 passengers. Princess Margaret launched her under the name RMS St Helena in 1978, one of very few vessels with the designation 'Royal Mail Ship'. She was used in a minesweeper support role by the Royal Navy during the 1982 Falklands conflict, and she continued to service the island until the middle of 1990, before being replaced by a larger vessel.

Just as the Austin Countryman was given away when we left Zambia, the Peugeot 404 was donated to the Bible Institute of South Africa in Kalk Bay at the end of their final road trip in Africa.

Sheena and James were already in England when we left Zambia and Kristi was staying on at UNZA, so just three of us made the journey. The Peugeot 404 was heavily packed, but our parents arranged a flat part of the large back of the estate car where I could lie down. To help me with the long drive one or two new science fiction novels were given to me as we set off, which was a real treat—though I do not remember any part of them. We had breakfast with Kristi on the Ridgeway Medical Campus of UNZA early on the 4th of December 1975, and then drove South and West to Choma, the capital of the Southern Province of Zambia. South clearly—the drive from Lusaka to Manzini would now be about 1700 km South and a little East. However the direct route involved crossing into Ian Smith's Rhodesia, which was inaccessible from Zambia. So we were headed West, in order to spent ten minutes or so driving through Botswana. Our mother, who had written and volunteered for many years with Africa Christian Press,[2] said she 'spent her last

[2]The Africa Christian Press was established in January 1964 as a new publisher aiming to provide Christian books relevant to the African situation. Our mother was involved in its founding, and worked with them for many years as editor and author [46]. She wrote several books published by the press, though for most of them we only have the titles, which include 'Newtown families' [235], 'Letters to Gabriel', 'Temptations of a nurse', 'Mr Mee escapes', 'What is Christian marriage?', and 'Letters to a student'.

Ngwee'[3] in the Choma Christian Bookshop. Then we drove on, West again, to the Kazungula crossing into Botswana by ferry.

At the time the crossing at Kazungula amounted to threading a political needle. The ferry crossed the wide Zambezi River just past its confluence with the Chobe River, aiming for a tiny landing ramp in Botswana. To the left, protected by barbed wire, was the hostile territory of Rhodesia (now Zimbabwe). To the right, also protected by barbed wire, was the hostile territory of the Caprivi strip which was part of South-West Africa (now Namibia) and under the control of apartheid South Africa. It felt like a tightrope strung between the sane countries of Zambia and Botswana, with madness to each side. After a few minutes drive in Botswana we entered Rhodesia, and had supper in the town of Victoria Falls at a Wimpy bar. The African customers were served food in small parcels to eat on the pavement as they were not allowed inside the building.

We took a last fond look at the Victoria Falls themselves, and drove through the night to Bulawayo and then on to Beitbridge through blazing heat and clouds of flying ants. The insects were so dense that we had to stop several times to clear the radiator grills to stop the car overheating. Once in South Africa we felt the relief of altitude as we drove up into the cooler and greener Soutpansberg and then down to Duiwelskloof (Afrikaans for 'Devil's ravine', and now Modjadjiskloof) where we spent the night at the Imp Inn Hotel.

The next day we drove on to Tzaneen and the Drakensberg mountains before joining the main road to Swaziland. The South African side of the border had two entrances labelled 'Blankes—Nie Blankes', one of many potent symbols of apartheid. The Reservation of Separate Amenities Act, Act Number 49 of 1953 was one of the legal pillars of the structure of apartheid: 'Act to provide for the reservation of public premises and vehicles or portions thereof for the exclusive use of persons of a particular race or class, for the interpretation of laws which provide for such reservation, and for matters incidental thereto.' It made legal the racial segregation of all public premises, vehicles, and services apart from public roads. Clauses inside it included Section 3a, allowing for the complete exclusion of people based on race from a facility, and Section 3b allowing provision made to different races to be unequal. It was repealed on the 15th of October 1990 by The Discriminatory Legislation regarding Public Amenities Repeal Act, 1990, Act No. 100 of 1990:

[3]Prior to independence in 1964, the colony known as Northern Rhodesia used British currency. By 1965 independent Zambia was using the Zambian pound, shilling and pence. The Currency Act of 1967 led to the introduction of the Zambian Kwacha comprising 100 ngwee in 1968. The Zambian pound circulated alongside the Kwacha at a fixed exchange rate until 1974.

Fig. 3.2 The family in Manzini in 1976, with the Peugeot 404 in the background

'Act to repeal or amend certain laws so as to do away with certain powers to differentiate between persons of different races; and to provide for matters connected therewith.' The Swaziland side of the border had a single channel, and we entered the country with a sense of return to normality despite it being entirely new to us. We had fish and chips in Mbabane and reached the University Campus of Luyengo very late that night, only to get lost in the dark and rain. Fortunately Frank Healey, who had been bursar at UNZA and was now bursar of the University of Botswana and Swaziland, happened to drive past, noticed the rather bedraggled car with an insect-splattered metal badge reading 'Z' for Zambia on the front, and rescued us. We stayed for the first few days with David Gooday, who had worked at the Natural Resources Development College (NRDC) in Lusaka, and this began our life in Swaziland (Fig. 3.2).

Manzini

We ended up driving several times between Swaziland and Zambia, but many of the journeys involved flying from Matsapha airport. Initially this meant travelling on the privately owned Swazi Air, which flew Douglas Dakotas dating from the Second World War (Fig. 3.3). In 1978 the Government created

Fig. 3.3 A Swazi Air Dakota in 1973 (Photograph taken by Will Blunt and reproduced with permission of John Austin-Williams of the Dakota Association of South Africa)

Royal Swazi National Airways Corporation, required Swazi Air to close its operations, and purchased a Fokker F.28 Mk.3000.

We started attending the Evangelical Bible Church in Manzini sporadically, with more frequent attendance at the Student sevice on the Kwaluseni campus of the University. We eventually settled into St. Georges and St. James Anglican Church in Manzini.

For some reason—perhaps because of experiences of friends in Zambia who had unhappy experiences at boarding schools in South Africa—I was anxious and uncertain about starting high school in Swaziland. The fact that life in Swaziland also meant more interaction with apartheid South Africa and Ian Smith's Rhodesia weighed heavily. At primary school we had learned very little about these places, though there was a distinct sense of their looming presence and, on occasion, violent intrusions into Zambia or its neighbours. Indeed, some of the maps we drew in order to learn the countries and capitals of Africa simply omitted them both, as if a gigantic inlet opened up at the Cape of Good Hope and extended as far North as the Victoria Falls.

Despite these anxieties, my patient parents managed to arrange for me to sit an entrance examination and then enrol in Year 2 at Waterford Kamhlaba School in Mbabane. They continued to invest huge effort into getting me to and from there. Waterford Kamhlaba was founded in 1963 by Michael Stern, who had a vision of a multiracial school in Swaziland. Waterford was established in clear and expressed opposition to the South African apartheid regime and its laws of racial segregation. It was given the name Kamhlaba (meaning 'of the world') in 1967 by King Sobhuza II Ngwenyama of Swaziland. It became the fourth member of the United World Colleges in 1981. A brief history of the early years of the school was written by one of the original teachers Tony Hatton [147].

My anxiety about the new school was much diminished when I discovered a friendly and familiar face: Chantal Krishnadasan, who had been at Woodlands Primary School in Lusaka with me, was now at Waterford and in my class. This at once made it seem less unfamiliar, and I quickly found my feet there and made several friends, including John Sheehy, who remained a close companion until he left the school in 1979. The university housing we were given was on Syringa Street in Manzini (Fig. 3.2). This was about 45 km away from the school, and for four years there was a complex mix of shared car lifts involving families living in Manzini, Malkerns, and further into the Ezulwini valley. It helped greatly, both financially and in managing all this logistically, that my mother taught Physics at the school several days a week. Fortunately for both of us she never taught me. As I look back, I fear all too often I was a rather difficult person to teach, opinionated and at times arrogant. It is a testament to their tolerance that I am still in friendly contact with several of my teachers at the time.

So it was that I started high school Mathematics at Waterford Kamhlaba in the class of Alan Wicks, and within a few weeks it was this subject that emerged as a favourite. We were following the 'School Mathematics Project' (SMP) textbooks, whose laboured modernity was reflected in abstract images of punched cards on their covers (Fig. 3.4). These grew from part of the 'new Mathematics' educational movement of the 1960s, influenced particularly by Bryan Thwaites [359]. He remains a controversial figure in education, and

Fig. 3.4 The school mathematics project textbooks (©University of Southampton; images taken by Prof. Christian Bokhove and reproduced with his permission)

in 1983 edited a collection of articles on the future of school education [297]. This relatively obscure document interested me because it in part articulates a view on a debate that continues. "[The] sheer pace of technological advance today necessitates change to which the world of education must respond" [297, p. 38] is a good example of a mindset that can place technological advances at the centre of the education picture rather than learning and the individual learner; I have been an advocate of viewing technological advances as simply adding to the toolbox we rummage in when trying to design educational experiences. This edited collection was the subject of a blistering review in the *British Journal of Educational Studies* which ended "Overall, it, but not the issues it raises, richly deserves the obscurity into which it seems likely to fall" [276].

The punched card images were already dated, though the household still had some stacks of punched cards in boxes from computers at the university. At some point in 1977 or 1978 a programmable Texas Instruments 59 calculator appeared, with the capability to store simple programs on magnetic strips [245] (Fig. 3.5). Arguably this TI 59 was the first 'real' programmable calculator in the world. It allowed for 960 program steps and up to 100 data registers (using almost a kilobyte of RAM), some built-in functions including basic statistical ones, solid-state ROM modules, and a built-in magnetic card reader. It allowed me to write some clunky programs, largely aimed at helping my father process marks from his teaching at the University. Around the same time my brother made a tiny programmable Science of Cambridge calculator and I laboriously assembled a Science of Cambridge MK14 computer, both from kits. The latter proved to be an immense hassle, because after many months and airmail letters to and fro, it turned out that the printed circuit board had an error in it. This was readily repaired by scoring through an errant strip of conducting metal once identified, but by then both my patient father and I had been bored to distraction checking and rechecking solder joints and integrated circuit pins. It did eventually work, and had the ability to store simple programs on audio cassettes. More exciting than all these were early signs of what was to come in the shape of some early computers at the Kwaluseni Campus of the university.

For some of the years I spent at Waterford Kamhlaba there were other Mathematics teachers, the names I remember being Chris Mvere, Rose Muir, and Jon Dickens. Each had their own eccentricities. Miss Muir had a penchant for a diet of windfall fruit with little to supplement it, and was I recall treated with great kindness by our mother, who took many troubled and unwell souls under her wing; Chris Mvere was a gentle and kind person who had a deserved reputation for helping anyone with what we might now call 'math anxiety'. The only other incident of note was that after a period of illness

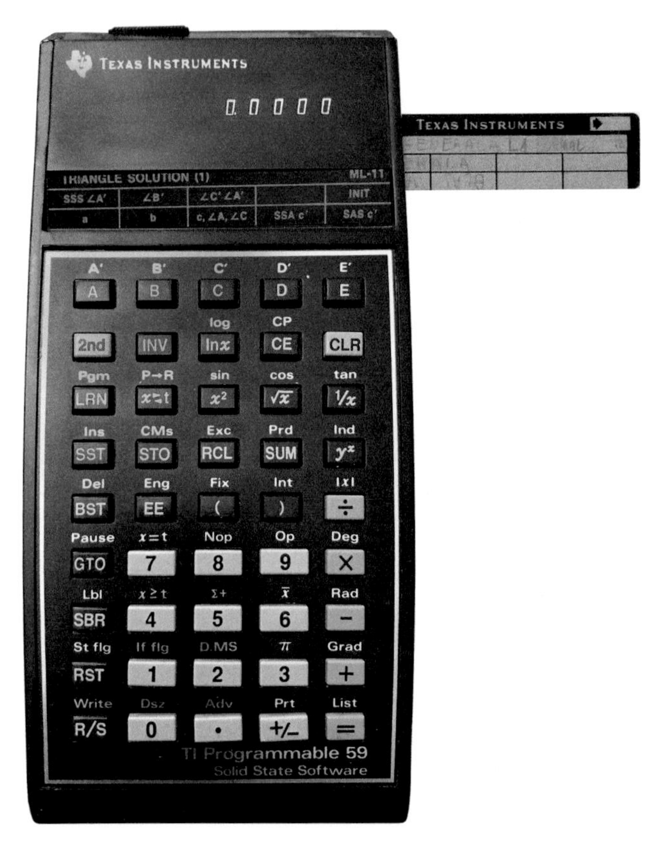

Fig. 3.5 The TI-59 programmable calculator with a magnetic card (Image: Pittigrilli, CC BY-SA 4.0, https://commons.wikimedia.org/w/index.php?curid=115064960)

meant I was not present when that year's SMP textbook was handed out and methodically numbered, Jon Dickens handed me my copy and numbered it 99X with a wink. At some stage another Mathematics teacher, Kip Sumner, ran an informal Mathematics club which added to the mix and further solidified my self-image as being 'a mathematician'. In the end it was the curriculum itself that I started to find interesting, and by the time choices had to be made about subjects it was clear that my route would be sciences, which entailed dropping both History and Geography far too early.

It was and remains a feature of school systems that find their origins in the English one that children are forced to make choices about subjects long before they can have any ability to do so. It seemed to be designed to produce scientists with no feeling for the humanities or social sciences and social scientists lacking any understanding of the magical biological, chemical, and physical principles that make the world the way it is.

More troubling still is the innate asymmetry created by the edifice of knowledge required to understand some of the scientific issues that matter so much to our lives. The resulting gulf—part of the 'Two cultures' identified by C. P. Snow [279]—grows ever wider, despite the many efforts at building inter-disciplinary capacity to address global problems.[4] From vaccine safety to climate change, it is now viewed as completely routine for public discussion to be dominated by commentators who have no conceptual framework for, let alone knowledge of, Bayes' theorem or the concept of thermal equilibrium respectively. A television panel discussing a recent play would normally reflect the voices of people who have read or written some plays, watched some theatre, and know, for example, that the audience generally faces the stage. In contrast a television panel on climate change routinely includes the voices of people with no particle of understanding of any of the concepts required to even formulate hypotheses about the phenomena being discussed. Challenges on this scale are inherently interdisciplinary, and require all perspectives and wisdom of all kinds—but interpreting that as an obligation to give a platform to more or less random nonsense has not been productive.

Wednesday June 16th 1976

Waterford Kamhlaba was founded explicitly in order to defy apartheid, by exhibiting a successful model of multi-racial education just a few kilometres from the border. The malign presence of the South African regime so nearby was a permanent backdrop to the experience of many pupils there. As a white person living outside South Africa I was both sheltered from and privileged within this context, but many school friends were not. Beyond the relentless brutality of the experience of simply living in that environment, some of them endured a litany of imprisoned parents, harassment, house arrests, teargassing

[4]In a follow-up to his essay in 1963 Snow wrote "A good many times I have been present at gatherings of people who, by the standards of the traditional culture, are thought highly educated and who have with considerable gusto been expressing their incredulity at the illiteracy of scientists. Once or twice I have been provoked and have asked the company how many of them could describe the Second Law of Thermodynamics. The response was cold: it was also negative. Yet I was asking something which is about the scientific equivalent of: Have you read a work of Shakespeare's?" I have had the pleasure of working with many academics from the Humanities and have never had such an experience—indeed my experience is that we are all trying to understand a complex world using different perspectives, language, and tools.

and beatings in protests, the endless humiliations of the regime, and targeted assassinations.

While ever present, this was sometimes in the background of life at Waterford. The events of the 16th of June 1976 pushed it to the front of every thought. Schoolchildren across the Johannesburg 'township' of Soweto, many of them younger than we were, began a series of protests and demonstrations in part against the Afrikaans Medium Decree of 1974 which forced 'black' schools to use Afrikaans and English equally as languages of instruction. Estimates vary, but the violent response from heavily armed police killed at least 176 children, and perhaps as many as 700, over the next three days. It was a shattering event, with pupils at Waterford losing loved ones. Even at this distance in time the shock of the images and stories that came out of that day remains.[5]

A House of Books

Among the extraordinary diversity of books at home there was a battered old copy of Silvanus P. Thompson's 'Calculus made easy', a trenchant, opinionated introduction to the practical techniques of the calculus. The full title of this extraordinary little book—first published in 1910 and still in print—is 'Calculus Made Easy: Being a Very-Simplest Introduction to Those Beautiful Methods of Reckoning which Are Generally Called by the Terrifying Names of the Differential Calculus and the Integral Calculus' [294]. It gave the impression in written form of a temporarily irritated person with a chip on their shoulder who deep down was possessed of great patience and skill. Contemporary reviews of the book reflected real umbrage taken at its tone.

> ... in encouraging his readers he continually jeers at the professional mathematician in what might be regarded as reckless nursery language. In spite of such faults, we have no doubt that the book will be useful to schoolboys who need the ideas of the calculus in their study of physical science. The young engineer or the clever schoolboy will think it illogical and slipshod [...] he will probably look upon the introduction of the expansion of $(1+1/n)^n$ when n is indefinitely great, as not quite playing the nursery game. (From the review of the first edition in *Nature* [2])

---

[5]There are multiple accounts of these events and the impact they had. The work of Elsabé Brink et al. tries to explain the context and trace the story from the viewpoint of 25 live witnesses [37].

I learned a great deal from this little book, and my mother was always happy to help explain anything that puzzled me. At times I would sit on the cool concrete counter in the kitchen of our white-walled house in Manzini while she was cooking, and we would work through some of the trickier examples. The approach was neither a simply formulaic presentation of the calculus nor a rigorous 'epsilon-delta' development. Instead it used a partially rigorous language of 'orders of magnitude', in something of the spirit of Gottfried Leibniz's development of the calculus.

Silvanus Thompson was a distinguished electrical engineer and physicist who became a Fellow of the Royal Society, and was a committed Quaker throughout his life. At the Manchester Conference in 1895, a seminal moment in liberal Quakerism, he spoke on the question 'Can a Scientific Man be a Sincere Friend?', culminating in a defence of the idea that not only could science and faith be compatible, they were able to be of mutual benefit.

> We have learned that there is no infallible man, no infallible church, no infallible book. We have learned that creed is not separable from conduct; that a man's religion is not that which he professes, but that which he lives; that our dealings with our fellow men must be judged from no lower standpoint than that of the springs which govern our inmost thoughts and actions. The habit of accurate thought and speech, of letting yea mean yea and no more, which is characteristic of Friends, is one that the scientific method tends ever to strengthen.
> (from the Manchester speech by Silvanus Thompson, 1895)

He died in 1916, like all Quakers appalled by the escalating war in Europe. In several regards, including a positive way to combine faith and science, and a greater interest in practical and applied Mathematics rather than the formal rigour and rhythm of definition–theorem–proof that has dominated parts of pure Mathematics, he seemed to have much in common with my father.

The household also had a good collection of popular science books, including George Gamow's 'One, two, three... infinity' and 'Mr Tompkins in Paperback' [133, 134], 'A random walk in science' edited by Weber [350], a collection of science quotations called 'The Harvest of a Quiet Eye' edited by Mackay [187], 'Flatland: A Romance of Many Dimensions' by Abbott [5], and many others including some of the Time Life Science Library books. Three more books more directly relevant to Mathematics stood out particularly: 'Mathematics for the Million' by Hogben [152], and the two famous personal accounts 'A mathematician's apology' by Hardy [143] and 'A mathematician's miscellany' by his collaborator Littlewood [183].

Fig. 3.6 The CRC handbook our parents used

Another intriguing book—something that makes no sense in the internet age—was a massive green volume that we all referred to as the 'rubber handbook' (Fig. 3.6). This was the CRC Handbook of Chemistry and Physics (originally a pamphlet given away with purchases of rubber aprons by the 'Chemical Rubber Company', with its own interesting history [38]; it is more commonly the 'rubber Bible' but the latter word was probably not used lightly in our household). It comprised an enormous collection of data for researchers across the physical sciences. Whether you sought the characteristic impedance of a vacuum, the molar Planck constant, the surface tension of liquid halogens, the diamagnetic susceptibilities of obscure organic compounds, or the melting points of the transition metals, this was the place to look. The copy we had was the 56th edition [349], and it ended up with me.

What interested me was not the fact that the boiling point of Gallium is 2403 °C—beguiling fact though that is—but the first section, 'Mathematical Tables', which included an extraordinary compilation of indefinite integrals. Long before we really encountered these at school, and certainly many years before I encountered anything like differential Galois theory which gives a precise language for 'types of functions', it fascinated me how functions that seemed of one type (rational, for example) could produce indefinite integrals of an entirely different type. Some of the limited number of definite integrals included in the book seemed to share the same slightly mysterious property, and the issue of understanding the values of specific families of definite integrals reappeared in my life when I encountered Mahler measures [189] and the extraordinary theory of Periods [171]. At the time one of these explicit formulas led me to a primitive series (that is, infinite sum) expansion for π, which Alan Wicks was able to place in some mathematical context. At the time I had no thought of the issues behind this, which start with being clear about how π and the trigonometric functions are defined. In hindsight the most interesting aspect is probably the terminology. When I showed my mathematics teacher the argument that—ignoring all subtleties of convergence and interchanging an infinite sum and an integral—showed that a certain infinite sum converged to π, he called it Gregory's formula. I would now know how to deal with the interchange, and the many ways in which it could go wrong—and would probably call it the Madhava–Leibniz series. What Alan Wicks also showed me was that this infinite sum converged extremely slowly—meaning that a huge number of terms would be needed to obtain a reasonable approximation. By raising the question of whether I could find an infinite sum that was more practical as an approximation he opened the door to a great deal of more sophisticated mathematics.

The rather intimidating weight of factual knowledge in the huge 'rubber handbook' did—despite my lack of interest in Gallium specifically—nonetheless contribute to a layer of adolescent certainty that the physical world was something that was ultimately knowable, like Charles Ryder's 'world of three dimensions [discernable] with the aid of my five senses'.

> 'I have left behind illusion,' I said to myself. 'Henceforth I live in a world of three dimensions—with the aid of my five senses.' I have since learned that there is no such world, but then, as the car turned out of sight of the house, I thought it took no finding, but lay all about me at the end of the avenue. (From 'Brideshead Revisited' by Evelyn Waugh)

The extent to which I was already showing interest in Mathematics before the O-level year is clear from my fifteenth birthday present, which was a Pan Reference Book optimistically entitled 'The Universal Encyclopedia of Mathematics' [3] (Fig. 3.7). The inscription reads 'To our dear mathematical son from Mum and Dad (with help from Aunt Vera and Uncle Douglas)'. The thanks were logistical—purchasing a book like this at the time required some planning, and the assistance of someone living outside Swaziland.

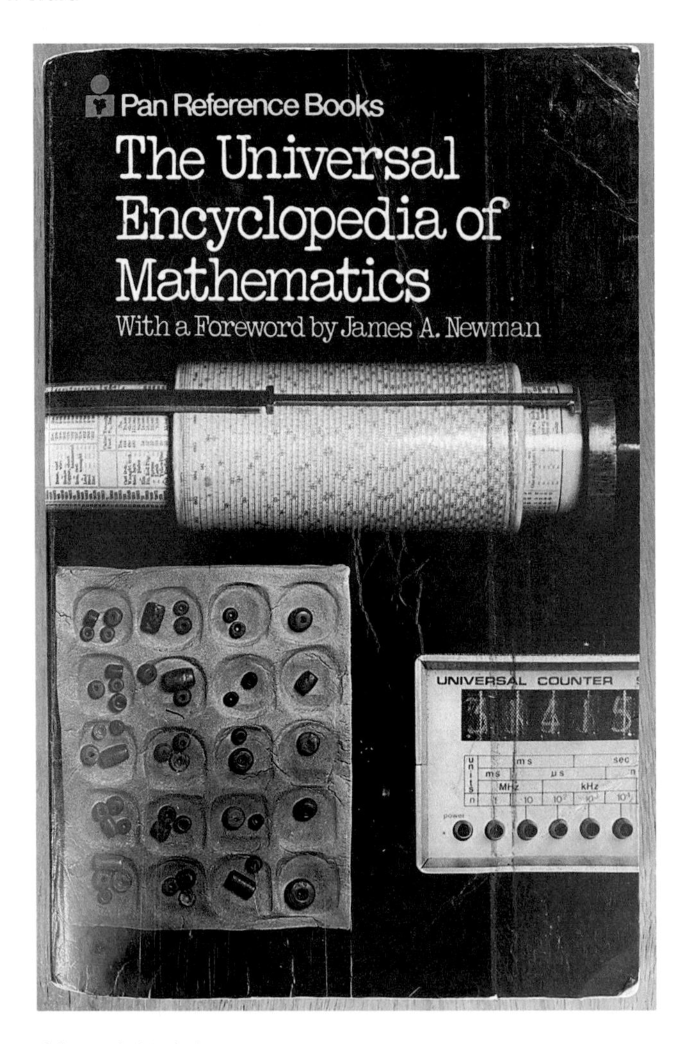

Fig. 3.7 My fifteenth birthday present

In addition to two parents with enormous knowledge of Physics and the Mathematics relevant to it, my older sister Sheena was studying Mathematics and Chemistry at the University of Keele, and showed me some of the things she had learned when home for vacations, including a first intimation of higher-dimensional differential calculus in the form of a book by Baxandall and Liebeck [17], and an elementary introduction to partial differential equations by Stephenson [285]. From the latter I first learned about the formal notions of partial differentiation and ideas like separation of variables for the wave and heat equations in a non-rigorous way.

I also spent a great deal of time reading fiction, and some of the characters and stories I encountered then remain with me. The ancient science fiction stories of Jules Verne and H. G. Wells, and the more recent ones of Ray Bradbury, Arthur C. Clarke, Isaac Asimov, Robert Heinlein, Philip K. Dick, Frederik Pohl, Clifford Simak, John Wyndham, and many others of that period. The Malory Towers stories of Enid Blyton, the Father Brown stories of G. K. Chesterton, the detective stories of Michael Innes, Dorothy Sayers, Ngaio Marsh, and Agatha Christie, the Rabbi David Small stories of Harry Kemelman, and much else became great favourites. Both the diversity and the volume of this reading came directly from our mother, who read constantly and had immensely wide interests.

The rate at which the reservoir of books that have become 'part of the furniture' of my mind is replenished has slowed since then, but never quite ceased. It has certainly always exceeded the rate at which it is being drained. The assumption that the mind could or should be furnished in this way is one of the many things our mother gave us, and it is something far more significant than any subset of the list of authors to which she introduced us.

> There are books that one reads over and over again, books that become part of the furniture of one's mind and alter one's whole attitude to life.
> (George Orwell, in 'Books v. Cigarettes' [234])

The library at home, compiled by our mother from second-hand bookshops in Ghana, Zambia, Swaziland, and England while on leave, was not without its quirks. While the Mallory Towers stories were welcomed, the stories of Noddy and Big Ears were strictly forbidden as our parents with good reason thought them racist. They have been sanitised since, and original editions would shock a modern reader. We had all of Georges Remi's Tintin books, but only in French. The reason for this—a family story that perhaps has grown in the memory over the years—is that it was a ruse to help us learn some French, with the existence of translations into English quietly suppressed. Every now and then an innocent teacher would be taken aback by some of the French we used at school, because it had been learned from the creative vitriolic invective of Captain Haddock.

One of the areas of special interest for our mother was Victorian literature, and the *Girl's Own Paper* in particular. Originally this was a weekly magazine starting in 1880 and costing one penny, published by the Religious Tract Society. Whenever we were on leave in England she slowly accumulated from secondhand bookshops around Dorset a complete run of the large bound annual volumes from 1880 to 1940 (Fig. 3.8). In retirement she built an index

Fig. 3.8 The collection of *Girl's Own Papers* 1880–1940 in our Durham house, Mishpocha

to the fiction and non-fiction in these 62 volumes, an enormous labour of love. The fiction index was published in book form [316], and I converted the underlying data into a simple database that could serve a web front end for the combined fiction and non-fiction index. The first version of the database was built in the HyperTalk language using the HyperCard development kit that was bundled with Macintosh computers at the time. This shared some features of the linking abilities of the World Wide Web as it emerged later. The website was eventually placed on the Lutterworth Press website [315] and I still get the occasional request from a researcher or interested reader through this.

Saturday July the 2nd 1977

Our father was awarded an 'Order of the British Empire' (OBE) in 1977 for his services to science education in the Commonwealth. The ceremony took place at the home of the British High Commissioner, Mr. J. E. A. Miles, in Mbabane. Kristi was in Zambia, but the three other children were able to be there. On the drive to Mbabane we stopped off at the entrance to the main University campus to take a group photograph among the sisal plants and the big concrete sign saying 'University of Botswana and Swaziland' (Fig. 3.9). It originally included 'Lesotho', and eventually said simply 'Swaziland' as the three branches matured and went their separate ways, with the university becoming the University of Swaziland (UNISWA) in 1982. The irony of the

Fig. 3.9 Our parents, Sheena, James, and me at the entrance to the Kwaluseni Campus of the University of Botswana and Swaziland on the 2nd of July 1977

terminology—the Order of the British Empire awarded for a lifetime of effort dedicated to science education in and for independent countries—is more evident in hindsight than it was at the time.

Last Gatherings in Lusaka

By 1976 Sheena was studying at Keele University in England, and we all gathered in Lusaka for Kristi's wedding on the 18th of December that year at Lusaka Baptist Church. Our mother flew up early to help out, and Sheena, James, our father, and I drove up a little later. At this stage—and for many years—the relationship between Ian Smith's Rhodesia and Zambia was close to a *de facto* state of war. This meant that the way to drive from Manzini to Lusaka was to travel West through South Africa, then North inside Botswana, and then use the Kazungula Ferry to cross into Zambia. An eery and partial memory from this long drive, fleshed out much later by my siblings, is of the car stopping in the middle of the night, waking James and me. Sheena had been driving and had stopped because an elephant was blocking the road. Nothing dramatic happened, the elephant moved on, and we proceeded with the long drive along deserted roads. Once in Lusaka we borrowed the old Austin Countryman from Mrs Whitehead to use as the wedding car.

On October the 3rd 1978 our father and I flew direct from Matsapha Airport to Lusaka in the new Fokker F28 of the recently launched Royal Swazi National Airways. There we joined up with my brother-in-law to attend Kristi's graduation from the UNZA medical school with an MBBS and the 'best medical student' award (Fig. 3.10). This was just thirteen years from the founding of the University, and Kristi had been a student there for more than half of its existence.

Sheena graduated from Keele in 1979, and came out to Swaziland with her fiancé for a wedding at St. Georges and St. James Anglican Church in Manzini on the 11th of August the same year.

Reports

School reports in the 1970s were probably more candid, and certainly more caustic, than would now be accepted. Sport—at this stage called Physical Training—did not come naturally to me, and the level of real skill in at least one of cricket, rugby, hockey, or football acquired in a typical South African primary school was at a completely different level to mine. One of the

Fig. 3.10 President Kenneth Kaunda awarding my older sister Kristina her medical degree from UNZA in 1978

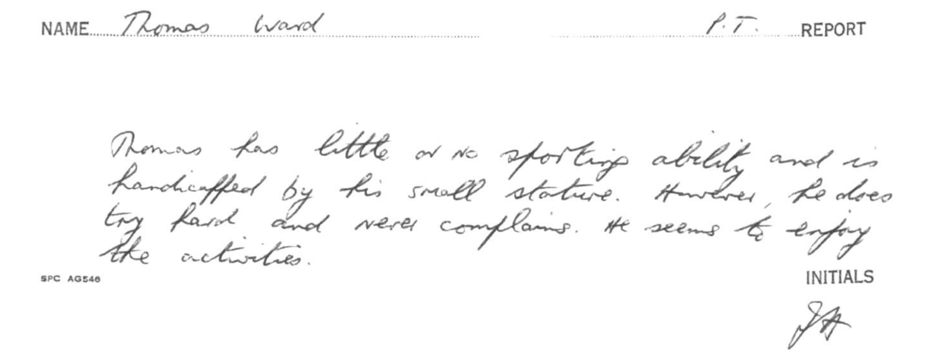

NAME Thomas Ward P.T. REPORT

Thomas has little or no sporting ability and is handicapped by his small stature. However, he does try hard and never complains. He seems to enjoy the activities.

SPC AG546

INITIALS

Fig. 3.11 A school report in physical training from 1976

reports for this subject in my first year at Waterford amused my parents greatly (Fig. 3.11). We all imagined that the teacher in question had conscientiously paused after the word 'little' and remembered his obligation to be scrupulously accurate—leading to the sentence continuing with 'or no sporting ability'. This effort at such a precise calibration of my sporting prowess made my mother in particular weep with laughter.

NAME Thomas Ward Art REPORT

Cycles:	Av. %
1st: C	58
2nd: C	63

Thomas calculatedly shows minimal interest in this non-scientific subject, and has more ability than he would like to admit. Despite his amusing additions to the attendance sheets (NAME: Prof | REASON: Sick of art) he has produced some good work, notably his well-observed onion which had a good sense of solid form to it. His Minotaur composition is flagging at present and he may be genuinely uninterested in the subject set.

SPC AB325 INITIALS CL

Fig. 3.12 An art report from 1979

Sadly a quite serious case of osteochondritis dissecans in both heels led to extreme discomfort during and long after physical activity. It was a relief when this was diagnosed in the form of fairly dramatic X-rays showing the crumbling outer part of the heel bone. This meant I had to sit out for all those lessons apart from swimming, which I much enjoyed, for some years. It was not really until the fifth form that I was able to enjoy sport—tennis and badminton in particular—again.

My pursuit of science disciplines did not preclude studying French and Art and Craft. The talented and patient Art teacher, Colin Lanham, endured what seems to have been my frequent complaints with good humour and managed to drag me to an adequate O-level result (Fig. 3.12).

The rich diet of literature at home was accompanied by studies of 'English Literature' at school. For the last few years before our O-levels I was taught by Andrew Forrest, who managed to get us through the instrumental steps needed to do well in the formal examinations without ever losing the excitement and opening up of worlds through reading (Fig. 3.13). The syllabus made some small gestures towards our location, and so we studied Chinua Achebe along with Shakespeare—but it remained staunchly the product of its history. Certainly the extraordinary incongruity of solemnly studying Edward Thomas' poem Adlestrop on a misty mountain in Swaziland is only clear in hindsight. Part of the enjoyment of the subject was the candid view Andrew Forrest took of the syllabus as an 'appalling grind'.

During 1979, the year I would take my O-level examinations, our parents were uncertain about whether Waterford was the best place for me to do the final two years of high school. They left in their files correspondence with various schools in England asking about the possibility of my joining them. In addition they sought advice from experts in Mathematics education. In 1971

NAME Thomas Ward

Literature REPORT

Cycles: 1st B 2nd B+

Exam/Test: %

Average: 51 %

Tom has a ready grasp of the texts and writes essays which are well-expressed and often very perceptive. I am pleased, also, that the appalling grind of the 'O'-level course has not undermined his enthusiasm or sense of humour.

INITIALS

Fig. 3.13 An english literature report from 1979

the Government of Swaziland and the United Nations Educational, Scientific and Cultural Organization (UNESCO) ran a project to develop in-service teacher training for primary school Mathematics teachers [229, 257]. The aim was to equip 600 primary school teachers who had been employed to teach only the first two or three years so that they could teach Mathematics across the whole primary school curriculum. The approach, built on learning from a similar earlier project in Botswana, involved three college courses of six weeks at William Pitcher College in Manzini, remote assignments, and radio broadcasts. One of the Mathematics education experts sent out to Swaziland to assess the project was Dr Robert Morris, and he suggested it might be worth writing to Douglas Quadling at the Cambridge Institute of Education for advice. This resulted in several long letters full of suggestions and advice for what to do with a 'mathematically precocious' 15 year old in our father's words. It was a real kindness to a complete stranger in another country. In fact Douglas Quadling had already played a role in my mathematical education, albeit indirectly, as he was also an influential figure in the SMP project that defined the syllabus we had followed. In the end I did stay at Waterford, and had exceptionally capable teachers across all the sciences in the 6th form.

Music

My experience of music bridged significant technological developments. Coasters Cottage contained a wind-up gramophone with thick shellac resin records. These included the Skye Boat Song, Shrimp Boats, some Beatrix Potter stories, and songs like The Three Billy Goats Gruff. In Lusaka we listened to a wider range of records, and my older sisters introduced some relatively up to date popular music. In Manzini the horizons expanded considerably with the ability to record onto cassette tapes. Most significantly, my older brother James returned from boarding school in England and introduced me to a huge range of music, which shaped many of my later musical interests. By the end of the 1970s friends at school further expanded what was available, and the era of home-produced—bootlegged—music cassettes really began.

Cinemas

My first encounter with real films, rather than the clattering 8mm silent movies on a home projector, was at the drive-in cinema in Lusaka. The most memorable of these were films like 'Those Magnificent Men in Their Flying Machines', 'My Fair Lady', and 'Chitty Chitty Bang Bang'. By the early 1970s we also started to see American comedies like 'The Love Bug', 'Bedknobs and Broomsticks', 'What's up Doc?', and 'The Hot Rock'. Somewhere among these wonderful memories lies the origin of my lifelong interest in films. In Manzini there was a more conventional cinema that I remember as being called Giulio's, though I can find no trace of it now under that name. There is a cinema there called Julio's which may be what it became. This often showed so-called 'spaghetti westerns', and every now and then something more glamorous like 'Star Wars'.

I must have seen a great many films there, but few stand out in memory in the way those earlier ones did. One that does—both for the ghastliness of what it depicted and for the glamour of Barbara Carrera—was 'The Island of Dr. Moreau', based on the H. G. Wells novel. Wells later talked of the novel being 'the response of an imaginative mind to the reminder that humanity is but animal [...] and in perpetual internal conflict between instinct and injunction', occasioned by the trial and imprisonment of Oscar Wilde, which he described as 'the graceless and piteous downfall of a man of genius'. The novel was published in 1896, and Wells was much influenced by the work of Charles Darwin; some writers have explored how the ferment of ideas of 'making humans out of nature' played a role [151]. It is not easy to shake some of the images—like the puma whose sounds are 'as if all the pain in the world had found a voice' for the narrator—from the memory.

At Waterford there were a small number of film screenings, the most memorable for me being Woody Allen's 'Interiors'. There were also some wonderful plays, ambitious in scope and execution. During my last year there two productions stood out particularly. One of these was a precisely orchestrated production of David Cregan's 'Arthur', in which Mark Eyeington, who became a great friend at Warwick, and Janine Bill who went on to marry my brother, played starring roles. The other was a bold choice of James Saunders' one-act play 'Barnstable' in which my dear friend Alice Stewart starred. The review in the school's annual publication 'Phoenix' by the teacher most interested in drama rightly described this as 'the performance of the play—the evening—the year, even'. I somehow always knew that I had no talent in this direction, but greatly admired the work of those who threw themselves into these productions. While I enjoy speaking in public, and am happy doing so extempore, it is always and entirely as myself.

Warwick University provided a film club, and the city of Coventry had several cinemas and theatres. Films and plays became a more frequent and in some ways less memorable part of my life. It seems strange now, with ready access to a vast catalogue of films on streaming services, to remember how much of an impact a new film could have when seeing it required planning and a trip to the cinema. This may simply be a consequence of scarcity entailing value, as for most of our childhood going to the cinema was a rare treat. It may also be related to the habit of intense engagement with the screen, because there were no other screens in ordinary life. Some films do stand out from the time I spent at Warwick for an extraneous reason, one above all. I met Tania in January of 1986, and our first outing together was to see 'My Beautiful Laundrette', a wonderful period piece about Britain during the Thatcher years based on a screenplay by Hanif Kureishi.

Live theatre, both at the impressive Warwick Arts Centre and the Belgrade Theatre in Coventry, produced many memorable moments. The two that stand out most in hindsight are a student production of Peter Shaffer's 'Equus' and a professional staging of 'Sizwe Banzi Is Dead' by Athol Fugard, John Kani, and Winston Ntshona at the Belgrade. My friend Mark Eyeington and I were able to talk to the English cast afterwards, sharing some of the phrasing and intonation that would be used in South Africa. It must have been an odd sight as we sat in the bar rehearsing how 'to hell and gone' would be said and used in South Africa.

Sixth Form and Universities

After sitting the O-level examinations in the form of the Cambridge Overseas School Certificate in 1979, I entered the sixth form at the start of 1980 (the school year coincided with the calendar year in Swaziland). For several reasons it made sense to board for the last two years at Waterford Kamhlaba. There was a strong sense that day pupils were missing out on much of the life of the school beyond the classroom, which was certainly true. More pressingly, the already difficult logistics involved in getting me to and from the school were going to be impossible as our mother had started working in the Science Pre-Entry Course at the Kwaluseni Campus of the University of Swaziland.

In the sixth form Alan Wicks returned—perhaps from a real absence, or just from my circle of teachers—and took a small group of interested Mathematics students through A-level Mathematics in the lower sixth year. For some of us this meant we could do a full four subjects for the final year: Physics, Chemistry, Pure Mathematics and Applied Mathematics as separate subjects in my case. Mathematics by this stage was unequivocally my natural home, and became very clearly the thing I wanted to 'do next'. Many years later it is still a thing I would like to 'do next', though the remaining time to do so is small in comparison to the time already spent, and too many other interesting and enjoyable things have intervened. When, many years later, I was one of the authors of a book aimed at interested students between high school and university [8], the three authors decided to dedicate it to teachers who had been particularly influential in each of our mathematical journeys. My dedication was to AW, and I had the pleasure of posting a copy to him, living in retirement in New Mexico.

Perhaps unsurprisingly, some subjects came more naturally for me. Reports in Mathematics and Physics in particular tended to be effusive (Fig. 3.14). In the sixth form Physics was taught by Mark Sylvester and Kevin Rosenberg, and I found every aspect of the subject both enjoyable and interesting. The basic conceptual machinery and language of Physics had been part of our lives from an early stage. Weight was not to be confused with mass at our breakfast table, nor power with energy at lunch—and incautious arguments using centrifugal forces were politely corrected to be framed using centripetal forces over supper. The theoretical side of Physics at Waterford was much the same as it would be anywhere, but the practical side had its own features. Some

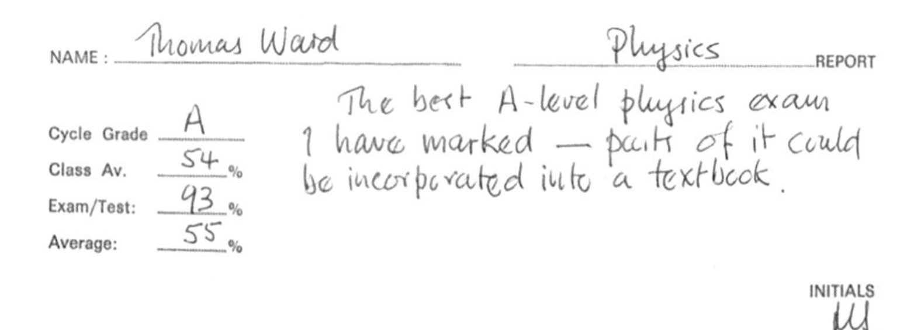
NAME: Thomas Ward Physics REPORT

Cycle Grade A
Class Av. 54 %
Exam/Test: 93 %
Average: 55 %

The best A-level physics exam I have marked — parts of it could be incorporated into a textbook.

INITIALS

Fig. 3.14 A physics report in the sixth form

of the fairly standard electrical experiments had not been designed for that climate, so we spent much time cleaning joints with sandpaper. In both Physics and Chemistry the commitment to enabling students to do real experiments was impressive, and dealing with the consequences of the climate is a good example of what may be lost in modern virtual experimentation platforms.

Mine was the last cohort that sat the A-level examinations at Waterford Kamhlaba, as it joined the United World College (UWC) movement in 1981. The first of these colleges, UWC Atlantic College in Wales, was founded by the German educator Kurt Hahn in 1962. His vision was for co-educational colleges for ages 16 to 19 that prepared students for life as well as further study, equipping them with resilience in particular. He had already founded the *Schule Schloss Salem* in Germany and Gordonstoun School in Scotland and played a part in the creation of the Outward Bound movement and the Duke of Edinburgh's Award Scheme. The UWC movement expanded steadily, adding colleges in Canada in 1974, Singapore in 1975, Swaziland in 1981, the USA and Italy in 1982, Venezuela in 1987, Hong Kong in 1992, Norway in 1995, India in 1997, Costa Rica and Bosnia-Herzegovina in 2006, the Netherlands in 2009, Germany in 2014, Armenia in 2014, China in 2015, Thailand in 2016, Japan in 2017, and Tanzania in 2019. Becoming one of the United World Colleges entailed changing from A-level qualifications in the final two years to the International Baccalaureate (IB), which offered a broader curriculum.

During my final year at Waterford Kamhlaba the question of what to do next grew in importance. The situation was ambiguous, as it was for most people who had lived in different countries. My place of birth was in England but I had never really lived there apart from brief vacations. On the other hand, for the whole family in no other place did we have any permanent right to live—a typical 'expat' experience. My older siblings had gone on to further studies, Kristi staying on in Zambia at UNZA, Sheena going to Keele University and James to the Cheshire College of Agriculture in England. Serendipity intervened in the form of Themba Gamedze. He was a former pupil at Waterford, engaged in post-graduate study in Mathematics at the University of Warwick. He spent a (Northern hemisphere) Summer back at Waterford, doing a bit of teaching, and suggested that Warwick was a good place to study Mathematics. So my parents wrote to the University of Cambridge and to the University of Warwick explaining my problematic residence status and interest in Mathematics.

The two responses were entirely different. Someone at Cambridge sent a stiffly formal letter saying that they would need a statement from a bank showing that my parents could afford to pay international fees for the entire course of study in order to consider an application, which was a complete impossibility. This was 1981, and the decision taken by the first Thatcher government to move to a 'full-cost policy' for international students [9, 362] had raised the stakes of all these questions. Someone at Warwick—a person I met many years later—wrote a long and friendly letter full of good suggestions. This contained a careful argument outlining why in their view I should be eligible for home fees and why they should be covered by Dorset County Council because of the family's home there dating back to the early 1950s. They went on to suggest that if we failed to make this argument successfully they would try and make it for us with the West Midlands. All this may have been an accident of two rather different individuals who opened our letters, or may have reflected something deeper about the history and self-image of the two institutions at the time. It was clear that the starting assumption at Warwick was that someone wanted to help us through the application process, had some insight into some of the specific difficulties I might face, and would do their best to be flexible if possible.

The welcoming response made Warwick the place I wanted to get to, which eventually happened—after a detour. This arose because of the mismatch between a Southern hemisphere end of the school year in December and the Northern hemisphere start of the academic year in October. I had some pleasant thought of spending this period reading in the garden and further extending an already burgeoning fondness for the excessive consumption of wine and John Player Special cigarettes in their slim elegant black boxes, but my parents thought it might be better spent adjusting to life in England and expanding my academic background.

Namibia

One of the more memorable of many road trips we made took place just after my two sixth form years at Waterford. Our mother's first cousin Robin Line, who had hosted us in Chingola in 1966 when we first arrived in Zambia, was working for the De Beers diamond company in Namibia by then, and the family were living in Oranjemund. Our mother, my brother James, and I visited them in December 1981, which involved driving more or less across the width of the continent. The distance—about 1,800 km each way—was not enormous, but it involved driving through some dramatically barren parts of the Karoo (Fig. 3.15).

In those days the town of Oranjemund (meaning mouth of the Orange [River]) was run by the diamond company, and the whole area and coastline

Fig. 3.15 James and me in front of the Peugeot 404 near Port Nolloth, December 1981

was closed to the public. The reason for this—what is now a national park entirely closed to the public—is the presence of diamonds. Some of these are close to the surface on the coast, and in places were simply lying around in the shifting sands. Part of the mining operation involved heavy equipment digging up the beaches at low tide and then getting up and off the beach before the tide turned. Rusting hulks of diggers that had been caught by the tide lay on some of the beaches. The wet sand passed through enormous sieves to extract the diamonds. Other parts of the operation involved sea-going vessels dredging and pumping up sand from the ocean floor near the coast, sieving the sand onboard and dumping the residue back into the ocean. The various names of the area reflect something of both the geology and the history: It is 'Diamond Area 1' for the company, the 'Sperrgebiet' (prohibited area in German), and the Tsau Khaeb National Park after Namibia liberated itself from South Africa.

Further North up the same coast lies the 'Skeleton Coast', an inhospitable long coast of dense sea fogs, constant heavy surf, and very little ability to get to safety across the desert for shipwrecked sailors who made land. The indigenous San peoples call it 'the land God made in anger', Portuguese sailors referred to it as 'the gates of Hell', and the name 'Skeleton Coast' was coined by John Henry Marsh for the title of his account of the shipwreck of the Dunedin Star [191].

The visit was full of extreme contrasts. Enormous wealth in diamonds close to parts of both South Africa and Namibia of great hardship. Warm Orange river water making for pleasant swimming out into the Atlantic ocean with extremely cold waters of the Benguela current on either side. Arid deserts surrounding a green oasis where some employees of the diamond mining company enjoyed green gardens and swimming pools. Dry heat and fierce sunlight close to dense ocean fog. Above all, the reason that the German colonial government had created the Sperrgebiet in the first place in 1908: A vast area, extending 100 km inland and 320 km along the coast, created to control who could collect tiny lucrative objects often a few millimetres across.

When, many years later, I read some of the words Antoine de Saint-Exupéry wrote about a much greater desert, they brought me back to those long straight roads across a deserted and hostile landscape.

> Everything is polarised. Each star shows a real direction. They are all the Magi's stars. They all serve their own God. This one marks a distant well, difficult to reach. And the distance to that well weighs like a rampart. That one denotes the direction of a dried-up well. And the star itself looks dry. And the space between the star and the dried well does not lessen. The other star is the sign-post to the unknown oasis which nomads have praised in songs, but which dissent forbids you. And the sand between you and the oasis is a lawn in a fairy tale. That other one shows the direction of a white city of the South, which seems as delicious as a fruit to munch. Another points to the sea. Lastly this desert is magnetised from afar by two unreal poles: a childhood home, remaining alive in the memory. A friend we know nothing about, except that he exists. (From 'Letter to a Hostage' by Antoine de Saint-Exupéry [66])

The alluvial diamonds at the heart of this extreme environment are themselves somewhat mysterious. The prevailing view seems to be that they originate in kimberlite pipes somewhere unknown in the central plateau of South Africa around 500 km inland, and that over the course of dozens of millions of years were carried into the South Atlantic by several river systems, the Orange river being the largest. Then, the theory goes, the Benguela current deposited them on the shallow ocean floor near the coast and along the coastline. Hearing all this at the time felt a bit like meeting characters from a novel by H. Rider Haggard.

It was a rare treat to spend some time again with our cousin Sarah Line—we could only see her and her siblings at odd moments, once or twice in a decade, and usually only if both families happened to be in England at the same time.

4

Dorchester

Thus I came to spend two terms at The Thomas Hardye School in Dorchester in 1982, which allowed me to do both A- and S-levels in Physics and in Further Mathematics, from a different examination board. It also meant a period of wearing a school uniform, something I had not done since primary school in Zambia. The S-levels originally were a tool used to help the Ministry of Education allocate 400 state scholarships [358]. State scholarships were abolished in 1962, and these examinations were rebadged as 'Special' papers. They were last set in 2001. The same idea was later expressed via the 'Advanced extension award' and 'Sixth term examination papers'.

This meant I had more exposure to a certain kind of Mathematics—techniques in calculus, some partially rigorous sequences and series, even a little finite group theory—than some of the cohort I started university with. Despite my leaning even then towards 'pure' Mathematics (based on a common misunderstanding: In the UK system 'pure' Mathematics at school consists largely of techniques that find most use in 'applied' Mathematics), it was impossible not to enjoy the feeling of satisfaction when a fiddly question in mechanics worked out neatly (Fig. 4.1).

Applied Mathematics at this level—mechanics in particular—is an imaginary world of perfectly elastic collisions, perfectly rough surfaces, massless rods, and inextensible strings. Arguably it is just as remote from the physical world as are the exotic things discussed in the ethereal realm of 'pure' Mathematics. It is easy to construct paradoxical mechanical problems in this world by arranging some of the objects incautiously. A simple example is a massless bicycle tyre balanced on a perfectly rough surface. If you attach a point

T. Ward, *People, Places, and Mathematics*, Springer Biographies,
https://doi.org/10.1007/978-3-031-39074-6_4

11. A uniform circular disc of mass m and radius a is suspended with its plane horizontal by four vertical light inextensible strings, each of length l ($> 2a$), attached symmetrically to the perimeter. The disc is twisted through an angle θ so that its centre rises vertically. Find the work done on the disc, and show that the couple required to maintain it in its displaced position is

$$\frac{mga^2 \sin\theta}{\{l^2 - 4a^2\sin^2(\frac{1}{2}\theta)\}^{\frac{1}{2}}}.$$

Fig. 4.1 One of the mechanics questions I answered on the Further Mathematics (Special) paper 0; June 29th 1982

mass to the edge of the tyre near its top and perturb it, then it will roll. When the point mass is at the bottom, it must be momentarily stationary, so the gravitational potential energy it possessed has simply vanished, as the massless wheel cannot have kinetic energy. Of course including some real parameters gives a realistic model, and the quantities involved do not converge as those parameters (the mass of the tyre and the coefficient of static friction between the tyre and the surface, for instance) tend to zero or infinity.

Yet again I encountered truly remarkable Mathematics teachers, Brian Savage being the one I remember most vividly. They expanded my knowledge considerably, and were always happy to explore things beyond the formal curriculum.

I did not find Hardye's School in Dorchester an easy place socially, though I met some wonderful people there. Visits to and from my brother James at Rycotewood College in Thame helped greatly with the feeling of being far from home. My sister Sheena was living in Glasgow at the time, and James and I drove up to greet a newly arrived niece early in 1982 in his blue Mini 850, a vehicle that required bailing out in heavy rain. In hindsight spending some months adjusting to life in the UK before starting the intensity of university life made things much easier. Even from a school background that was not only anglophone but essentially an old-fashioned version of an English education, there was much to adjust myself to. Partly as a result of my own experience, I now view with immense admiration the courage, tenacity, and flexibility of international students who come to study in the UK in a strange language and culture.

I shared a room in the attic space of a small boarding house attached to the school. The boarding house was a legacy of the close links the school had to the armed forces, in part because of the proximity of the Royal Naval Air Station Yeovilton, the Allenby Barracks, Blandford Camp, and the Bovington Training Area. Sadly an early formative experience came in the form of the Falklands war in the Spring of 1982. One of the younger boys in the house lost a father in the conflict, bringing the tragedy vividly to our door.

The boarding house was hardly a brutal place, though I counted myself fortunate to become a resident there aged 18 rather than 12 or 13. Here I learned to enjoy beer, snooker, and Lambert & Butler cigarettes, and was exposed to colour television and shows like Top of the Pops on a Thursday evening. The low camera angles and short skirts worn by some of the performers on Top of the Pops are one of the things I remember vividly from this period. It all seemed rather sophisticated and glamorous. I also began to learn to navigate the peculiar terminology of meals in England. 'Dinner' was an early and weighty meal comprising two stolid courses, while 'supper' consisted of white toast with jam and large metal jugs of cocoa, all consumed while standing around in the kitchen. I learned later that in England all these matters are subtly freighted signifiers of both class and location, and have long since abandoned any attempt to keep track of it all.

There was also a small and somewhat chaotic library in the boarding house. Two books stand out from this period: 'Darkness at Noon' by Arthur Koestler and 'A Portrait of the Artist as a Young Man' by James Joyce. Each in its own way moved me considerably, and triggered interests both in those writers and in the periods of time they were writing about that remain with me. There was much else to read, because my parents had arranged for me to receive the *Guardian Weekly* and the *New Scientist* in the post, and there was a long letter from my mother most weeks. These letters, carefully typed with multiple carbon copies, were maintained throughout the period any of her four children were living in another country starting in the early 1970s. They kept the whole family connected to a remarkable extent. Something of the imprint of those carefully typed letters remains in the fact that her four children started a Sunday evening catch-up on Zoom during the first Covid lockdown, and that has been retained to this day.

Alongside the interest of studying new things in both Mathematics and Physics, there was a small computer room at Hardye's School with some BBC (British Broadcasting Corporation) Micros (Fig. 4.2). These machines were part of a conscious national effort to raise the level of knowledge about computers in the country. We were taught—or somehow gathered—how to write simple programmes using the 'BBC BASIC' programming language interpreter that was built into its read-only memory. Two of us in particular took to this readily, in ways that perhaps presaged something. A friend—whose name has long since faded from memory—wrote a computer game of the 'PacMan' variety, which required keys to be pressed in sequence at high speed, producing a distinct clatter in the computer laboratory. I wrote simple programmes to plot Fourier series of square wave functions, triggered probably by one of the books my father had bought for me when I left Swaziland. These

Fig. 4.2 A BBC micro from the early 1980s (Image: Public Domain, https://commons.wikimedia.org/w/index.php?curid=11672213)

books were the Schaum outline series volumes on Fourier Analysis, Laplace Transforms, and Calculus. The series developed mathematical material with an emphasis on a large number of concrete problems and worked examples. The graphs I produced were finite approximations to Fourier series, and so illustrated the so-called Gibbs phenomenon, a visually distinctive aspect of elementary Fourier analysis. The ability to add more terms and see the small spike near each discontinuity move closer and closer to the discontinuity without shrinking away, and to see this happening almost immediately, was striking—and provoked mathematical questions that were beyond all of us.

My parents had generously ordered several of the Warwick Mathematics department's suggested reading list of books to be sent to Coasters Cottage. They had leave in England during the Summer of 1982, and we were able to spend some time together in Dorset before I went off to University. Three of these books were particularly influential.

Spivak's 'Calculus' [281], a large square green hardback, began my introduction to rigorous or 'epsilon-delta' analysis. There were interesting exercises throughout, developing little nuggets like Viete's infinite product for $\frac{2}{\pi}$, Cavalieri's argument for integrals of powers, and one of the 'elementary' proofs of the irrationality of π. The book on foundations [287] by Ian Stewart and David Tall introduced the basics of set theory and axiomatic Mathematics in a careful way. If those two comprised the main course and starter of a substantial meal, then the third of these books, 'Concepts of modern Mathematics' [288], was all exotic confectionary. It included at a superficial level enticing topics, including the extraordinary work of Matijasevič on Diophantine representation of primes [194], and the work of Jones, Sato, and Wada [161] giving a polynomial in 26 variables whose positive values coincide with the primes.

5

Coventry

In the Autumn of 1982 I enrolled at the University of Warwick, living for the first year in 'International House', a small modern residence near the centre of the campus, intended for international students. We had somehow arranged for a small tea chest to be left with distant relatives who lived nearby, so I started with more than a single suitcase of belongings. Included in this box was a basic soldering gun and an electric meter, both of them still in use. These were of course gifts from our father; our mother with a different sense of what is important practically gave me a raincoat and a small suitcase when I first left Swaziland.

Despite the benefits of those extra months of pre-university study in Dorchester, the aura and seeming confidence of the other Mathematics students proved to be intimidating. I felt unsophisticated and unsure of social conventions in comparison to the home students, and naturally gravitated to the company of other students who also found themselves in a liminal setting there for one reason or another.

It was only much later that I learned about the fiercely contested political climate surrounding the early years of Warwick University. E. P. Thompson's account of the student occupation of the Vice-Chancellor's office and the Registry in 1970, and the resulting discovery of what became known as 'The Warwick Registry Files' [293] is a fascinating insight into some of the attitudes of the time. The files referred to individual students, but the wider issues raised about the role of the University, how it should balance the interests of different parts of society, and who ultimately should have authority over its destiny are still deeply contested.

T. Ward, *People, Places, and Mathematics*, Springer Biographies,
https://doi.org/10.1007/978-3-031-39074-6_5

I started buying a discounted copy of the *Guardian* newspaper each day at the campus newsagent, which began a lifelong liking for political cartoons. Part of the attraction for me is related to a similarity between mathematics and political cartoons: In both cases there is reliance on an extraordinary efficiency or compression of meaning. In mathematics a simple phrase like 'Let X denote a Kähler manifold' is laden with meaning that would take a considerable length of time to fully explain. Similarly a cartoon like Will Dyson's prescient 'Peace and Future Canon Fodder' from the 17th of May 1919 with its enigmatic caption 'Tiger: Curious! I seem to hear a child weeping!' does not make sense unless the reader recognises French Prime Minister Georges Clemenceau, President of the United States Woodrow Wilson, Italian Prime Minister Vittorio Emanuele Orlando, and British Prime Minister David Lloyd George leaving the Versailles Conference—and knows something of the context.

At the time the *Guardian* carried both the syndicated 'Doonesbury' cartoon by Garry Trudeau and Steve Bell's 'If…' cartoon. Two different examples of the fine tradition of political satire, the former a rapier and the latter at times a scatological bludgeon. Each year I either bought or was given the collection of Steve Bell cartoons in book form, and they are an intensely evocative reminder of the political climate of the early 1980s.

At one of the first lectures—possibly the first one—Prof. Tony Pritchard ended by saying it was a Warwick tradition for the students to stand and applaud at the end of each lecture. This we sheepishly did, victims of a very English form of gentle practical joke delivered deadpan. Many years later I looked him up online in order to send him a copy of our functional analysis book [88], and was sad to discover that he had passed away rather suddenly in 2007 [62].

1982–1983

Starting from the first year there was a considerable amount of choice within the Mathematics degree programme.[1] As undergraduate students we slowly learned that organisational complexity was not something that the department ever feared (Fig. 5.1). The structure of routes through their degree programmes were laid out with great clarity in a series of 'Plan Your Degree Course'

[1] There is no universal terminology for the components of higher education even within some institutions. I have chosen to ignore the actual words used at the time and arbitrarily chosen 'programme' or 'course' for an entire degree and 'module' for a specific topic component.

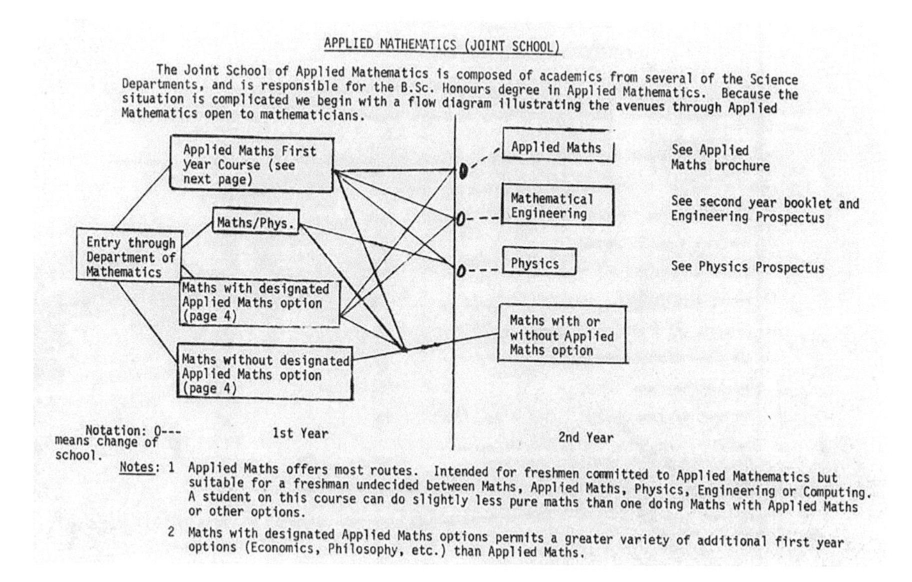

APPLIED MATHEMATICS (JOINT SCHOOL)

The Joint School of Applied Mathematics is composed of academics from several of the Science Departments, and is responsible for the B.Sc. Honours degree in Applied Mathematics. Because the situation is complicated we begin with a flow diagram illustrating the avenues through Applied Mathematics open to mathematicians.

Notes: 1 Applied Maths offers most routes. Intended for freshmen committed to Applied Mathematics but suitable for a freshman undecided between Maths, Applied Maths, Physics, Engineering or Computing. A student on this course can do slightly less pure maths than one doing Maths with Applied Maths or other options.

2 Maths with designated Applied Maths options permits a greater variety of additional first year options (Economics, Philosophy, etc.) than Applied Maths.

Fig. 5.1 The 'Applied Mathematics' routes on offer in the 1982–1983 first year 'Plan Your Degree Course' booklet (With permission from the Warwick Mathematics Institute.)

booklets, and in fact one of the biggest problems during my years at Warwick was having to choose between too many attractive modules.

Despite an early interest in Astronomy and a fondness for school-level Physics, my interests lay in what I understood to be 'pure' Mathematics. In the end the modules I chose comprised Discrete and Continuous Systems I and II, taught by Tony Pritchard, Foundations by Christopher Zeeman, Analysis I by Roger Carter, Mathematical Techniques by Trevor Hawkes, Group Theory by Stewart Stonehewer, Analysis II by David Elworthy, and Linear Algebra; from the Physics department I studied Mechanics, Quantum Phenomena, Electricity, and Probability. This was long before the dead hand of things like the national Teaching Quality Assessment or the systematic shift to a mass higher education system had undermined much of the flexibility and ambition possible in the undergraduate curriculum. The teaching was largely didactic and traditional, generally with well-organised lectures on a chalkboard solemnly transcribed by the audience.

Even at this stage there were indications of some distinct features of the Warwick approach to Mathematics. One of these was weekly supervisions, in which a pair of undergraduates spent an hour with a post-graduate student going through feedback on homework in detail. The second was a specific piece of work, developing an iterative approach to learning how to write clear Mathematics. This involved a written proof of something non-trivial being read several times by a personal tutor, with feedback for improvement at each stage. Mine was something to do with word cancellation in free groups, and had its own frustrations because several steps were somehow easy to 'see' but difficult to write down clearly and explicitly. I learned a great deal from the careful reading of each attempt by my rather precise personal tutor, Anthony Manning. The lessons in converting an idea that is relatively easy to believe into something written in formal language was valuable, a form of apprenticeship in the art of mathematical writing.

We often think of key conceptual steps in mathematics in technical terms like understanding set-theoretic notation or using existential quantifiers and predicate logic, but those are simply pieces of the language of mathematics. That mathematics is expressed in a language, written in sentences constrained by a rich and subtle grammar and carrying a narrative to a conclusion, is a deeper truth that lies behind all of it.

Exploring more deeply the journey from intuition to rigour is a lively and important part of the subject, with significant errors coming to light in some parts of the Mathematics literature. There is a distinct feel that the Mathematics of various—for some people all—past eras is viewed by some as lacking rigour, while the current golden age has supposedly rid itself of such problems. It has not been easy for mathematicians to confront the many problems created by a naive belief that just because the subject is presented as a network of paths that proceed from axioms to conclusions via logically certain steps does not make it so in practice. One of the recent moves to redevelop the subject using computerised theorem provers under the banner of 'Formalising Mathematics' (a particularly persuasive exponent of this is Kevin Buzzard [41]) is starting to bear fruit. It is conceivable that Mathematics may become a domain in which artificial intelligences lead the way, but what seems more likely to happen within the foreseeable future is for the habit of developing Mathematics within a formalised environment using a 'proof assistant' to become commonplace. What is undeniable is that the subject does need to address some problems: There are important statements in number theory (for example) that are believed in some communities but not universally accepted. Being unable to definitely answer the question of whether the published proof of a definite statement is correct or not says much about the sheer complexity of parts of the subject—and is clearly unsustainable.[2]

I returned to Swaziland for the Christmas holidays in 1982, and something did not work out in the timings of my return journey via Johannesburg. This meant that I missed a small examination in January 1983, and the first formal mark recorded against my name that would actually count towards my degree at Warwick was zero.

At some stage I also attended a module called 'Introduction to Quantitative Economics'. This was engaging and entertaining, but in the end unsatisfying. Even with the modest experience of life at my disposal, the attempts to link real human behaviours and motivations to utility graphs and the like seemed implausible and unenlightening, even in aggregate. The phrase was not I think used, but the idea of building a theory predicated on the existence of *Homo economicus* seemed problematic and naive, particularly without much consideration of whether this interesting creature existed in the first place. I was intrigued to read much later about a more evidence-led version of that nagging anxiety about the performative nature of some economics teaching. Roughly speaking, graduates of economics degrees of this sort are more likely to behave in accordance with the ideas of rational choice theory, but the general population are inclined to make decisions on much more complex grounds, and in particular to be more altruistic than the theory suggests they should [15, 135, 242].

The cost of travel to Swaziland meant that I stayed at Warwick for the Easter vacation throughout my time there. These sustained periods of quiet revision with mainly international students around me each year became a particularly enjoyable part of the year. A University campus outside term-time has its own feel for a student, and the small numbers of us there developed close friendships.

Outside my formal studies and a great many hours spent in the Student Union bars playing pool and drinking beer, my time was taken up with voluntary organisations like the Anti-Apartheid Movement and some desultory political engagements. The formal political life of the Students Union was

[2] I am neither competent nor eager to add more words to the contested territory surrounding the so-called 'ABC conjecture'. A non-expert view is something like this. ABC is a simple to state conjecture in number theory with enormous implications and many important consequences. There is a published proof in a special issue of *Publications of the Research Institute for Mathematical Studies* of Kyoto University. Serious questions have been raised about parts of the argument, and the community of experts in anabelian geometry is split on the question of whether they have been addressed adequately. An anonymous commentator called HM on one of the blogs discussing this expressed the state of the issue particularly neatly: 'Does it become a political statement whether one assumes ABC?'

both noisy and exciting, as it was a period in which differing visions of society found particularly clear expression in party political terms. Elections for the student sabbatical officers who would lead the Student Union for the year were dominated by political 'slates' aligned with national political parties rather than cohering around issues directly related to higher education.

The 1983 general election took place at the end of my first undergraduate year, and it was a febrile time of considerable polarisation. The Labour Party was led by Michael Foot, and its manifesto 'The New Hope for Britain' was described as 'The longest suicide note in history' by the Labour Member of Parliament Gerald Kaufman. The manifesto grew in part from Michael Foot's determination that it would comprise all the resolutions of the party conference. This meant that it committed a possible Labour government to unilateral nuclear disarmament, more progressive personal taxation policies, abolition of the House of Lords, withdrawal from the European Economic Community, and the return to public ownership of the recently privatized British Aerospace and British Shipbuilders Corporation. The left-leaning vote was diluted by an alliance between the Social Democratic Party (SDP), founded by breakaway Labour MPs, and the Liberal Party. The SDP had been founded in 1981 by four prominent Labour party member who had held cabinet posts before the Conservative election win of 1979. Their concerns with the direction that the Labour party was taking found explicit expression over nuclear disarmament and withdrawal from the European Economic Community, but was part of wider set of tensions about the role of an entryist Trotskyist group called Militant Tendency in the Labour party. The public announcement of the departures was made by Roy Jenkins, David Owen, Bill Rodgers, and Shirley Williams near the home of David Owens in Limehouse, and became known as the 'Limehouse Declaration'. The tensions between those who view the Labour party as the vehicle for the creation of a social democratic society closely integrated into the European Union running a mixed economy and those who see it as the route to a socialist economy with widespread public ownership moving away from a neo-liberal European Union are still with us. The Conservative Party had a difficult hand to play economically, but presented a simple vision and possessed the strident clarity of the Falkland Islands conflict. The outcome was always expected to be a victory for the incumbent Prime Minister Margaret Thatcher, but the scale of the 144 seat majority was decisive. This enormous parliamentary power influenced much of the next decade and beyond.

I was more interested in the individual than the collective, and perhaps that is what made me sign up as a volunteer with 'Nightline' (a campus version of the Samaritans) in my first term in 1982. This remained a significant part of my life through all of my seven years at Warwick, and it had a huge impact on me in several ways. Tania and I met through Nightline, when she also volunteered a few years later, and I was the trainer of her group. Nightline gave me a direct insight into the prolonged and profound suffering that mental health problems can cause, and how much of an impact a little human compassion, a modest amount of training, and some truly terrible instant coffee can have. Less positively, it led to many years of disrupted sleep patterns. It also demanded, and largely achieved, scrupulous anonymity on a relatively small campus.

1983–1984

The wide range of choices expanded in the second year, and I had some difficulty in deciding quite what to do. In the end the modules I chose were Introduction to Linear Systems Theory, Automata and Formal Languages, Differential Equations, Variational Principles taught by Klaus Schmidt, Topics in Algebra by Colin Rourke, Applied Sources of Pure Mathematics by John Rawnsley, Metric Spaces by Dr Lima, Galois Theory by Charudatta Hajarnavis, Measure and Integration by Bill Parry, Calculus in Higher Dimensions by Dai Evans, Geometry and Topology by John Jones, Advanced Calculus by George Rowlands from the Physics department, and a module on ecology called Science and the Natural Environment.

Three of these modules stood out particularly. The basic machinery of rings and modules and canonical forms for various notions of similarity of matrices over rings in the module taught by Colin Rourke was compelling and useful to me later. His lectures were full of flair and entertainment, and we much enjoyed several not entirely successful attempts to carry out non-trivial calculations of normal forms live in the room. The module on Geometry and Topology developed fundamental ideas like Klein's Erlangen program and the classification of surfaces; many years later John Jones and I served together on the London Mathematical Society Publications Committee. The third module that stands out in memory was Bill Parry's module on Measure and Integration. This was my first real encounter with early twentieth century analysis, the notion of Lebesgue measure in particular, and the measure-theoretic underpinnings of probability. Measure Theory is one of those parts of mathematics that really develops a new way of thinking, and becoming confident about initially strange concepts like statements that are true 'almost everywhere' or sets that can be large viewed in one way and of size zero viewed in another was both enjoyable and important for me later.

In addition to the taught modules there was a compulsory essay. Mine was on the fundamental group in topology, with some simple examples of homotopy arguments. The essay was handwritten, an almost inconceivable idea now. It seemed extraordinary to me that so much meaning—and such powerful tools leading to such important results—could be extracted from the superficially floppy notion of homotopy equivalence.

At some stage I also briefly attended short modules on Astronomy. They did not rekindle my childhood enthusiasm—the magic of looking through an eyepiece on a warm Zambian evening was not readily matched by calculations of the mean free path of photons generated in the core of a star, interesting though that may be.

The UK miner's strike began early in 1984, and grew to become a defining conflict between the Government and the National Union of Mineworkers. My view was not nuanced: The proposed pit closures, and the wider feeling of a conflict deliberately sought out to rebalance society away from the power of the collective, seemed entirely wrong. My anorak, as a result, bore several 'Dig deep for the miners' badges and I joined in various protests about the issue. Many Student Unions tried to find ways to support the strike, but were sharply constrained by the legal limits prescribing their activities. We all learned the meaning of the doctrine of *ultra vires*, and many Student organisations were on the receiving end of threatening letters from the Attorney General, Sir Michael Havers. Dreadful things happened during the strike, and the divisions within communities and families created at that time remained a powerful presence in the lives of some of the people we got to know when we moved to the North-East of England many years later. In hindsight, I would simply say that while the issues at play were immensely complex, the fact that income inequality has grown more or less relentlessly ever since, and that it is now viewed as unremarkable to have foodbanks in every town in this relatively wealthy country, is not unrelated to the decisive political shifts of that time.

1984–1985

My choices for the final undergraduate year at Warwick in 1984 were as follows: Algebraic Topology taught by Brian Sanderson, Applied Functional Analysis by Anthony Pritchard, Differential Geometry of Curves and Surfaces by Jim Eells, Fourier Analysis by Klaus Schmidt, Group Theory by Roger Carter, Linear Analysis by David Evans, Modern Control Theory by Diedrich Hinrichsen, Qualitative Theory of Differential Equations by David Rand, Representation Theory by Sandy Green, and a module in Complex Analysis

Fig. 5.2 The Google books Ngram chart for the frequency of occurrences of the phrase 'Catastrophe theory' between the year 1880 and 2020 in the literature visible to Google (Data source: https://books.google.com/ngrams/.)

taught as a reading course led by Ian Stewart. Strictly speaking this was two more than required, but even getting down to this list had already involved painful choices. I started but dropped an introductory module in Algebraic Geometry taught by Miles Reid, and could not fit in Catastrophe Theory with Christopher Zeeman, Topics in Mathematical Physics with John Rawnsley, nor the option of an extended essay.

I did buy the large orange paperback book on Catastrophe Theory, as it was so much talked about at the time [368]. This consisted mainly of a collection of papers from the 1970s covering both the Mathematics and some of the ways in which the theory had been used as a model for various physical, biological, sociological, psychological, or economic phenomena, edited and collated by Christopher Zeeman. It was my first real encounter with a mathematical theory that went through a period of being intensely fashionable and then seemed to become less relevant (Fig. 5.2). Indeed the mathematical theory came under some criticism and the many applications in other disciplines generated some negative reactions.

Missing out on even an introduction to Algebraic Geometry left a real gap in my knowledge. Even at the most naive level—I am doubtless viewed with pity and suspicion by 'real' algebraic geometers, thinking as I do of points and zero sets of polynomials rather than more elevated concepts—the basic tools and language of the commutative algebra involved proved important to me later. Happily I am in extremely illustrious company in holding this naive view of the magnificent edifice that is algebraic geometry.

> In support of his view of Zariski as a geometer for whom modern algebra could provide, at best, only a "temporary accommodation," Weil recalled a well-known anecdote about a conversation at a party between Zariski and Claude Chevalley,

a French algebraist who had moved to Princeton from France just before the war. The two men were discussing algebra versus geometry in algebraic geometry when at long last, exasperated, Zariski exclaimed, "But when someone says 'an algebraic curve', surely you see something!" "Yes of course," Chevalley quietly replied. "I see this: $f(x, y) = 0$." (From Carol Parikh's biography of Oscar Zariski [240, p. 80])

The Fourier Analysis module taught by Klaus Schmidt was remarkable for me in several ways. It explained and gave a wider context to my early efforts to visualise the Gibbs phenomenon using a BBC micro-computer in Dorchester years earlier, and it provided some insight into genuinely subtle problems in analysis of great importance in the history of Mathematics. More importantly, while Klaus dutifully did a thorough job of developing some of the classical theory, every now and then his lectures started to illuminate a broader canvas and some twentieth century mathematical ideas. In particular, it was in this module that I first encountered things like harmonic analysis on locally compact abelian groups, the orthogonality relations for characters, maximal ideals in function algebras, and the mysterious objects called 'solenoids' that were to occupy much of my later mathematical life. This material also began a real appreciation in me for aesthetic aspects of mathematics. The English mathematician G. H. Hardy famously said 'Beauty is the first test: There is no permanent place in this world for ugly mathematics', and some have suggested that functional magnetic resonance imaging experiments show that the subjective experience of mathematical beauty engages the same parts of the brain as does appreciation of fine art or music [371]. It is impossible to single out any one thing, but in hindsight it was this type of mathematics that is closest to the impact that Emerson had on Whitman for my life as a mathematician.

I was simmering, simmering, simmering; Emerson brought me to a boil. (Attributed to Walt Whitman by John Townsend Trowbridge [300])

The range of options available was remarkable—beyond my own interests there was a rich offering of other modules in Mathematics, as well as ones in engineering, computing, social sciences, arts and humanities, and much else besides. Few Mathematics departments would now be able to offer such a replete menu of choices.

My brother James was with me when I walked up to a display board with the degree classifications pinned up outside the Arts Centre in the centre of the Warwick campus on the 27th of June 1985, and he said he could tell from a distance that I was pleased with the outcome. My degree performance in fact led to a modest departmental prize of £25. I put this towards the purchase of my first camera, a Praktica MTL 5B. This was a manual single lens reflex camera, with the sophistication of 'through the lens' exposure metering. The lens mount was a simple M42 screw thread, and I slowly accumulated some other lenses. It seemed to be a piece of almost indestructible East German engineering, and was only retired to a box in the attic, still working perfectly, when I bought a fully automatic Minolta SLR in 2000. The Minolta also proved to be remarkably robust—despite its complexity it is still in use, in the hands of our son-in-law, who retains a liking for film cameras and vinyl long-playing records.

My links to Swaziland were shrinking more and more as life in England became more familiar, and as the friendship group from Waterford Kamhlaba scattered to many different countries. Our parents were starting to plan for retirement and a return to the UK, but were both still active at UNISWA (Fig. 5.3).

Students attending a lecture at the University of Swaziland.

UNIVERSITY OF SWAZILAND

EDUCATION

Tertiary Education

The University of Swaziland (UNISWA) which came into being in 1982, has two campuses, one at Kwaluseni and the other at Luyengo. The University was formerly part of the University of Botswana and Swaziland, which was formed after the breakup of the University of Botswana, Lesotho and Swaziland, founded in 1964.

The University of Swaziland, as an institute of higher learning in the country, is committed to:

- The development of high-and-middle-level manpower required for the developmental needs of Swaziland. To this end, a number of suitable study programmes are offered.
- The promulgation of development-orientated research, with particular reference to Swaziland. To this end, a number of research institutes and units have been established in the University.

Fig. 5.3 A page from the Swaziland Review of Commerce and Industry 1985 showing our father lecturing on the Kwaluseni Campus of UNISWA. By this stage both 'Lesotho' and 'Botswana' had been removed from the concrete sign at the entrance to the campus

After Graduation

I had no coherent ideas about what to do next, but had set my heart on a doctorate. When I raised this possibility with my personal tutor Anthony Manning he was not encouraging, more in the spirit of general advice than being critical in any way. This was sensible really: There are many stages when the role of the wiser head should be all encouragement, but nobody should make the decision to start a doctorate unless they have an inner drive to do so capable of overcoming much discouragement.

I knew little about how to apply for a place for doctoral studies—the sophistication of some current undergraduates applying to multiple institutions, sometimes in several countries, with well-developed ideas for doctoral research and sometimes a serious research record themselves, impresses me greatly. I also had little awareness of any wider mathematical world, so applied to do a PhD at Warwick with chosen topic 'Linear Analysis'. There was no science behind this really—it was probably triggered by a feeling that the classification of 'approximately finite-dimensional' algebras using K_0 groups in the eponymous undergraduate module seemed exciting and sophisticated. In those days the Mathematics Institute at Warwick had a block of PhD grants from the Science and Engineering Research Council (SERC), and these would be allocated through a competitive process between all eligible applicants. Happily I had done sufficiently well in my final exams to be awarded one of the coveted funded places.

Two unexpected things then happened, one major and one minor. In the early Summer of 1985 it seemed the world was my oyster. I had won a funded place for postgraduate study, giving me confidence about what I would be doing for the next three or four years. My then girlfriend and I were making plans and exchanging currencies for a holiday on Inter-rail around Europe for a few weeks of the Summer. Shortly before we were due to set off, a disaster arrived in the form of a letter from SERC on the 2nd of August. Having grown up in several different countries, there had been some difficult moments persuading various arms of the UK government that I met the criteria for home fees. This had in the end all worked out for my undergraduate studies, but SERC decided I was not sufficiently British to be eligible for one of their awards. We reluctantly decided that I simply had to stay in England to try and sort this out, so my girlfriend set off on our planned holiday with her sister—and I started writing letters and talking to the Mathematics Institute. The head of the Institute at the time was Bill Parry, and he and the Institute Secretary May Taylor gave as much help as they could, both morally and practically.

My parents—at the time in Swaziland—did what they could remotely. This all proceeded at modest speed, with physical letters to and fro. In great anxiety and some desperation I even contacted the local Member of Parliament, John Butcher. As far as I can tell, his views and mine would be diametrically opposed on every policy question the country faced then or since—but his response was energetic and exemplary. The experience left me with an abiding respect for the great effort that lies behind the phrase 'a good constituency MP'. I was one of his constituents, so he did all that he could. It was never clear which letter, person, or telephone call was the last straw for SERC, but in the end they relented in the form of a letter on the 19th of August saying they were 'able to process' my case and as a result I would indeed be allowed to take up one of Warwick's PhD grants. So I stayed in Coventry and began my post-graduate life—somewhat dishevelled and tired.

A few years later at a dinner involving some of the faculty from the Mathematics Institute I was taken to task by one of their partners, who thought it quite right that someone with so tenuous a connection to the UK should not have been allowed to take up such an opportunity. Perhaps they had a point—I have certainly felt the obligation to pay something back ever since.

The second incident was much more minor, but in hindsight represented a fork in the road of great significance for me. For reasons that never quite became clear to me I was initially assigned to Klaus Schmidt rather than David Evans as a supervisor, while my friend David Pask was assigned to David Evans. At one point someone suggested this was because David Evans had too many students at that time and Klaus too few, but it never seemed important to find the exact reason. Also starting at the same time with Klaus was Zaqueu Coelho, who worked on subshifts of finite type. Klaus seemed to tolerate my blundering early mathematical efforts with good cheer, and I had enjoyed his module on Fourier Analysis just as much as the operator algebras David Evans had shown us in his Linear Analysis module, so I was quite relaxed about all this.

Some time later the three of us, David Pask, Zaqueu Coelho, and I all ended up as students of Klaus, studying in three entirely different areas of dynamics (Fig. 5.4). David worked on ergodicity of skew products over irrational rotations, but returned to his first interest of operator algebras after completing his doctorate. Zaqueu worked on central limit theorems for shifts of finite type, and continued to work at that kind of interface between probability and dynamics. My area followed naturally from my Masters dissertation, namely the dynamical properties of compact group automorphisms. During the same period Klaus was working with Bruce Kitchens on a far more systematic

Fig. 5.4 Klaus Schmidt with his three students in 1987: Zaqueu Coelho, David Pask, and me (from right to left) (With permission of Jacqui Ramagge.)

development of the link between these systems and commutative algebra, with a significant publication [167] appearing in 1989.

The system at Warwick then was—and perhaps still is—in some ways a hybrid of a North American doctoral program with taught examined modules and a British one, which can involve plunging into research. There was a good range of taught modules offered each year, and a multi-stage route through the program which brought in its own complexities (Fig. 5.5).

The approach helped bridge the gap between the BSc, which by most international standards was too short and had simply too little material, and becoming an effective research student. These curricular problems were particularly acute in pure Mathematics which, to a greater extent than most other disciplines, is precariously (and at times vertiginously) reliant at each stage on the solidity of earlier stages. Algebraic Geometry is a perplexing mystery if your grounding in Commutative Algebra is shaky, Functional Analysis relies on basic Real and Complex Analysis, much of the statements of Ergodic Theory cannot really be formulated without some Measure Theory and Functional Analysis, and so on. There was a significant price to pay though. At the time the doctoral awards from the research councils like SERC funded three years of post-graduate study, and the structured Masters degree in the first year left too little time for most students. This has improved

-18-

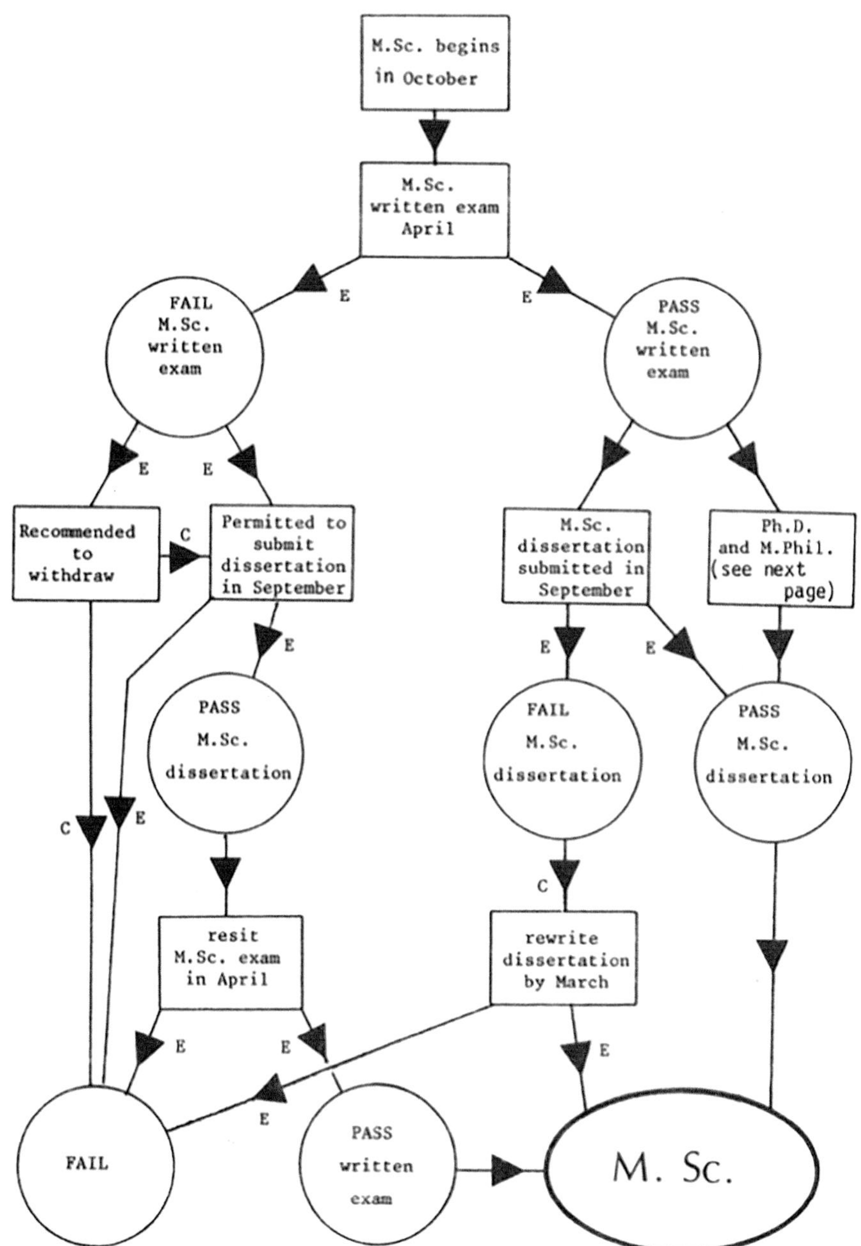

Fig. 5.5 The route to an MSc at Warwick (from 'Plan your higher degree course' 1985–1986) (With permission from the Warwick Mathematics Institute.)

considerably since then, but the canonical route to a doctorate in the UK remains at the lower end of time in full-time study by international standards (Fig. 5.6).

As had been the case in the third year of the BSc, there were too many interesting and attractive modules to choose from in the first year of postgraduate study. Influenced perhaps by the fact that Klaus was my supervisor, I decided to do the two paired modules he planned to teach, and it is not much of an exaggeration to say that a great deal of my later professional life relied in an essential way on ideas from those two modules. My final selection was this: Ergodic Theory taught by Peter Walters, Locally Compact Fields by Klaus Schmidt, Complex Dynamics by Anthony Manning in the first term, and Number Theory taught by Klaus Schmidt in the second term. I also attended parts of a module on Manifolds by Caroline Series and one on Numbers and Dynamics by Bill Parry.

Anthony Manning had been my undergraduate personal tutor, and had been the source of sound advice on several occasions. Most memorably, when I first tentatively suggested I was interested in pursuing a doctorate, his measured and sobering response was "Don't assume that everyone who gets a first [class degree] will get funding"—long pause—"and don't assume that everyone who embarks on a PhD completes one"—even longer pause—"and certainly don't assume that everyone who gets a PhD will get a post-doctoral position"—another pause—"nor that anyone with post-doctoral experience will land an academic job." Sage words certainly, then and now. Arguably the likelihood of translating a PhD into an academic job are even slimmer now. In most countries post-graduate volumes have grown far more than the number of academic posts available. On the other hand, we were given very few opportunities to develop skills or confidence for any future other than an academic one as post-graduate students in the 1980s. A modern doctoral centre is usually designed to equip the students with both the skills and the expectation of a wider range of future professions.

Once again I am struck in hindsight by the rich diversity of taught modules and their level—some were in effect third year undergraduate modules with some additional reading, but many were specifically aimed at masters students. The module in Ergodic Theory taught by Peter Walters followed his book [303] closely, and these meticulous lectures began my life as 'an ergodic theorist'. It was in the two linked modules taught by Klaus that I first learned about Pontryagin duality, global fields and local fields, adeles and

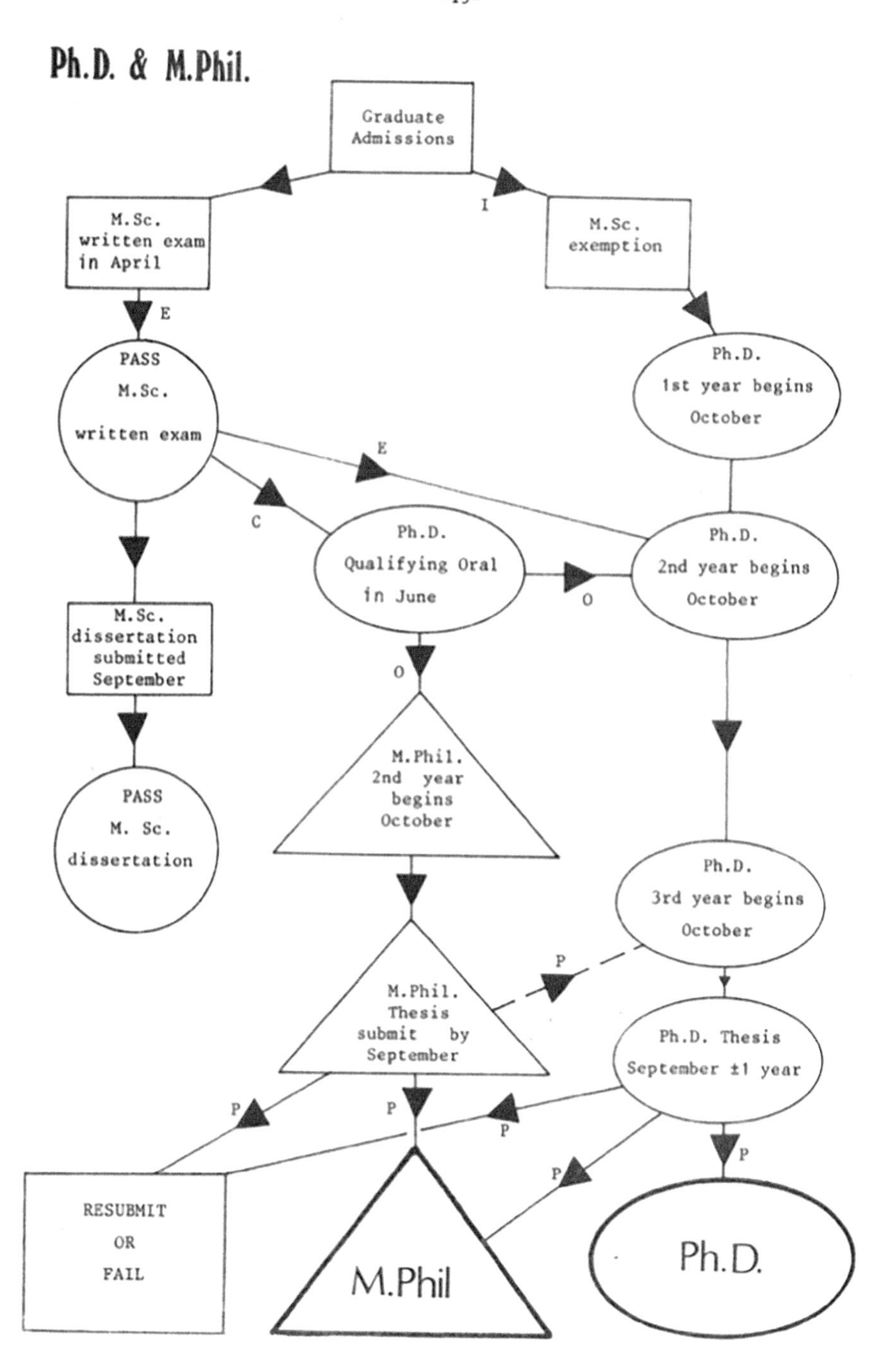

Fig. 5.6 The route to a PhD at Warwick (from 'Plan your higher degree course' 1985–1986). (With permission from the Warwick Mathematics Institute.)

ideles.[3] This particular slice of twentieth century number theory influenced my mathematical interests greatly, and is ancient lore that will remain with me to my last hour.

Where possible, some of the masters modules offered at Warwick were aligned with research interests of faculty members or planned research activities. At the time I did not really understand quite what a centre of mathematical activity Warwick was—and had been for decades. Klaus was at the time interested in André Weil's elegant book 'Basic Number Theory' [351] for his own reasons, and there was a large conference in Smooth Ergodic Theory planned for the Summer of 1986 as part of a longer symposium running from January to August of that year. Outside that, there were regular seminars, including one on a Tuesday afternoon on the topic of Ergodic Theory and Dynamical Systems.

Wednesday July 16th 1986

During the Smooth Ergodic Theory conference Klaus said he wanted to introduce me to someone who had an interesting idea to discuss. Looking back, this was the first substantial instance of the generosity of others that has so shaped my mathematical life.

The person was Doug Lind from the University of Washington in Seattle, and we had a conversation at one of the low tables in the coffee room of the Mathematics Institute at Warwick (Fig. 5.7). The idea was indeed interesting, and I was ideally placed to make some progress with it—as was Klaus. Seeing the attractive problem, the insight that the machinery needed for it could be found in the number theory he had been teaching, and having an existing long-standing friendship with Doug Lind, Klaus handed this wrapped gift to his young student.

Doug wanted to expand on an idea whose origins lay with his own earlier work on exponential recurrence for compact group automorphisms and an insight from Hillel Furstenberg into the geometry of group automorphisms.

He sketched out the idea: Entropy calculations for compact group automorphisms reduce to the case of solenoids, and previous approaches worked

[3]Class field theory is a branch of number theory that describes certain extensions of fields using objects connected to the field. In 'local' class field theory the group of units of the local field plays a central role. In 'global' class field theory, the idele class group takes on the role of the group of units. The terminology of idele (or idèle) for an 'ideal element' (or id.el.), and 'adele' (adèle) for 'additive idele' was coined by the French mathematician Claude Chevalley in the 1930s [48]. The power of these ideas is that they enable all possible completions of the field to be considered simultaneously and on an equal footing.

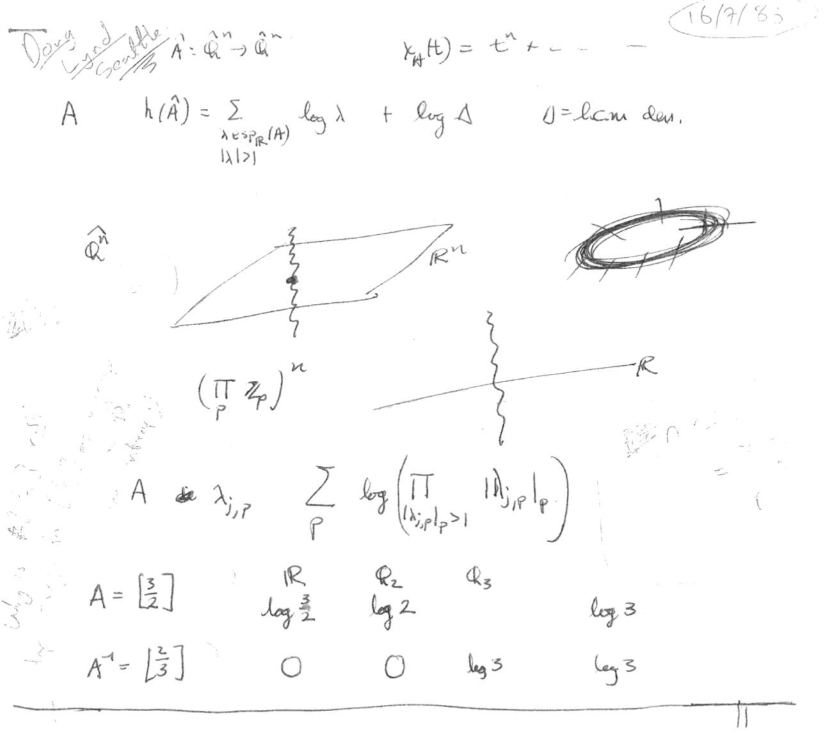

Fig. 5.7 My first meeting with Doug

directly inside the complicated geometry of the solenoid. He had already used some of these ideas—that the way in which a compact group automorphism dilates distances locally may be invisible viewed in the familiar notion of distance in directions that look like the real numbers, but visible in directions that look like Cantor sets—in several ways, including proving exponential recurrence in this setting [179]. On the other hand, a slightly flexible approach from the 1950 thesis of John Tate [292] to using adeles as covering spaces for a certain type of solenoid suggested a simpler approach. This idea of 'localizing' or 'lifting' a messy calculation into a linear one all worked out, resulting in my Masters dissertation [319] and my first publication [181] (Fig. 5.8).

This first research collaboration marked a transition in multiple ways. A change from student of Mathematics to practitioner in a modest way certainly, but how the collaboration was carried out marked something more significant. It began with the two of us sat at a table in Coventry, and continued by air-mailed letters. In one of them Doug suggested we switch to electronic mail,

The length of the dissertation should normally be not more than 30 pages double-spaced on A4 paper. The more original the dissertation, the shorter it could be.

It is essential that the dissertation be legible. It should normally be typed; but poor typing can be harder to read than neat handwriting.

Fig. 5.8 Sound advice on the presentation of the MSc dissertation (from 'Plan your higher degree course' 1985–1986) (With permission from the Warwick Mathematics Institute.)

tally, if you have access to the computers there, I can be reached by electronic mail at the above address. This is quick*, and would vastly simplify communications.

Best regards

* $\leq$ 1 day.

Doug

Let me strongly urge you to try getting electronic mail working. A colleague here is collaborating with someone at Cambridge this way, and it works fine. To get you started, from Cambridge this fellow uses the address:

"bass@entropy.ms.washington.edu AT rl.earn"

↑ spelled out.

Fig. 5.9 Footnote to a letter from Doug in April 1987 (With permission from Doug Lind.)

which the more adventurous and knowledgeable students were already using (Fig. 5.9).

My Masters dissertation was typed on an electric typewriter, with mathematical symbols added in black ink using a ruler and ball-point pen (Fig. 5.8). Our first exchanges of letters were hand-written. Doug quickly started writing his letter in the mathematical typesetting program TEX, and the joint paper was written using this. For some reason I persevered with bundled software on the Macintosh computers available in the Mathematics Institute, eventually writing my dissertation using long-obsolete software called WriteNow.

Doctoral Studies at Warwick

My mathematical time at Warwick was spent in the old Mathematics Institute at Gibbet Hill. This shared a site with Biological Sciences a short walk from the main campus. As an undergraduate it was simply a university building much like any other, though it was notable in hindsight that it had a great many photographs on the walls of the attendees at conferences spanning many different fields of Mathematics.

For a post-graduate student it had a stronger identity, with its own rich culture and habits of life as we navigated a way through our studies (Fig. 5.6). At the time life in the Institute included coffee served in China cups with saucers each morning from 10 to 11 a.m., tea served each afternoon from 3 to 4 p.m., and a croquet set for the use of postgraduate students. I was never consciously a scholar of the nature of the Institute itself, nor of its history—but much about it bore the clear imprint of the people who founded it. Most influential of these was Professor Sir Christopher Zeeman, FRS. Others who know more, and can express themselves more eloquently, have written extensively about the ways in which his vision shaped the Mathematics Institute [367]: As a research centre which grew to have global influence certainly, but also in a multitude of other ways, some going well beyond what is normally thought of as part of the role of a founding head of department. Starting in 1965, the Mathematics Research Centre (MRC) began a series of year-long research symposia. It was one of these that enabled me to meet Doug Lind, for example, a tiny piece of an enormous story of international collaborations and research initiatives that were started, supported, or enriched by conversations at Warwick. Many of these resulted in influential conference proceedings, those in dynamical systems being familiar sources for some influential publications. At one point Zaqueu Coelho, David Pask, and I, along with a few other PhD students, were placed in a large office that had once been used by Prof. Zeeman. The boxes on the floor and behind desks still contained copies of some of those conference proceedings particularly close to his diverse interests [49, 63, 190, 252].

The Institute also reflected some of Zeeman's ideas as a physical space, with a substantial library, a large common room, small cubicles for 'supervisions' and, most famously of all, a set of five houses and two flats just behind the Institute, for longer-term research visitors. These were built in the late 1960s, designed by Bill Howell. They are now listed[4] and won the RIBA Architecture Award in 1970. Many years later Tania and I stayed in one of these with our young daughter and one of my PhD students, and enjoyed the distinctive features which included blackboards running around all the walls of the bedrooms.

It is a third aspect of Zeeman's legacy that mattered most for me. The second half of my career has been spent holding responsibility on various executives for the education a university enables, and everything I have read about his vision shows that his passion for education was serious, long-lived, and deeply influential. I suspect he did not make the lives of people in my sort of role particularly easy, because there were doubtless many courteous battles fought over how his vision did or did not fit with university policies, or the various waves of infuriating and largely irrelevant external regulatory scrutiny. His requirements for optional module choices in the undergraduate mathematics curriculum gives a flavour both of the breadth of his own intellectual landscape, and of a certain unequivocal style.

> Rule 1: mathematicians must spend at least 50% of their time in the Mathematics department.
> Rule 2: they can do anything they like with the other 50%.
> Rule 3: no more rules.
> (From [367])

Part of his commitment to the centrality of education was expressed in his determination to teach every Mathematics undergraduate. For me this arose in the first year, where he taught a module called Foundations. A module with that title aimed at first year undergraduate Mathematics students could mean almost anything, ranging from an arid but precise articulation of elementary logic to a vague meandering over various issues at a superficial level. In the hands of Prof. Zeeman this was instead one of the highlights of a long period as a student of Mathematics. Topics included some basic logic, a careful construction of the real numbers from the rational numbers using Dedekind cuts, and an account of Newton's argument showing how the inverse square law for gravitation implied Kepler's laws of planetary motion. For a young Mathematics student feeling the lure of 'pure' Mathematics but not then—nor since—fully recovered from an early infatuation with astronomy and physics,

[4]A 'listed' building in England means one that has been placed on the statutory list maintained by a non-departmental public body of the Government called Historic England. This is done for buildings of particular architectural or historic interest, and means that it may not be demolished, extended, or altered without special permission.

it is hard to think of anything that could be more interesting at that early stage. Contemplating many years later the way in which teaching was the first thing rather than the last thing squeezed out of my life by the relentless demands of various university roles, I am in awe of how tangibly real Zeeman's commitment to education remained over long periods of time.

The Mathematics Institute was also a sociable space in multiple ways. Friendships and relationships formed readily among the postgraduate students, and the fact that the Institute was a little distance away from most of the campus helped create a sense of community. Klaus and his wife Annelise were extremely hospitable, and hosted regular convivial dinners for his students and their partners.

Klaus initially suggested I read Zimmer's elegant account of rigidity phenomena in actions of Lie groups [372], but it quickly became apparent to both of us that my background in Lie theory was inadequate—absent would perhaps be a more accurate word. So I started to study the systems he had been working on with Bruce Kitchens, looking at growth rates of periodic points in a certain type of 'algebraic $\mathbb{Z}^d$ action', meaning a dynamical system in which several different maps that commute with each other all act. Klaus spent the first six months of 1988 (during my third year as a postgraduate student) at the Institute of Advanced Study in Princeton. We had by then become interested in an entropy problem, calculating the entropy of actions of $\mathbb{Z}^d$ by automorphisms of compact groups. My Masters dissertation was concerned with a specific calculation for the case $d = 1$. However, entirely new ideas were needed for $d > 1$, and for some time we did not even have an expectation of how this would work out. At various points I had erroneous arguments suggesting that the answer for a non-trivial system was always zero, or always infinite, and it took me some time to really understand what the issues were. By then it was routine to communicate by email, and around Easter of 1988 Klaus emailed from Princeton to say that he and Doug thought they had made some real progress. The simplest of the systems we were all looking at were determined by a single polynomial with integer coefficients in several variables, and they had the sketch of an argument that suggested the value of the entropy was given by something called the logarithmic Mahler measure of the polynomial, an expression introduced by Kurt Mahler in connection with inequalities between various notions of size or measure for polynomials in several variables [188, 189].

Proving this result and dealing with the general case involved a great deal of work really carried out by Doug and Klaus. I was receiving updates by email, and at times struggled to keep up because so many different ideas in commutative algebra, perturbation theory, and measure theory came into the picture at high speed. The result, along with many related ideas, questions, and examples appeared in print after I had started my first post-doctoral year at Maryland [178]. The identification of entropy for this class of dynamical systems as a Mahler measure has been influential in multiple directions, and it was a significant step in our understanding of such actions. Mahler measures already had a rich history, and in particular computing the Mahler measure for families of polynomials brings in many different ideas in Mathematics.

We use the word 'computing' in many different ways in Mathematics, and what was of interest here were 'closed formulas', how these changed in families of polynomials, and insight into the polynomials whose Mahler measure was zero. It is easy (for the right sort of numerical analyst—in fact like every branch of Mathematics doing this well is a sophisticated business) to 'compute' the Mahler measure of a given polynomial to 50 or 100 decimal places. It is a different type of argument to 'compute' one in the sense of showing that it is equal to an expression involving other known constants like π, or values of some classical function at specific points. Some of the arguments involved are sufficiently difficult that the former numerical calculations helped inform some of the latter exact ones.

The idea here is something close to a mathematical belief in a benign deity. If a 'small' polynomial (for example, in three variables and having no variable appear to degree higher than two and coefficients no larger in absolute value than 10) has a numerically computed Mahler measure that agrees to 50 decimal places with the decimal expansion of some simple expression involving a classical object (an 'L-function') attached to an elliptic curve with small coefficients up to a rational factor with rather small numerator and denominator, it is certainly enough evidence to try and prove the closed formula that the numerical arguments have hinted at. There are few numerical coincidences of this sort that are deceptive, but they do exist. An extraordinary aspect of the Mahler measure is that pursuing this idea brings in many beautiful and seemingly unrelated parts of Mathematics.[5]

At the time we called on work of Chris Smyth [278] for some explicit examples in which he was able to compute the Mahler measure in terms of certain associated Dirichlet L-series of odd characters, objects of importance in Number Theory. Even the question of when the Mahler measure vanishes—a classical lemma due to Kronecker in the case $d = 1$ of a single

[5]An introduction to some of these surprising identities and directions may be found in work of David Boyd [32] and Christopher Deninger [68].

Fig. 5.10 Lind, someone, and Schmidt at the Lorentz Center in Leiden in September 2014 (With permission from Evgeny Verbitskiy)

variable—becomes a non-trivial result, proved by David Boyd [31] and Chris Smyth [277].

I was much amused many years later at a conference, also at Warwick, when Shmuel Friedland cited the Mahler measure entropy result as being due to 'Lind, Schmidt,... and someone'. We all hope to be someone at some stage of our lives, and this was my unexpected route to the accolade (Fig. 5.10). During this conference I found myself by chance with the four people who had been most significant in helping my journey from undergraduate to academic and managed to record the moment (Fig. 5.11).

One strand of this work was a use of knowledge of periodic point data to determine when the systems being studied had an extremely strong mixing property called 'completely positive entropy' or 'K' (for Kolmogorov). This was important because one of the seminal examples in the subject due to François Ledrappier [175] was a two-dimensional action of an algebraic sort that was mixing but not mixing of all orders, and had zero entropy. In some sense—and with the considerable benefit of hindsight—it was this example, contained in a note of just over two pages, that contained the seeds of many

Fig. 5.11 Klaus Schmidt (my PhD supervisor), me, Peter Walters (who first taught me ergodic theory), Doug Lind (my first research collaborator), and Anthony Manning (my undergraduate tutor) at an Ergodic Theory & Number Theory conference at Warwick in April 2011

later developments. As presented originally it comprised a simple and elegant construction of a dynamical system with some unexpected properties. Viewed in a different and superficially more complicated way, it was a special case determined by some data from commutative algebra. Changing that data produced a huge diversity of different systems and phenomena, and this was the domain of research mapped out in the work of Kitchens and Schmidt. Two of the main questions raised in [178] were in this circle of ideas.

The first question was this. Is 'completely positive entropy' for this class of systems equivalent to a stronger property called 'Bernoullicity', essentially amounting to the strongest possible form of randomness? This was to be expected—having the former property without the latter was certainly possible to find in the vast *terra incognita* of measure-preserving systems in general, but would not be expected in such a constrained and algebraic setting. Needless to say, the gap between what is expected and what is known comprises the beautiful and sometimes hostile landscape of research in Mathematics.

The second question—not explicitly raised in the paper, but raised earlier in work of Klaus [269]—was to determine when these systems had a weaker property, called 'mixing of all orders'.

6

Seattle, Shuffleboard, Vitaly

During my last year as a doctoral student the opportunity arose to attend a conference at the University of Washington in Seattle. This was an 'NSF-CBMS Regional Conference in the Mathematical Sciences' event entitled 'Algebraic Ideas in Ergodic Theory', comprising a series of lectures by my supervisor Klaus Schmidt. The event had been organised by Doug Lind, my first collaborator, and Selim Tuncel, who had also been a doctoral student in dynamics at Warwick a few years earlier. Part of the structure of these events is an expectation that the principal speaker produce a set of lecture notes [270], and these lecture notes helped me understand something of the wider world my own work lived in.

There was no real funding available, but the Warwick Mathematics Institute agreed to help with a small sum in return for my marking the first year analysis examination. This would all be viewed as dreadful exploitation now, but at the time I was grateful for both the experience and the opportunity.

Tania and I were living in Bordesley Green, Birmingham at the time and the journey began with a taxi ride. The whole event was heavy with meaning for us, because we already knew that I was going to spend the next academic year in Maryland while Tania stayed in Birmingham teaching at Shaw Hill Primary. I explained all this to the friendly taxi driver, who refused any payment and wished us well for the future.

The conference introduced me to a good number of people who have remained significant figures in my life (Fig. 6.1). It was both scientifically informative and convivial, perhaps too much so. One of the evenings was particularly significant for my future. A group of us ended up at a bar which

T. Ward, *People, Places, and Mathematics*, Springer Biographies,
https://doi.org/10.1007/978-3-031-39074-6_6

Fig. 6.1 My first experience of travelling to a conference: 'Algebraic ideas in ergodic theory' in Seattle, July 1989 (With permission of the University of Washington Mathematics Department.)

sold pitchers of locally made beer, where we were able to play shuffleboard. Even then there was a large gap between bland, barely identifiable corporate product and characterful locally produced American beer.

As the evening progressed the more sensible people departed. I eventually found myself at one end of a long wooden table silently toasting someone I had not yet met. By this stage the pitchers of beer had been abandoned, and we were drinking more or less randomly, with much more enthusiasm than wisdom.

My companion turned out to be Vitaly Bergelson, a name already familiar because of his beautiful 'weakly mixing PET' paper [21]. We talked about the Mathematics I was working on, and Vitaly suggested I visit him in Columbus and apply for a temporary position there. These 'Research Instructor' positions had already been held by many young ergodic theorists, and this would be joining a long list of people I would later meet at conferences and workshops.

I honestly do not know how much weight this conversation carried for later events, but I did indeed work at Ohio State later. I remain troubled by the ease with which habits and channels of informal contacts in academic life repeatedly default to settings that are not readily accessible to young researchers who feel unwelcome in such a setting, for whatever reason. Alcohol is often the unintended mechanism through which the social side of academic life becomes less welcoming to women, or to people of specific faith or national backgrounds. I take no pleasure in remembering that I have been an active part of this culture to a much greater degree than I have been a challenger of it. Certainly all the evidence available points to enormous and entrenched problems with enabling everyone to access opportunities to work in higher education, and there is much that needs addressing. A recent striking study on some of the mechanics at work in the nearby discipline of Physics deserves to be widely read and acted on [365].

7

College Park, Maryland

Doubtless because of Klaus' influence, I was appointed to a post-doctoral year with the dynamics group in College Park, Maryland for the academic year 1989–1990. At the time the group with the National Science Foundation funding that made this possible comprised Mike Boyle, Misha Brin, and Dan Rudolph. College Park had been a centre of activity in dynamics for many years, and there were other faculty members working in the area. Among these were Joe Auslander and Nelson Markley, who worked in abstract topological dynamics, Ken Berg, one of whose early results on measures of maximal entropy had been generalised in the work with Doug and Klaus, Hsin Chu, who had largely worked in the area of topological groups but had done early work on ergodic properties of affine transformations of compact groups, and several others. It was a lively and friendly place, with a large number of visitors and regular weekend workshops in dynamical systems held jointly with Pennsylvania State University. Given the problem that arose during my doctorate of the relationship between completely positive entropy and Bernoullicity, my primary post-doctoral adviser was Dan Rudolph. Nonetheless, the whole group were actively helping and welcoming.

Our move to the USA began several parallels to my father's early career in my life, as his first post-doctoral position was also in another country. Despite the fact that my own life had by then been in four different countries, our mother's view was that "for the first time I think one of you is going somewhere really foreign". She also sent me the sensible advice that the Church Historian Frederick John Foakes Jackson gave to a young academic in Cambridge at the

T. Ward, *People, Places, and Mathematics*, Springer Biographies,
https://doi.org/10.1007/978-3-031-39074-6_7

turn of the century: “It’s no use trying to be clever. We are all clever here; just try to be kind. A little kind.”

Before I moved to Washington someone in the dynamics group had, as a favour, moved my teaching assignment from a module in the usual ‘calculus’ sequence to something more interesting. This turned out to be a higher level Statistics module, and I spent the flight hastily learning some from the book specified, a text by Peggy Tang Strait [289]. I had learned a little Probability as an undergraduate and stationary stochastic processes arise naturally in ergodic theory—but I knew almost nothing about Statistics. This was my first experience of the powerful mechanism of learning by teaching, and it was a particularly interesting one because I had no feel for the way in which theory and practice fit together in statistical reasoning. It also involved frequent use of the Greek letters β and θ, and the way I pronounced these produced a wonderful, confused, cross-cultural moment. For me these are /'bi:tə/ (bee-ta in the International Phonetic Alphabet) and /'θi:tə/, while for the students they were /'beɪtə/ (bay-ta) and /'θeɪtə/. The member of faculty in charge of teaching called me in to go over some feedback on my lectures, and said that one of the students was particularly complimentary, writing: “this new German lecturer speaks very good English”.

Joe Auslander and Ken Berg met me off the flight from London at Dulles airport on the 21st of August 1989. Washington felt blazing hot and humid to me, and my English clothing was not really appropriate—but my J-1 visa let me in, and my post-doctoral life began. The logistics of life in a new country were much aided by the kindness of others as well. Misha Brin drove me a considerable distance to obtain a social security number; Mike Boyle had already helped me find a shared house on Lewisdale Drive in Hyattsville with two post-graduate Mathematics students, Joe Wissmann and David Pierce. Later in that year both Joe and David helped me obtain a Maryland driving licence, which involved driving David’s pick-up truck around an artificial driving course laid out in a parking lot. After completing a PhD in Maryland, David moved to Turkey where we later visited him and his wife Ayşe Berkman, also a mathematician.

The three of us lived an austere life, with no air conditioning and little in the way of heating. On hot days this provided a strong incentive to travel onto the campus a few miles away on a free shuttle bus early in the morning. I was in a shared office with metal desks in a basement, and enjoyed wandering the large building with its cool corridors and spaces. Seminars were frequent, as were various workshops and weekend conferences. Tania made a brief visit around Easter of 1990, and met some of the friends I had made there. One of these was Robert Miner, a PhD student in computational geometry. Robert was so fascinated by Tania's description of a 'chip buttie' that he insisted on setting out to make one on the spot. So the three of us piled into his car, found suitable sliced white bread and butter in a local shop, collected some French fries from a drive-through MacDonalds, and came home to manufacture chip butties. A modest moment in the annals of cultural exchanges, but an enjoyable and memorable one. Sadly Robert passed away in 2011 in Minneapolis.

Early on at Maryland I decided I had to learn how to use the mathematical typesetting program TEX, so bought a copy of the grey ring-bound book 'The Joy of TEX' by Michael Spivak [282] in the campus bookshop. In advising a student now one would point them at existing examples on the internet, and would use LATEX, with its automated cross-referencing, but in 1989 the internet was a very different thing. This habit of copying and modifying existing ways to do what one needs to in LATEX is so widespread that there is something like a 'phylogenetic tree' of code fragments. It would have multiple points of origin of life, but the bulk of it would represent LATEX documents containing many fragments of other, earlier, pieces of code, often vestigial in the sense that their purpose has long been forgotten or obviated by packages developed later.

So my first efforts began with the early pages of Spivak's book and an empty text file. From the moment I first successfully compiled a TEX document, every word of Mathematics I have written has used it or one of its derivatives.

1989–1990

My first post-doctoral year happened to coincide with an extraordinary time for the world. The Soviet Union was in rapid disintegration, and a few months after I started work in Maryland the Berlin Wall fell. Berlin had been the home of my mother-in-law until she escaped as a *Kindertransport* in 1939, and the city retained a powerful hold on our family ever since. The fall presaged enormous changes across the Soviet Union, and an extraordinary feeling that the threat of nuclear conflict that had shaded the world since 1945 was lifting slightly. By February of 1990 Nelson Mandela had been released from the Victor Verster Prison in Cape Town and the long journey to some kind of democratic government in South Africa had started. There was a distinct feeling that we could look at each other at celebrations on New Year's Eve of 1989 in the knowledge that freedom, democracy, and self-determination for many people had expanded or been envisioned with credibility in its reach across the world. There have been very few years since where the same could be said.

Dan

The experience of working with Dan Rudolph for a year was extraordinary. Rankings and 'best at' statements about something so vast and multi-faceted as even a tiny branch of Mathematics are absurd: We are all humbled by the discipline and the challenges it brings. Nonetheless, I am comfortable joining many others in saying that Dan was one of the most profound thinkers, and one of the most powerful, creative practitioners in measurable dynamics who ever graced the subject (Fig. 7.1). Many people better placed than me to do so have tried to capture something of what was extraordinary about Dan and his Mathematics, including Mike Boyle and Benjamin Weiss in an obituary note [34]. It is tempting to quote a great deal of their brief account—with some effort I will edit this down to a handful of sentences. The first is about Dan the mathematician, the second about Dan the person.

> His relentless positivity was humbling and inspiring. In a problematic colleague, he would see the part to admire; with a problematic student, he would find some path to success. Dan did not write people off. He brought out the best in the people he knew.

Sadly Dan became ill with amyotrophic lateral sclerosis (ALS) a few years later. The University of Maryland organised a workshop on dynamical systems in honour of his 60th birthday in 2010, but he passed away shortly before it

Fig. 7.1 With Dan Rudolph on the College Park campus in 2005

started. As a result it became a memorial to a remarkable life. The range of participants in research area, in age, in national origin, and in number reflected the scale of Dan's contribution both as a mathematician and as an encourager of others.

> At one ALS website, a member posted a list of 100 things that made his life worth living, which he was losing one by one; he said he would kill himself when only 50 remained. Dan said, when he lost one, he would find another one to put on the list. (Also from [34])

I learned a little of what is sometimes called 'Ornstein theory' from Dan, but perhaps my experience was more about encountering an 'Ornstein–Rudolph approach'. A way of thinking about the sometimes elusive structures that remain visible in dynamical systems when you have lost the spectacles that allow you to see geometry, or algebra, or even points.

Dan's Mathematics as expressed in the form of formal research publications, while always creative and at times dazzling in its brilliance, could be dense and technically demanding. His Mathematics as expressed in person in seminars or privately was fluent, dramatic, compelling, and persuasive. To a greater extent than anyone else I encountered, he brought his whole self—a once in a generation brain, a beautiful physical presence honed by dedicated effort in modern dance, and a spirit of exceptional generosity and size—into the process of doing or explaining Mathematics.

The knowledge that he was taken from the world so young makes it painful even many years later to recall some of our conversations. Many concerned the Mathematics I was trying to do as a stumbling young post-doctoral researcher, where he was a notably generous guide. The ones I remember most vividly were about his broader vision, and in particular how much enjoyment he found in the idea that people of many different kinds could all contribute to Mathematics. The problems thrown up in the subject are so vast and varied that there is ample space for the wild-eyed visionary, the intrepid explorer, the pedantic and cautious botanist, the methodical builder, the technician, the expositor, the artist, the archivist. After some time with Dan, anyone would feel confident that they too had a place, a role, or a contribution to make in Mathematics if they wished, be they fox or hedgehog,[1] bird or

[1] The poet Archilocus wrote 'The fox knows many things, but the hedgehog knows one big thing', and this distinction was used in a light-hearted essay by Isaiah Berlin [22], who reasonably said 'Every classification throws light on something'.

frog.[2] A remarkable gift—made all the more so by the fact that this building up of confidence was independent of the many confusions, mis-steps, and failures on my part that Dan patiently helped with.

My own work was primarily about special cases of the Bernoullicity question raised in the work with Doug Lind and Klaus Schmidt [178]. With Dan's help I made some progress on a key example and one more general case [321, 322], but there were two problems with my work. First, the argument was frankly a bit woolly in one key step. Second, the special case was far more effective at shining a light on the difficulties in the general case than it was in illuminating a path through them. I moved away from this kind of measurable dynamics, but Klaus and Dan went on to resolve all these difficulties [262], building in part on an elegant use of the product formula in number theory to revisit a much earlier alternative proof of Bernoullicity for toral automorphisms by Doug and Klaus [182].

Bill

In the Spring of 1990 I attended a module on Low Dimensional Dynamical Systems taught by Michael Yakobson, introducing the basic machinery of Markov partitions, renormalization theory and so on motivated by standard families of examples like quadratic maps, circle maps, and the Hénon maps. This was just one of several interesting modules and reading seminars on offer, and it was difficult to avoid filling my mind with too many of them.

A single year post-doctoral position meant I was applying for my next role within a few weeks of arrival. By happy chance Bill Parry, who had taught me measure theory as an undergraduate, was spending that semester at College Park. This was enormously helpful in several ways. He was already the natural person to act as my 'internal' examiner, and Dan Rudolph was one of the obvious choices for 'external' examiner. So on the 13th of November of 1989 the formal thesis defence took place in Dan's tiny office in College Park, saving me a long and expensive journey back to England. Bill and Dan were kindly examiners, though both asked penetrating and insightful questions.

The second great kindness of Bill Parry was to spend an hour or two over coffee making suggestions as to where I should apply for the next position. His knowledge and network were vast, and his advice invaluable.

[2]Freeman Dyson was scheduled to give the AMS Einstein Lecture in October 2008. The event itself had to be cancelled, but the notes were written up and comprise an interesting overview of a broad sweep of Mathematics, particularly as it pertains to Physics [69].

The third kindness was the way in which Bill and his wife Benita somehow embraced me as a real person not an obscure post-doc. On arrival at one of their parties in a rented apartment in Washington DC, several things were thrust upon me: Filterless French cigarettes, whisky, and the expectation that I would participate as an equal in debating the political questions of the day. At the time Ronald Reagan was still President, and I had recently joined the large 'Housing Now' march on the Washington Mall protesting about the lack of affordable decent housing, and we had several political interests in common.

After stripping away the fierce brilliance they both possessed, this all reminded me of the deeper attribute that Dan Rudolph shared with Bill: A determined assumption that there was something of value in everyone they met.[3]

Several other more established mathematicians offered advice at this early stage, including Eli Glasner when he visited College Park for a conference. Eli took me aside and in an almost conspiratorial manner suggested that I should 'prove good theorems'. Kindly meant, but at first glance this is a little like revealing the idea of scoring more goals than the opposition as a tactical masterstroke for a football team. In fact it was not only kindly meant but wise: Over the course of a career, appropriately balancing the easy wins against the more difficult longer projects is important. The vast growth in the volume of published literature across most disciplines suggests that we are not collectively heeding this advice effectively. I certainly can make no claim to have avoided the lure of the easy result and the cheap publication in my time.

> An extrapolation of its present rate of growth reveals that in the not too distant future Physical Reviews will fill bookshelves at a speed exceeding that of light. This is not forbidden by general relativity since no information is being conveyed. (Attributed to Sir Rudolf Peierls and quoted by David Mermin [199, p. 57])

[3]Bill Parry passed away in August of 2006, and his remarkable life both as a mathematician and as a person is recorded in a short obituary by his colleague David Epstein and former student Mark Pollicott [99]. A fuller account of his many mathematical contributions appears in a volume of the journal *Ergodic Theory and Dynamical Systems* [246]; he was one of the founding editors.

8

Columbus

Towards the end of my year at College Park I bought my first car, a Honda Civic with 80,000 miles on the clock—later dubbed the 'red hot tamale' by Aimee Johnson, one of Dan Rudolph's students—and arranged a rental apartment in Columbus. Aimee's husband Todd Drumm had kindly spent a morning driving me around car dealerships, and helped arrange insurance and so on. Aimee and Todd treated me with great kindness during that year, and remained good friends who visited us in the UK several times. Sadly the kind and relentlessly good-natured Todd passed away unexpectedly in 2020.

I returned to England in the Summer of 1990, and a week later Tania and I were married, in two stages. The first was the legal ceremony at a registry office in Twickenham on the 17th of August, the second was a service we had written ourselves and wedding reception at the Grim's Dyke Hotel in Harrow the next day. We had to invent the service because our different faith backgrounds meant neither Church nor Synagogue made much sense. We flew to Dulles via New York on the 10th of September and picked up the red hot tamale from Hyattsville. After a few days staying with Joe Auslander and his wife Barbara Meeker, we drove to Columbus to really begin our married life. Our apartment in Wake Robin Apartments meant we were near neighbours to Vitaly and Valia Bergelson and their family, whose kindness and warmth of welcome was boundless.

Ohio State University is enormous by UK standards, and the Mathematics department was on the same scale. It had a vibrant research culture, with too many interesting seminars to choose from. I felt at home in the busy multinational department, but it was a difficult adjustment for Tania in a new city far from home.

T. Ward, *People, Places, and Mathematics*, Springer Biographies,
https://doi.org/10.1007/978-3-031-39074-6_8

I worked on finishing up the Bernoullicity questions [321, 322] in one seminal case dubbed the 'three dot' system and for a class of systems under the hypothesis of expansiveness, and looked at a periodic point issue that also arose specifically in the $\mathbb{Z}^d$ setting for $d > 1$. Under some hypotheses concerned with the 'size' of the action, the periodic points distributed themselves uniformly around the space. However the fact that the acting group was large for $d > 1$ meant there could be complicated dynamics in a lower dimension, and here the periodic points behaved in strange ways depending on arithmetic phenomena. This resulted in a proof that the uniform distribution was related to the system having completely positive entropy [323]. In this work once again I relied on the hypothesis of expansiveness, and the essentially Diophantine problems that arise without this assumption are still actively studied.

Working in such a large department brought many benefits. The library was extraordinary, with an outstanding collection and back runs of a huge array of journals. There were frequent visitors and interesting seminars—and there were taught modules aimed at postgraduate students. So I was able to expand my basic Mathematics knowledge in several directions. I attended a module that Peter March taught on Brownian Motion and Stochastic Calculus, and one that Alice Silverberg taught on Elliptic Curves and Shimura Varieties. The latter was my first real introduction to the extraordinary connections between geometry and arithmetic.

Qing

At some point in 1991 Qing Zhang, who was at the time a post-graduate student of Vitaly Bergelson, knocked at the door of the large windowless office I shared with David Ginzburg and asked if something called the 'Abramov–Rokhlin entropy formula' held for amenable group actions. I had recently looked—with limited understanding, despite the patient instruction in this kind of Mathematics I had received at College Park from Dan Rudolph—at a monumental work by Don Ornstein and Benjamin Weiss on entropy and isomorphism theorems in exactly this setting [233]. So I felt confident enough to say "Yes of course it does—let's work out the details." We started to do so more or less immediately, standing at a blackboard and seeing what needed to be changed from the original lucid argument of Abramov and Rokhlin [6]. It is hard to imagine the luxury of being able to do this now—conversations or video conferences about Mathematics have to be scheduled weeks or months in advance to work around diary commitments.

We set to and quickly realised how this would go. Roughly speaking, at each step an equality in the argument used by Abramov and Rokhlin in the late 1950s became an approximation in our setting—replacing, for example, exact tilings of the integers by intervals with Følner sets that nearly tiled. So the whole thing became a matter of controlling the size of errors, and needed some of the more general machinery that had been developed for amenable group actions. We quickly convinced ourselves that this was possible, and over the next few weeks had the opportunity to discuss some of the places where we could not quite control some error with two great experts: Dan Rudolph and Benjy Weiss. With their kind assistance the last pieces of the proof came together—a key step was to use a deep result of Rosenthal [260] guaranteeing the existence of a finite generator under the assumption that the action was free, ergodic, and of finite entropy—and so we then needed an additional argument to deduce the general case from this one. By the time we had finished I was back in the UK, and so my affiliation on this paper reads University of East Anglia [348]. In a modest way, this result turned out to be useful and quite widely cited, though I was amused by one of the early citations, which appeared in a remarkable paper of Rudolph and Weiss [263].

> We point out that J. Kieffer [5] constructed a version of the Shannon McMillan Theorem for amenable group actions predating the work of [7] by the introduction of a generalized tail-field. T. Ward and Q. Zhang in [9] use this method to obtain a conditional entropy theory for such actions. This approach does not appear able to directly give Theorem 8 via this generalized tail-field. The work of Ward and Zhang does give essentially all the results we need for conditional entropy. We give an alternative approach to these results anyway building on the methods of [7] rather than those of [5].

It was an early and relatively trivial insight into some of the issues that might arise in paying too much attention to citation metrics, which later solidified into real scepticism about their naive use. Indeed the wider use of metrics and the relentless growth in the pressure to measure and count outputs in the form of research papers and their citations and outcomes in the form of graduate earnings has changed the experience of working as an academic in fundamental ways. Slow scholarship and bold creativity in both research and study risks being steadily ironed out of the system as a result.

Library Discards

During 1991 we were notified that some duplicate journals in the holdings of the Ohio State University library were being discarded. My formal title was 'Research Instructor', and I was doing some teaching, mainly in the elementary calculus sequence for non-mathematicians. However I certainly had plenty of time for research, and was actively looking around for new ideas. So I went down to the basement and sorted through some of the journals stacked up in recycling bins. If I came across an article that looked as if it might be interesting or useful I tore it out, returning to my shared office with a small pile of articles. I left these on my desk and left shortly afterwards to attend a 'Colloque de Théorie Ergodique' at the Centre International de Rencontres Mathématiques (CIRM) in Luminy from the 27th to the 31st of May 1991. Both Doug and Klaus were there, and Klaus reminded us both of the mixing of all order problem, which by then had been distilled down to a single question: If a $\mathbb{Z}^d$ action by automorphisms of a compact connected group was mixing, was it mixing of all orders?

'Mixing' has a meaning in Mathematics close to what one might mean by mixing in more familiar settings. One of the standard illustrations of the concept goes as follows. If two ingredients of a cocktail are poured into a glass, then the action of a stirrer is called mixing if after some time every mouthful tastes almost exactly the same. That is, every mouthful contains the two ingredients in essentially the same proportions. Mixing of all orders is more or less the same idea, but allowing for more than two ingredients.

While still at the conference I was sure I had an argument for the mixing problem, but Klaus gently pointed out that my approach applied equally to known mixing examples which were certainly not mixing of all orders. So we went our separate ways after the conference with the problem firmly in my head. On the flight back to Columbus I played with the problem, trying to understand what it was we really needed to know. Viewed from one angle it came down to a property of mixed multiplicative and additive Diophantine equations in number fields. After a few weeks back in Columbus, I got round to going through the little pile of papers I had left on my desk, and one of them was a recent paper of Schlickewei [266] with a form of 'S-unit theorem'.

It took me a moment to realise that this deep paper proved something that was exactly what we needed as the reduction step for a simple and beautiful backwards induction argument. In fact our mixing of all order result and Schlickewei's S-unit theorem (in the case of a finite-dimensional compact group at any rate) were really so closely connected that they amounted to the same statement in two different languages, with Pontryagin duality as the dictionary to translate between them. I sent an email to Klaus in some excitement, and we quickly set to work. As usual there were some additional steps needed to sort out the details, but we quite soon had the proof of the full case sorted out and published it together [268]. For technical reasons we used a slightly different formulation of Schlickewei's result [267] in the final version. It is a beautiful connection between dynamics and a certain part of number theory, and many years later I wrote about how this all happened when I was invited to contribute a 'Trip to the proof' column for the *Nieuw Archief voor Wiskunde* [340].

The fact that this mixing of all orders result and the entropy paper both appeared in *Inventiones Mathematicæ*, a highly-regarded journal, probably made life easier during the precarious period of academic life between a doctorate and a first 'permanent' position.

Graham

Columbus in the early 1990s was a comfortable and easy place to live, though it felt somewhat isolated from the wider world. It was easier for me than for Tania, because I was in an outward-looking, large, international Mathematics department that had much in common with any other Mathematics research department. Some aspects of life in the United States never stopped feeling strange for us nonetheless. One of these was how the difficult legacies of race played out, sufficiently problematic in the UK and the countries of my childhood. Our apartment complex was almost but not quite entirely 'white', though for some of the time there our immediate neighbours were a friendly African–American family. The apartment complex across the road seemed to be entirely African–American. Despite the proximal shadow of apartheid throughout my childhood, neither of us had ever seen such overt division on these lines in residential neighbourhoods, and after living in Birmingham for a few years it seemed odd and unnatural. There were other features of mid-Western life we found unfamiliar. The overwhelming ubiquity of the car, which ranged from drive-through liquor stores to shops and restaurants being widely scattered islands in huge car park seas was hard to push back against. I tried to resist this in modest ways, including developing the habit of lining up on foot in a queue of cars at the bank, to the amusement and consternation of the other customers.

Our return to England was closely related to a subscription to the *Guardian Weekly*. A strong theme of randomness permeates my career, and this was an example. Nowadays a young academic looking for their first relatively secure role would be signed up on organised websites like `jobs.ac.uk` and `www.mathjobs.org`, but in 1991 the internet was nothing like the pervasive presence it is now. The choice to subscribe to the *Guardian Weekly* was driven by two things: Our own feeling of being isolated while living in Columbus, and my memories of reading it as a child.

Knowledge of the wider world during my childhood very largely came from listening to the World Service of the BBC on crackling short-wave radio, and subscriptions to *New Scientist*, *National Geographic* magazine, and the *Guardian Weekly*. The words "This is London" followed by the lively Irish jig Lilliburlero drifting in and out of precise tune on a short wave radio transports me back to warm Zambian evenings and the family all listening to a short-wave radio on a large gramophone.

I read each *New Scientist* cover to cover, giving me a wide but now profoundly outdated sense of science in the round. Indeed my first experience of seeing my name in print occurred early in 1979 as a 'letter to the editor' in this magazine: Our mother was often the first to read the *New Scientist* when it arrived and while doing so one day called out "Who's T. Ward, Swaziland?" I had written a letter responding to an article on the role of specific terminology in science teaching, and there it was in the magazine. The *Guardian Weekly* was printed on thin airmail paper, with a distinctive blue background to the header. It has a long history, originally launched by C. P. Scott in 1919 as a way to project the best of the *Guardian*'s reporting to the world. During my childhood it included weighty articles from the *Washington Post* and *Le Monde*, and gave a somewhat rounded—albeit unabashedly Western—viewpoint. It evolved later into a magazine format, and now reflects the growth in the Australian and American sections of the *Guardian* newspaper. I was amused many years later to encounter *Le Monde* again in a quite unexpected setting: Oliver Babish, the fifth legal counsel to the White House in the television series 'The West Wing', read *Le Monde* exclusively.

So the *Guardian Weekly* was a regular and welcome arrival in the flat metal mailbox of our apartment in the Wake Robin complex. Tania and I more or less had an implicit understanding that I would apply for any reasonably plausible Mathematics job back in the UK, and a small box advertisement for a Lecturer in Mathematics at the University of East Anglia (UEA) was the first such opportunity to come up. Had we not subscribed to the *Guardian Weekly*, I would certainly never have seen the advertisement. I kept this copy on a shelf in my office for a few weeks, because something made me reluctant to apply. I am not sure what this was, but certainly I had something of an assumption that only candidates with an Oxbridge PhD would be considered seriously. I was also reluctant to shorten my time at Ohio State and the attractive National Science Foundation funding there.

Eventually I plucked up the courage to send in an application to UEA, and early in December of 1991 received an email from Prof. John Johnson, the head of the School of Mathematics and Physics there, saying that a letter was on its way which would invite me to an interview in Norwich—the following Tuesday. The letter did indeed eventually arrive, but would have been too late, so this email proved to be significant. Arranging flights from Columbus to London at short notice involved faxes and scans of credit cards, but it all worked out until the last leg of the journey. With perfect timing, the train from London to Norwich ground to a halt on the morning of the interview, with little available information. I talked to the guard, and as soon as I said I was on my way to an interview he did all he could to help, ending up with my ringing UEA on some device and asking them if they could kindly arrange for me to be interviewed a little later in the day.

So it came about that I was interviewed at UEA, by a large panel in the Council chamber, on December 10th of 1991. I only remember two questions. Alec Fisher, a mathematical philosopher, asked me what 'Fourier analysis', a phrase that I had used in another answer, meant. Like many mathematicians before me, and doubtless many after, I bent the truth somewhat and deviated from the fact that I meant it in the context of locally compact abelian groups, solenoids in particular. Instead I held forth on how one might use Fourier analysis to reduce the problem of designing a concert hall to have good acoustics to the problem of a point source of a single frequency perfect sine pressure wave. After sitting the other side of the table on several hundred occasions, I would of course now say that producing a not ridiculous answer that made some sense to a non-expert was the point, rather than anything of substance. The rest was routine I think—certainly not memorable—until the chair of the panel handed over to Graham Everest, a number theorist.

Here another strange random choice played a role. Because I had very few publications at the time, I had included the title of a preprint—the paper on mixing of all orders with Klaus—in the modest list of publications for my application. Nowadays this would have been placed as a preprint on the `arXiv`[1] as a matter of course, and anyone who wanted to could have looked

[1]The arXiv is an open-access repository for many science disciplines including Mathematics. The astrophysicist Joanne Cohn began emailing preprints in string theory around 1990 in TeX format on a large scale, but rapidly encountered capacity problems [125]. What became the arXiv was started by Paul Ginsparg in August 1991 as a Physics preprint server at the Los Alamos National Laboratory; in 2001 both its creator and the archive moved to Cornell University where it changed to the name `arXiv.org`.

at it. It was not easy then to disseminate preprints, and we could easily have chosen a different title for 'Mixing automorphisms of compact groups and a theorem of Schlickewei', which appeared a little later [268]. Graham was fascinated—as it turned out, he had used something called the 'subspace theorem', a profound result in Diophantine analysis and essential technical precursor to the result of Schlickewei we had used, to make progress on some problems in number theory. He wanted to know how this kind of Diophantine analysis played a role in questions like mixing in ergodic theory, starting with what was meant by 'mixing'. Sadly, all too few years later I had to describe this moment in an obituary for Graham [338], which I did as follows.

> I first met Graham across an interview table in 1991. Despite the constraints of the setting, it was quickly apparent that we shared an interest in the remarkable, then recent, work of Schlickewei and others on the subspace theorem, and the resulting insights into the solutions of S-unit equations in fields of characteristic zero (from [338])

The fact that the origins of our paper involved more chance than science made this all the more remarkable.

The day of the interview also involved having lunch with some of the pure mathematicians, in my case David Evans and Johannes Siemons. They helped round out the picture of the department and its modest size—between the interview panel and the lunch I certainly met almost all the pure mathematicians that day. I flew straight back to Columbus, and on the 12th of December a job offer was faxed to me. John Johnson emailed to say he had persuaded the registry to add one increment to the salary, increasing it to £15,688. This was an enormous drop from what I had been earning at Ohio State, but that aspect of moving back to the UK never concerned us.

Tania and I went to Blendon Woods Metro Park in Columbus and walked around in the snow, weighing up what to do. There were some other possibilities in the USA, and neither of us knew Norwich at all. We had lunch with a colleague who listened carefully to how we both talked about this, and pointed out the power of how we both used the word 'home'. This was an astute observation—he felt he was now too old and too senior to return to the country he considered home, and urged us to think carefully about the longer term question of where we wanted to live. A trusted external mentor slightly poured cold water in the form of an email saying "you could always move somewhere better later", an early foretaste of a certain viewpoint that was to be later captured rather too effectively when the Russell Group came into being. The group is named for the location of the first informal meetings in the Russell Hotel in London before wider gatherings of the Committee of Vice-Chancellors and Principals (now Universities UK or UUK) in Tavistock Square. Its remit "to help ensure that our universities have the optimum conditions in which to flourish and continue to make social, economic and cultural impacts through their world-leading research and teaching" was explicitly about the members not the sector. It initially comprised the 17 research universities Birmingham, Bristol, Cambridge, Edinburgh, Glasgow, Imperial College London, Leeds, Liverpool, London School of Economics, Manchester, Newcastle, Nottingham, Oxford, Sheffield, Southampton, University College London and Warwick; Cardiff and King's College London joined in 1998, and Durham, Exeter, Queen Mary University of London, and York joined in 2012. The legitimacy with which it is often seen as the 'high quality' part of the sector is fiercely contested and only weakly supported by the data, with the exception of their combined research power.

Nonetheless, the wish to return to England and the lure of a notionally permanent position made the eventual decision an easy one.

I had raised at the interview the question of whether a way could be found to make use of the third year of my position at Ohio State, and the relevant members of the panel agreed to look into this. The arrangements were largely made by fax, and in the end worked as follows. I would take an unpaid leave of absence from Ohio State for the academic year 1992–1993, and we would move to Norwich in June of 1992. The start date meant I could contribute to UEA's submission to the 'Research Assessment Exercise' (RAE) that year. I would then take an unpaid leave of absence from UEA for the year 1993–1994, allowing us to return to Columbus for the third year of my position there. Both departments did their best to be helpful, but all this took a little arranging of course, as we also needed to change my H-1 visa to allow us to return to Columbus. This was all arranged by fax, on devices that now feature in museums of technology. A happy and unplanned by-product of all this was the initial three years of NSF funding I had access to from a joint award with Vitaly Bergelson and Joe Rosenblatt provided some research support through my first year back in England, and the three of us were fortunate enough to win another three year award which provided some support right up to 1996 while I was back in England.

9

Norwich and Graham

We arrived in Norwich in June 1992, and moved into a small house on Caernarvon Road rented from my new colleague Johannes Siemons. Because of an unfortunate piece of timing, we had committed to a year rental on the Columbus apartment ending in October, and we were also trying to sell Tania's little house in Birmingham which had been rented out for some of our time in Columbus. The two rents and one mortgage together exceeded my salary, so this was a difficult period financially. Tania did some supply teaching, and we used up our savings for some months. Gradually this all resolved itself: The apartment in Columbus was taken by someone else, the house in Birmingham sold for a large but one-off loss, and some sense of normality resumed.

The School of Mathematics and Physics (MAPS) at UEA was a friendly and lively place. The 'pure' Mathematics group was a small subset of a small school, so it took considerable effort to provide a good curricular offering. Despite being a small unit, the research atmosphere was lively, and there were regular seminars.

Graham Everest was enormously helpful professionally. We started as joint supervisors of Vijay Chothi[1] in the Autumn of 1992, and quite quickly the following idea emerged as being of interest. The specific dynamical systems that Doug Lind and I had worked on, and that featured in my masters dissertation, were parameterised in part by subgroups of the rational numbers (and products of such groups), and in that first publication we had included the cryptic remark shown in Fig. 9.1. A step in our entropy calculation

[1] Vijay graduated in 1996; the external examiner was Prof. Peter Walters from Warwick [51].

T. Ward, *People, Places, and Mathematics*, Springer Biographies,
https://doi.org/10.1007/978-3-031-39074-6_9

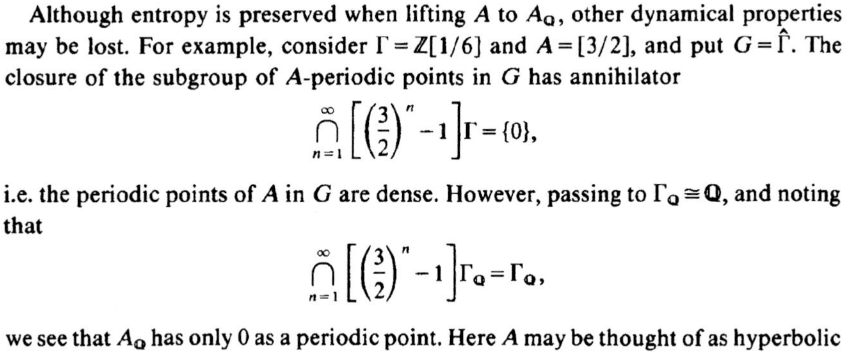

Although entropy is preserved when lifting A to $A_{\mathbb{Q}}$, other dynamical properties may be lost. For example, consider $\Gamma = \mathbb{Z}[1/6]$ and $A = [3/2]$, and put $G = \hat{\Gamma}$. The closure of the subgroup of A-periodic points in G has annihilator

$$\bigcap_{n=1}^{\infty} \left[\left(\frac{3}{2}\right)^n - 1 \right] \Gamma = \{0\},$$

i.e. the periodic points of A in G are dense. However, passing to $\Gamma_{\mathbb{Q}} \cong \mathbb{Q}$, and noting that

$$\bigcap_{n=1}^{\infty} \left[\left(\frac{3}{2}\right)^n - 1 \right] \Gamma_{\mathbb{Q}} = \Gamma_{\mathbb{Q}},$$

we see that $A_{\mathbb{Q}}$ has only 0 as a periodic point. Here A may be thought of as hyperbolic

Fig. 9.1 A remark from the paper [181] pointing towards understanding how dynamical properties change as you move in the parameter space of one-dimensional solenoids (©Cambridge University Press 1988; reproduced from *Ergodic Theory and Dynamical Systems*, Volume 8(3), 411–419 with permission.)

involved simplifying the problem by making the groups involved much more complicated, and the meaning of the remark is that this process kills all the non-trivial periodic orbits in the system.

While looking for a tractable problem for our new student to look into, it struck me that the simplification process could be done in multiple—indeed, in infinitely many—steps, and it might be interesting to understand how these intermediate steps influenced the way in which the periodic points were killed. This did indeed turn out to be interesting, and Graham, Vijay, and I ended up able to exhibit an enormous range of behaviours, some of which continue to be studied [50]. This type of question—what periodic orbit growth is possible for compact group automorphisms, and how the structure of the compact group influenced these possibilities—dominated much of my research for many years.

Alongside this project, the three of us tried to make sense of the way in which the entropy and its decomposition into p-adic components from my work with Doug [181] varied as the direction changed in a $\mathbb{Z}^2$-action defined by a pair of commuting automorphisms of a solenoid. This was only partially successful, but the ideas of entropy rank, directional stability of dynamical phenomena, and several other things that were implicit in our efforts cropped up in later work [52].

Jointly supervising Vijay really began a collaboration with Graham that lasted twenty years. Through all the turbulence and pressures of an academic career, and alongside our own work, Graham and I always had someone in the department interested in discussing research ideas. Graham also was a great source of encouragement and solace during some difficult times, particularly when I became head of department.

In the middle of the academic year 1992–1993 our daughter Adele was born at the Norfolk and Norwich Hospital. We had bought a semi-detached house in Hellesdon for £53,000 by then, and I had started a pattern of cycling to work at UEA from it along the Norwich ring road. With an eye to the future, we had chosen to buy a house in a far less fashionable district than the part of Norwich known as the 'golden triangle' where many university colleagues lived. Doing this allowed us to afford a garden and a bit more space. Over the years we added an upstairs toilet and a garage, and this was the house where our two children grew up. Norwich was not immune to the wider forces of growth in the number of households, inadequate volume of house building, and the rapid decoupling of the exchange value of houses and land from their use value. The expansion of areas where a new lecturer with a modest deposit could not afford to buy a house continued relentlessly, and eventually swallowed up most of Hellesdon.

We arranged to rent the house to five students for the 1993–1994 academic year, and left our battered Ford Mondeo with a UEA colleague, Alan Camina. He wisely said we should actually transfer ownership to keep the paperwork in order, and this proved to matter as the car was written off in an accident while parked outside their house a few months later.

Tania already had an F-1 visa allowing multiple entries to the United States until June 1996, based on her studying for a masters degree at Ohio State University. Adele was added to Tania's passport in March of 1993. We had checked as carefully as possible what needed to be done about this before we left Columbus, and naively assumed things would proceed smoothly.

10

Columbus Revisited

The return to Columbus on June 16th of 1993 proved to be a little more exciting than we would have hoped. Miraculously the details of coaches and trains between Norwich and Gatwick Airport meant that we would either be extremely early for our flight or nerve-shreddingly tight, and we had opted for the former. Even in 1993 flying to the United States often involved additional security and immigration checks, and I was and remain something of an anxious traveller.

We had the idea of checking in, enjoying a leisurely meal and then having a wander round the airport showing the six month old Adele the sights. At the first part of check-in it came to light that Adele could travel on her mother's passport but needed her own visa within it. We had been advised differently before we left Columbus, and either the rules had changed or this was an error in the first place. The next few hours are something of a blur in memory, but must have looked as if we were filming an action movie. I did some rapid mental arithmetic and decided we should simply travel as quickly as possible to the US Embassy in central London and hope for the best. We took the fast train into London, pleaded with a taxi to get us there as quickly as possible, and then boldly walked past the immense queue of people lined up for their booked appointments and found the largest and most impressively armed guard we could find.

US Embassies the world over can be difficult places to deal with for non-citizens, and are often associated with lengthy delays and understandably intense concerns about security. On this occasion—doubtless influenced by our wild eyes and the presence of a small baby—they were extraordinarily

T. Ward, *People, Places, and Mathematics*, Springer Biographies,
https://doi.org/10.1007/978-3-031-39074-6_10

helpful. As soon as the huge guard had grasped the situation, and because we were able to produce not only our visas but the blue approval forms for them that seemed to have magical significance, we were ushered into another door, taken to an office for a brief interview, and then told to wait. During the interview we had to attest on her behalf that Adele was not under indictment for any war crimes and had no plans to overthrow the government of the United States. Many stories and myths attach to these questions. One is that the diminutive and courteous mathematician Shizuo Kakutani interpreted the confusing double negative way in which he was asked them with English that was more logical than colloquial, and ended up giving the wrong answer to all of them—to the considerable surprise of the immigration officer. Gilbert Harding—once dubbed the rudest man in Britain—is supposed to have answered 'sole purpose of visit' to the question about overthrowing the government. The latter anecdote has been attached to multiple people, including Oscar Wilde, Michael Foot, Evelyn Waugh, and Peter Ustinov. I have no direct evidence that either story is true, but the latter is mentioned in a biography of Harding [254].

After an extremely anxious wait, Tania's passport was handed to us with a visa for Adele, and we raced back to the airport. The taxi driver was bold, the fast train to Gatwick was on the platform waiting for us, and we made the original flight as planned.

Boarding an aeroplane often feels like something of a relief, as it means all the hurdles of getting to the airport, navigating multiple queues, producing the right document for the right person, reassembling shoes and belts after security, and so on have been navigated—and all subsequent activities in support of your journey are now entirely out of your hands. On this occasion it was a little more than that: Both Tania and I were drained, and Adele had doubtless picked up enough of the emotional energy to also be exhausted.

At our destination we were met by Valia Bergelson in an estate car with a boot full of the equipment needed for a small baby. Bags of clothes, stroller, high chair all produced by the Jewish community of Columbus and all returned to them when we finally left. Both logistically and financially this was a huge help to us, and we have tried to help other young parents in similar practical ways ever since.

Interacting with the complexities of the medical system in Columbus was not easy for people used to the UK's National Health Service (NHS). The insurance scheme provided by Ohio State University was excellent, but that did not translate into being easy to navigate. Two incidents stand out in memory. One was an interaction with a dentist that seemed to change in emphasis from routine to exciting when it came to light that I worked at Ohio State. The suggestion to come back in a year for a check-up suddenly became certainty that half my teeth needed to be replaced with crowns. This was unsettling not in itself—for all I know he was quite right, though I do still retain all but one of those teeth, stubbornly uncrowned, more than thirty years later—but because of the realisation that I was laid far back in a dentist's chair and had suddenly lost all confidence in them as a disinterested clinical practitioner with my best interests at heart.

The second striking interaction with the system concerned our new daughter Adele. Our naive assumption was that we would simply enter her details into a form, add her to the family health insurance, and pay a bit more each month—but this immediately hit a problem. We were allowed to *transfer* her from a previous health insurance scheme to the Ohio State one, but could not *add* her unless she had been born within its purview. The fact that she had done something so unimaginable as to arrive in this world ex nihilo without benefit of private health insurance seemed to paralyse the system. We never worked out whether this was a meaningful and carefully constructed barrier intended to protect the scheme from undeclared or unknown pre-existing conditions, or if it was simply a scenario that had not been planned for. After multiple attempts to explain why she had no health insurance while we were in the UK we were on the point of giving up and returning to England when a kindly and forceful administrator in the Ohio State Mathematics department finally persuaded the scheme to accept her.

This unsettling experience helped to quash any lingering regrets about leaving the United States and solidified our own political views about what was called 'socialized medicine' there. In a poignant detail, the same departmental administrator burst into tears when I went in to finally say goodbye. I was taken aback, because while she had certainly been immensely helpful to us we had not interacted much, and certainly not socially. When I enquired further, she simply that both Tania and I were among the people who always said 'please' and 'thank you', and she would miss that. It was a sobering insight into the complex cultures that sometimes facilitate inappropriate behaviour on university campuses.

The year back in Columbus with a small baby was more than hectic. The Mathematics department were kind enough to arrange my teaching to comprise repeated classes in pre-calculus on Monday, Wednesday, and Friday at 7.30 and 9 am. This made it possible for me to get to work early, and then do an exchange with Tania, swapping car key for baby at the door in the middle of the morning. In this way Tania managed to complete a masters degree at Ohio State, and we both were able to spend huge amounts of time with our new daughter. I became adept at working with her on my lap, though her

Fig. 10.1 Our first computer at home (Photo by Danamania - Own work, CC BY-SA 3.0, http://commons.wikimedia.org/w/index.php?curid=2105185)

random contributions to my TEX documents were not always constructive. By this stage we had a Mac Powerbook 100 at home, bought at huge expense from the Ohio State bookshop (Fig. 10.1). We had added a modem, and managed to squeeze an immense amount of activity through its tiny monochrome screen and 20MB hard disk.

I'm Not Faculty, I Work Here

From the start of the year back at Ohio State we knew we would be leaving soon, so had done our best to accumulate as little as possible. In this we were not especially successful, and we ended up shipping some goods back to England. This included a second-hand sprung rocking horse and a traditional red cart for Adele. The latter resembled a real 'Radio Flyer' cart, but was I think an imitation brand. Nonetheless it was a properly made wooden cart capable of carrying two children easily, and thirty years later it is still in service.

The bulk of our modest belongings needed to be disposed of somehow. Many baby clothes and toys were returned to Valia Bergelson for the next family with a small child to use. Some things were given to the Goodwill Thrift Store on North High Street. Several items of furniture were rejected by the charity shops as being too battered or old-fashioned.

For some small items we used an email list server at Ohio State to offer them round, and one of these produced an inadvertently striking phrase which has stayed with me. Someone I did not know rang my office number to say they wanted one of the objects, and I suggested they come over and collect it right away before I went home. "I can't do that—I'm not Faculty, I work here" was the response. The intended meaning was of course that they could not leave their desk, but it is a striking phrase that has been at the back of my mind many times since. Not in connection with who does or does not work hard on university campuses, but in relation to the complexities of affiliation and who one's tribe really is. For many academics the tribe whose validation really matters is the research community which might be scattered all over the world, while for many professional services colleagues the tribe might be located within the University. Some of the moments of most profound mutual misunderstanding within higher education arise from this difference.

11

Norwich Revisited

Our friends Greg and Anni took us to Columbus Airport for our final departure on the 1st of July 1994. I had first met them in the Student Union at Warwick in 1986, resulting in a memorable evening that ended with the consumption of a bottle of Dutch Genever in their flat on Gibbet Hill. Through a strange sequence of events we were to live in the same city several times: In Coventry in the late 1980s (indeed we shared a house, 3 Lollard Croft in Cheylesmore, for some years), in Columbus in the early 1990s, and years later in Norwich when Greg came to work at the John Innes Centre. Some juggling of heavier items between suitcases and hand luggage was required to meet the weight limits, and the final weeks of packing and cleaning had been tiring. It was a real relief to finally board.

This journey marked the end of life as 'resident aliens', the end of dealing with private health care, and a return to a city that had already started to feel like home. We were nonetheless both immensely grateful for the opportunities afforded to us by the USA and its openness to visitors from other countries. Two of its many great public Universities had supported me across the perilous journey from doctoral student to relatively secure academic employment, and Ohio State University had enabled Tania to complete a Master's degree.

I settled down to work at UEA, and picked up the mathematical conversations with Graham Everest. I taught a mixture of real and complex analysis across the undergraduate years one and two, and each year a third or fourth year module. The latter allowed great flexibility, and over the years these included Topology, Functional Analysis, Coding Theory, Symbolic Dynamics, and other enjoyable things. I enjoyed lecturing and the role of personal

T. Ward, *People, Places, and Mathematics*, Springer Biographies,
https://doi.org/10.1007/978-3-031-39074-6_11

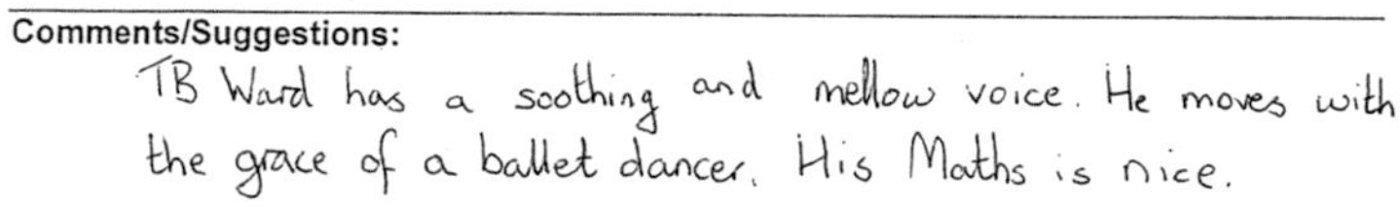

Comments/Suggestions:
TB Ward has a soothing and mellow voice. He moves with the grace of a ballet dancer. His Maths is nice.

Fig. 11.1 A comment on a student evaluation from second year Complex Analysis in 2006, and from a more innocent time—one to treasure

tutor greatly, and am still in touch with many of the students from that era. At the time the ratio of staff to students was manageable, and both lectures and seminars could be genuinely interactive. It was also a human environment, where staff and students got to know each other and to some extent both worked and played together. Our approach to things like student evaluations and lecture observations would look amateurish nowadays, but seemed effective. These too were more human than they have become in some Universities now, where it is all rather grim and metric (Fig. 11.1). I have tried where possible to avoid dreadful practices like allowing crude metric assessment of teaching evaluations as a criterion for promotions, but the world of good-humoured humanity at times seems rather remote.

Much like the so-called Teaching Excellence Framework (TEF) that was to come later, there are two difficulties with the habit of using metrics from teaching evaluations to assess the capabilities of lecturers. The first is simply muddle: If you want to use any kind of survey for a purpose, that purpose needs to be known when the survey methodology is designed. There are researchers who know how to design an approach that could be effective in helping a lecturer to improve their practice, for example. It is extremely difficult to design an approach that really measures, for example, the 'learning gain' for the students, but is not entirely impossible. What is impossible—and unethical—is to just muddle along and let data from surveys gathered for one purpose (or indeed for no purpose beyond compliance with an edict that they should be run) be used for another. The second is that even where it is explicit that surveys are going to be used for evaluation of teaching effectiveness by lecturers, and some attempt has been made to design the surveys to reflect that purpose, all the biases present in society find their way into the results. Women and people from other countries particularly tend to receive lower grades to such an extent that the practice of using these numbers for promotion decisions would not stand up to legal scrutiny. There is a substantial body of research that shows at the very least how carefully a system of evaluation needs to be designed to be effective; a useful overview appears in work of Boring et al. [29]. Troy Heffernan and Paul Harpur have studied how University use of student evaluations might end up with legal challenge [148].

Three things in particular have made it more difficult to maintain that human and collaborative atmosphere in Universities. A great many dedicated staff do their best, and many students still experience a genuine community in their academic department of study at some institutions, but it is difficult to avoid feeling that for many there has been a diminution in the quality of the relationships.

The first is simply weight of numbers: Expansion where possible has been the primary mechanism used to try and maintain the financial stability of Universities. This has created the Darwinian environment that policy makers sought, with some institutions ballooning in size and others struggling to survive. For the individual student, the role of personal tutor is qualitatively different if the number of students for each tutor is too high. Technology has supported ingenious ways to interact with large audiences, but some of these (displaying a live survey result on screen, for example) inherently group together the voice of the students—in a small lecture the individual student can express things and ask questions more easily.

The second is the huge growth in intrusive regulation of the sector—and a fair amount of self-imposed additional regulation and scrutiny of educational processes from the powerful institutional superego. The balance between trusting professionals to act appropriately and checking that they are doing so seems to have shifted more and more to the latter. This is not unique to higher education at all, and it is a difficult tilt to nudge back again. Part of the problem is that the sector has not done a good enough job at making sure it does indeed maintain its own professional standards. The knee-jerk response of fighting to preserve the principle of institutional autonomy has too often masked some poor practice. The solution of increasingly aggressive regulation may have done some good, but has certainly done some damage as well.

The biggest factor of all concerns how the funding of Universities has been reframed. The principle of a balanced contribution by both taxpayer and student to higher education is reasonable, reflecting the fact that both society and the individual benefit from it. The extent to which the way these ideas are framed influences perception is manifest in the fact that there always has been a mixed contribution model, but it felt very different when it was current taxpayers and future ones sharing the burden without much in the way of explicit debate. The political determination to position the student as a consumer and the educational experience as a commercial transaction framed in terms of consumer rights expressed in the Higher Education and Research Act 2017 has had its effect. It is difficult to maintain the idea of a shared responsibility for learning, in which student and academic are allies working together to explore a subject deeply, in such an atmosphere. Students are well aware that real education is a much more complex process than a consumer transaction, but this has had a slow corrosive effect on the quality of the experience.

I also started to learn some more about the history of the department, which had not always been easy. The historic strengths had been in applied Mathematics, particularly classical fluid dynamics. The pure Mathematics group were mainly interested in algebra and nearby disciplines, and at the time had no professor. The group was really too small to be sustainable, but maintained a highly active research and educational culture nonetheless. The absence of a professor in pure Mathematics was related to battles and personal conflicts in the 1970s that had taken place between some members of staff long since retired. While this was all opaque to a junior lecturer, there was an impression that behind the scenes pure Mathematics at UEA was slightly on the institutional naughty step, had been for many years, and would remain so. Indeed it had taken considerable efforts for the group to obtain permission for a new lecturer, and the subject area of the new hire had been the subject of multiple strongly held opinions, understandable given the small size of the unit. The fact that my own work involved both analysis and algebra may have made it easier to hire me.

Church in Hellesdon

We had both been thinking about the challenge of bringing some structured spiritual meaning into the lives of our young children in a way that made sense of our different faith backgrounds. We spent a little time in a welcoming 'liberal and progressive' Jewish community that met in the Octagon Chapel in Norwich, and our daughter Adele had a baby blessing there. This was a lovely and welcoming community, but we both struggled to find spiritual significance in its services. Something prompted us to attend a midnight carol service on Christmas Eve 1995 with Adele and her newly arrived brother Raphael at the local Anglican Church, St Paul's Hellesdon. The vicar followed up and visited a few days later, and that began a lengthy process for all of us. By the end of 1996 we had joined St Paul's, Tania had been baptised into the Anglican Church at a ceremony with the Bishop where I—having been baptised as a child—was also blessed. This was a much more complicated journey for Tania than it was for me, and for many years afterwards we also interacted with various 'Messianic' Jewish communities. We stayed at St Paul's for thirteen years, and formed many lasting friendships there. In the end we were pulled away to Norwich Central Baptist Church but have very fond memories of the role that St Paul's Hellesdon played in both our lives.

Graham and Integer Sequences

Graham and I started to exchange ideas again, and began to work on various questions in elliptic curves, arithmetic of linear and bilinear sequences, and dynamics of group automorphisms. We quickly arrived at a way of working together that was relaxed and convivial, and was only ended by his untimely death. Graham was creative and optimistic in his Mathematics, and many of our exchanges involved real scepticism on my part about what felt like slapdash summing of error terms, interchanges of limits, and other mathematical elephant traps on his part. For some time I kept a printed page on my notice board with hand-written annotations showing where Graham had replaced my word 'troubling' with 'interesting', which I had changed to 'disturbing'—only for him to change this to 'exciting'. This was about a certain analogy we were trying to develop between linear recurrence sequences and bilinear ones for which the natural analogue of one particular concept definitely did not hold. He was a constant encouragement to me professionally, in all aspects of academic life. A steady stream of PhD students added to a sense of a group, with a mixture of formal and informal supervisions by Graham, me, or both of us.

Graham encouraged me to take on PhD students on my own, and the first of these was Gary Morris, who had studied at UEA some years earlier.[1] We worked on some ideas motivated in part by work of Mark Shereshevsky [273] on the 'size' of a group needed for it to be capable of acting with prescribed dynamical properties. The type of question Gary looked at was to ask if a group of polynomial growth (for example) could act expansively on the circle. In another direction, we explored how the 'size' (in a different sense) of a dynamical system constrained dynamical properties of its symmetries. The latter work led to a proof that a mixing endomorphism commuting with an algebraic $\mathbb{Z}^2$-action of completely positive entropy must have infinite entropy [221]. This proved a conjecture of Shereshevsky under the (extremely strong) additional condition that the dynamical system be algebraic. Ten years later Tom Meyerovitch from Tel-Aviv University used remarkable aperiodic

[1]Gary graduated in 1998; the external examiner for this thesis defence was Prof. Peter Walters from Warwick [222].

tiling ideas to produce an enormous family of counterexamples to the conjecture in its full form, showing that our modest contribution certainly could not work without the algebraic condition [202]. Gary and I also clarified a minor aspect of the distinction between actions of a semigroup subset of a group and their invertible extensions to actions of the group [220]. This had its origins in a conversation triggered by my being asked to review a note by Yves Lacroix on mixing [173] for *Mathematical Reviews*. Among the many things we seem to have lost in the headlong rush towards a market-driven mass higher education system is the space to talk about papers we are reviewing over a coffee.

Vienna

In the Spring of 1997 I spent a period of research leave visiting Klaus Schmidt at the Erwin Schrödinger Institute in Vienna. One day he suggested I go over to the University of Vienna as one of his postgraduate students was giving a seminar on a topic that I would find interesting. I wish now that I had kept any notes I took of this seminar, or at least had recorded the date. The speaker was Manfred Einsiedler, talking about some work on generalizations of the Mahler measure as it arose in the context of algebraic dynamics, which later appeared in print [92]. My impression was that this seminar was aimed largely at other post-graduate students, and was taking place in German. I arrived a little late, and Manfred politely switched to English on the arrival of an unfamiliar face. We talked briefly after the seminar, and that began a mathematical collaboration and personal friendship that continues to this day.

Random Examples

Around the same period I became interested in the following kind of question. The space of one-dimensional solenoids could be parameterised fairly easily, and the essential part of this description had the structure of a probability space in multiple ways. The reason this space can be readily parameterised is that it corresponds to the subgroups of the rational numbers, and a good description of this space has been available for many years. In particular, one could think about dynamical properties of automorphisms of solenoids and how they change as you move around in the parameterising space. This led later to work on dynamical properties that vary smoothly, but the starting point for me was the realization that one could ask for 'typical' properties in the sense defined by the probability space structure on the parameter space.

These ‘random’ group automorphisms and their dynamical properties lead naturally to several interesting but unsolved problems in number theory.[2] As a result we do not really know the growth rate of periodic points in a ‘random’ group automorphism. These questions about random examples led to a short note [328] and a longer paper exploring these ideas in the totally disconnected case as well [330].

A surprising footnote to all these questions concerned higher-dimensional solenoids. These correspond to subgroups of $\mathbb{Q}^d$ for $d > 1$, and here no real ‘classification’ of the possible subgroups is available. Nonetheless, Richard Miles showed later that there is in principle a ‘formula’ for the count of periodic points in all cases—but the arithmetic complexity of this formula remains an issue in understanding questions concerning growth rates [215].

Richard and Yash

My second singly supervised PhD student was also a UEA graduate who had left to do something else first, in this case teaching at high school. Richard Miles[3] became interested in using more sophisticated algebraic approaches to study group automorphisms, and he quickly established a separate and successful research track. I have learned from all my PhD students, but in the case of Richard I was learning things and exchanging ideas as research partners almost immediately. One of the things I had been groping around for was a way to find the appropriate generalization of the S-integer dynamical systems that Graham and I had studied with Vijay Chothi to higher rank (that is, for several commuting automorphisms). Richard quickly brought in a more sophisticated use of ring theory to make sense of this, and develop a more coherent approach to some of my messy ideas [214]. He also became part of the ongoing conversations about number theory and dynamical systems, publishing work with one of Graham’s students [65] and joining in collaborations with Manfred Einsiedler, about which I will say more later.

---

[2]The simplest of these is a soft version of the ancient problem of Mersenne primes as follows: How big is the infinite sum $A = \sum_{n=1}^{\infty} \frac{1}{2^{\omega(2^n-1)}}$? Here $\omega(2^n - 1)$ denotes the number of distinct prime factors of $2^n - 1$, so for example $\omega(12) = 2$. If there are infinitely many Mersenne primes, that is primes of the form $2^n - 1$, then infinitely often we have $\omega(2^n - 1) = 1$ and the sum A contains infinitely many terms of size $\frac{1}{2}$. The expected outcome is indeed that the sum ‘diverges’ (that is, is bigger than any number you choose), and much weaker statements than the conjecture that there are infinitely many Mersenne primes would suffice to prove this—but we know nothing about A.

[3]Richard graduated in 2000, with Prof. Franco Vivaldi from Queen Mary University of London as external examiner [213].

Alongside Richard, I was also supervising Yash Puri.[4] Yash and I started thinking more carefully about the real import of a completely trivial observation: If a map from a set to itself has one fixed point, then the set of points that are fixed if the map is applied twice must have an odd number of elements (if it is finite). The reason is simply that the set of points fixed by the map applied twice comprises the one fixed point, and a certain number of pairs of points that are swapped by the map. That is, a sequence of integers that comprise the count of periodic points under iteration of a map—any map—must obey certain arithmetic conditions. These ideas are both old and elementary, and have been used and thought about in multiple ways for many years. Our first real result was that sequences of Fibonacci type, meaning those of the form $(1, a, 1 + a, 1 + 2a, 2 + 3a, \dots)$ where the next term is the sum of the previous two, could only count periodic points in this abstract sense if $a = 3$, adding another special attribute to the well-known Lucas sequence that begins $(1, 3, 4, 7, 11, \dots)$. We published this as a short note in the *Fibonacci Quarterly* [250]. In his thesis Yash pushed this further, and in particular completely characterised the binary linear recurrence sequences that are 'realizable' in that they count periodic points for some map [251]. We also developed some more general results in this direction, and found some integer sequences arising in other contexts with the unexpected property of realizability [249]. Yash was in my car when I collected his external examiner from Norwich Railway Station, and I had to bite my lip when he explained that his thesis was trivial. In fact it was elementary in the way mathematicians use the word—meaning it used no sophisticated machinery—but it was ingenious, and Yash had done great work pushing through difficulties. Learning and doing Mathematics can at times feel like that. Some problems become easy—after you have solved them.

I visited Manfred Einsiedler at Pennsylvania State University in October 2001, and gave a graduate student seminar in which I described some of this work. I raised the question of the existence of smooth models on manifolds for sequences that are abstractly realizable and satisfy the constraints coming from global topology (the 'Lefshetz conditions', which relate global topological properties of a space to periodic point counting properties of any continuous map on that space). Alastair Windsor was one of the PhD students in the audience, and a few years later he solved this problem completely in the setting

[4]Yash graduated in 2000, with external examiner Prof. Sanju Velani, who was then at Queen Mary University of London [251].

of no global constraints, by showing that any realizable sequence could be realised by an infinitely differentiable ('smooth') map of the annulus [364].

Two relatively obscure papers were particularly useful to me as I started to think about the properties of those sequences that could arise as counts of periodic points in this abstract sense. The first was by Bowen[5] and Lanford [30] and contained several interesting ideas on the interactions between arithmetic, analysis, and zeta functions. The second is a genuinely obscure note by England and Smith [98] about zeta functions of solenoids with a specific example of an irrational zeta function that seemed impossible. Indeed it was an error, and understanding how and why their example was impossible was useful as I started to explore the arithmetic of these sequences. Specifically, in their example a realizable sequence was modified in finitely many initial terms only, in a way that was clearly incompatible with the arithmetic constraints on realizable sequences.

Elliptic Curves

While working at Ohio State I had attended a postgraduate module taught by Silverberg on elliptic curves, and bought my own copy of the outstanding book on the arithmetic of elliptic curves by Joe Silverman [274], so was not coming entirely new to the subject. Nonetheless, there was much to learn and Graham was a patient guide. In various combinations with PhD students we made some progress on the arithmetic of elliptic divisibility sequences. Graham was very interested in finding parallels between objects in the linear recurrence world and analogues in the bilinear or elliptic world. The aspect of this that turned out well involved the Mahler measure, which arose naturally in many places including the question of how rapidly a linear recurrence sequence grows. Graham and his student Bríd Ní Fhlathúin found an appropriate notion of elliptic Mahler measure, which proved to be conceptually useful [102].

All this provoked Graham to teach a year four module (that is, for the final year of the four year integrated masters programme) about Mahler measure and heights of polynomials. The lectures covered things like Lehmer's problem (which remains open, and is of significance both in number theory and in ergodic theory), and we decided to write the notes up together. Writing a book

[5] I never had the chance to meet Rufus Bowen, a brilliant mathematician who passed away very young in 1978 [290] having made enormous contributions to ergodic theory and dynamical systems. The paper on zeta functions is relatively obscure, but his creativity, technical power, and clarity of exposition comes across clearly in those few pages.

was new territory for both of us, and we sent out a few enquiries to publishers. Springer Verlag emerged from this, partly because of a lunch in Norwich with Susan Hezlet, who at the time was Mathematics Editor for Springer in the UK. The agreement was for us to provide camera-ready copy in the end, but there was some editing of the electronic files needed. The contracts were physical documents to be signed and posted back, and much of the details were sorted out using fax messages. I learned a great deal through this process. There were particular issues with the way in which I had produced the encapsulated postscript figures, as my work-around to ensure the fonts and symbols matched the rest of the text was not to their liking. This was all resolved eventually, and the book appeared in 1999 in the Springer 'Universitext' series. It proved to be quite widely cited and used [113].

Through this process I started to become more interested in—and sometimes a little obsessive about—the appearance of mathematical text on a page: Line and page breaks, widows and orphans, and more generally the way in which typesetting choices can make life easier or more difficult for the reader. Every now and then something would come my way that reminded me of this. Someone—I cannot recall who—persuaded me that commutative algebra is simply a conceptually easier subject if the rings, modules, and ideals in it are denoted with italic computer modern type like R, M, I rather than the older tradition of Fraktur typefaces like $\mathfrak{R}$, $\mathfrak{M}$, and $\mathfrak{I}$. This stuck in my mind as an interesting instance of the interaction between typography and effective communication of mathematical ideas. Many years later a conversation with Gunther Cornelissen in Utrecht encouraged me to put some of these thoughts down on paper, which resulted in a short and slightly facetious note [344] about the impact of typesetting choices on the readability of Mathematics.

> Most people think typography is about fonts. Most designers think typography is about fonts. Typography is more than that, it's expressing language through type. Placement, composition, typechoice. (Mark Boulton, retrieved on the 20th of May 2022 from the theysaidso.com web site.)

Graham and I also wrote up a more generalist undergraduate module in number theory built around the concept of factorization a few years later, which appeared in the Springer Graduate Texts in Mathematics series [116].

EPSRC and Manfred

Up until 1965, the Minister for Science and their advisers had made the main decisions on governmental support for scientific research, putting cases direct to the Treasury for funding. The Report of the Committee of Enquiry into the Organisation of Civil Science [299] (known as the Trend Committee, after its chair the civil servant Sir Burke Trend, later Baron Trend) led to the Science and Technology Act 1965, which created two new research councils: The Science Research Council, to take over from the Government Department of Scientific and Industrial Research and The Natural Environment Research Council. These joined the existing Medical Research Council and the Agricultural Research Council. The SRC became the Science and Engineering Research Council (SERC) in 1981. In 1994 the Director General of Research Councils UK (RCUK) led a reorganization of the four research councils, which *inter alia* resulted in SERC being split into three more specialised bodies: the Particle Physics and Astronomy Research Council (PPARC); the Engineering and Physical Sciences Research Council (EPSRC); and the Biotechnology and Biological Sciences Research Council (BBSRC).

Graham and I had made a few unsuccessful applications to the Engineering and Physical Sciences Research Council (EPSRC) over the years. This is an essential and sometimes frustrating part of academic life, particularly when an application gets excellent reviews but fails to make the cut in a given funding cycle. We had several successful applications for individual PhD studentships, but had been less successful with applications for post-doctoral research funding.

In 1998 we were thinking about some computational number theory problems, and perhaps it was this angle that led to a successful funding application. The grant was not tied to any one candidate, though we had one of Graham's students in mind initially. In the event she had already started working for an Irish bank. Meanwhile Manfred Einsiedler had completed his PhD [93] in 1999, and the timings worked out to mean he could take up the role (Fig. 11.2).

In this way Manfred spent the academic year 1999–2000 at UEA, formally working on the EPSRC funded project 'Practical and theoretical aspects of elliptic divisibility sequences'. The idea behind this project had its origins in two different chains of ideas.

The first concerned linear recurrence sequences, going back to the work of Marin Mersenne in the early seventeenth Century. He studied the occurrence of primes in the sequence $(1,3,7,15,31,63,\ldots)$ whose nth term is 2^n-1 for $n \geqslant 1$. Mersenne noticed that the values of this sequence are primes for n equal

Fig. 11.2 Manfred, Graham, and me in the garden of Graham's house in Reedham in 2000 (With permission of Tania Barnett)

to 2, 3, 5, 7, 13, 17, 19, 31, erroneously claimed the same for the indexes 67 and 257, and correctly claimed the same for 127. It is elementary to notice that $2^n - 1$ can only be a prime if n is a prime, but the properties of those indices n for which $2^n - 1$ is prime remain something of a mystery. Indeed, it is not known if this happens infinitely often or not. However there is a network of standard conjectured behaviours in this setting, roughly saying that any such sequence will be prime (or have a bounded number of prime factors) infinitely often. The same questions make sense for any integer linear recurrence, and this type of question was part of the motivation of some influential work of Lehmer [176]. He was interested in using linear recurrence sequences of a certain form to generate large primes, and so became interested in sequences that do not grow too rapidly. These sequences, now called Lehmer–Pierce sequences, grow at a characteristic rate determined by (what we now call) the Mahler measure of an associated polynomial. In a remarkable computational *tour de force*, Lehmer noticed that the polynomial

$$x^{10} + x^9 - x^7 - x^6 - x^5 - x^4 - x^3 + x + 1$$

had the smallest positive growth rate among the integer polynomials he looked at, and asked if there could be an integer polynomial with smaller growth rate. This question, now called Lehmer's problem, remains open. Extensive computational experiments and theoretical developments since then have shown that among all integer polynomials of degree smaller than 55 no smaller example exists.[6] The 'linear' part of the project was to explore the question of where primes appear in Lehmer–Pierce sequences with small growth rate. The growth rate is given by the Mahler measure of the characteristic polynomial associated to the recurrence, so the starting point was to use the computational work of others that produced a list of polynomials with small measure.

The second concerned 'bilinear' recurrence sequences, which arise naturally in the part of arithmetic and geometry covered by the phrase 'elliptic curves'. Specifically, we wanted to study the appearance of primes in elliptic divisibility sequences. Here even the conjectural landscape was less clear—some arguments of the Chudnovsky brothers [53] pointed at the possibility that sequences like this might (after controlling a small number of prime factors appearing to high powers) be prime infinitely often. Our heuristic arguments suggested the opposite—that for a given elliptic divisibility sequence there would only be finitely many prime incidences. In this setting the growth is quadratic–exponential, and the rate is determined by so-called height data from the underlying elliptic curve. The arithmetic of the sequences is also more complex, with badly behaved primes again coming from properties of the curve.

Both projects were interesting in that the heuristic arguments brought in interesting ideas, and the computational approach would require us to learn a bit about using computers in number theory. They were also unsatisfactory, in that the chasm between any amount of finite evidence and what is true or not true about any statement in which a parameter can take on infinitely many values is unbridgeable.[7] For questions in this area—whether $2^n - 1$ is a prime

[6]The authors Mossinghoff et al. [225] estimate that almost six years of CPU time on workstations at Simon Fraser University were required for this search. Reducing the search in given degree to a finite one requires non-trivial number-theoretic arguments.

[7]Viewed from the lofty grandeur of the infinite, all numbers are negligible and all calculations irrelevant. Number theory has thrown up many questions in which the behaviour of the first few terms is deceptive, for startlingly large values of 'few'. A simple example is given by Mertens' function, given by $M(n) = \mu(1) + \mu(2) + \cdots + \mu(n)$. Here $\mu(n)$ is the Möbius function, which is defined to be 1 if n has an even number of prime factors and no square factor, -1 if n has an odd number of prime factors and no square factor, and 0 if n has a square factor. Deep questions in number theory are connected to how much or how little cancellation in the values of μ occurs when its values are added. Around the end of the nineteenth Century Stieltjes and Mertens [200] conjectured that the absolute value of $M(n)$ could never exceed $\sqrt{n}$. This easily stated conjecture would, for example, imply the famous Riemann Hypothesis. In 1985 Odlyzko and te Riele [231] showed that the conjecture of Stieltjes and Mertens was false, and that there must be some value of n smaller than approximately $10^{1.39\times 10^{64}}$ but larger than 10^{16} for which the absolute value of $M(n)$ is larger than $\sqrt{n}$. No explicit counterexample is known, though there are more 'effective' versions of the result due to Pintz [244]. There is every indication that if all the resources of the earth were devoted to building computers to search for such a thing, none would be found.

for infinitely many values of n, for example—the journey can feel circular. Some heuristic arguments are assembled, vast computer searches carried out, and at the end of the process we have moved from not knowing the answer to not knowing the answer, but perhaps in a more informed way.

> What we call the beginning is often the end
> And to make an end is to make a beginning.
> The end is where we start from [...]
> We shall not cease from exploration
> And the end of all our exploring
> Will be to arrive where we started
> And know the place for the first time.
> (From 'Little Gidding' by T. S. Eliot)

We combined existing knowledge of Lehmer–Pierce sequences of small growth rate (arising from integer polynomials of small Mahler measure), standard tables of points on elliptic curves with small Néron–Tate height, and some elegant programming by Manfred to explore the two questions numerically, and published the results together with some context and heuristic arguments [74, 76]. The calculations were carried out on the network of desktop Linux computers in the Mathematics department, by making them run 'nicely' (that is, with lower priority than anything the person using the computer in their own office wanted to do). In later work, and rather unexpectedly, Graham was able to prove that many elliptic divisibility sequences do indeed take on only finitely many prime values [118–121].

Alongside the work we had committed to doing on the funded project, we explored other research areas and started some other projects. Manfred and I had already started some work on presentations of modules in commutative algebra, using Auslander–Buchsbaum theory to work out properties of the associated dynamical system from a presentation of the module defining it [81].

Together with Graham we explored the idea of building a sort of dictionary between on the one side linear recurrence sequences, periodic points and entropy for compact group automorphisms, Mahler measures and on the other elliptic divisibility sequences, putative periodic points and entropy in some unknown system, elliptic Mahler measures [75]. This project generated more questions than answers, but the questions led to interesting developments.[8]

[8]For example, Joseph Silverman and Nelson Stephens went on to show that (the absolute value of) an elliptic divisibility sequence can never be a count of periodic points [275] and Florian Luca and I showed that the terms of an elliptic divisibility sequence cannot arise by sampling a linear recurrence sequence along the squares [184].

The idea of elliptic analogues of linear phenomena was not new, and has some reasonable linkages in arithmetic.

The three of us also started looking at periodic points for rational maps on curves in the area now called 'arithmetic dynamics', leading to some publications a few years later [77, 78].

The most significant side project started during Manfred's year at UEA concerned directional dynamics. Manfred, Richard, and I were interested in the recent work of Boyle and Lind [33] on 'subdynamics'. This explored the idea that within an action of $\mathbb{Z}^d$ for $d > 1$ there are rich structures visible in lower-dimensional sub-actions. Their viewpoint was primarily topological, and we wanted to understand how this all worked in the special setting of commuting group automorphisms. We made partial progress, in particular enjoying a moment at a whiteboard in my office at UEA when a key coding lemma we called the 'Lego argument' (because of the way it could be understood using aligned rectangular blocks of coordinates) became clear. Manfred discussed one of the examples we used, dubbed the 'space helmet' (because of the shape of its non-expansive set) with Doug when they both were in Vienna, and Doug joined the project (Fig. 11.3).

The space helmet gave Manfred the insight that the complete portrait of non-expansive behaviours could be arrived at by combining infinitely many objects called amoebas—logarithmic images of varieties—together, one for each prime and one 'at infinity'. The resulting 'adelic amoebas' featured in an intense period of work using 'Microsoft NetMeeting', a video conferencing platform that was bundled with versions of Windows prior to Windows XP. NetMeeting provided an audio link and—crucially—a shared whiteboard. By modern standards this may have looked rather primitive, but really it was ahead

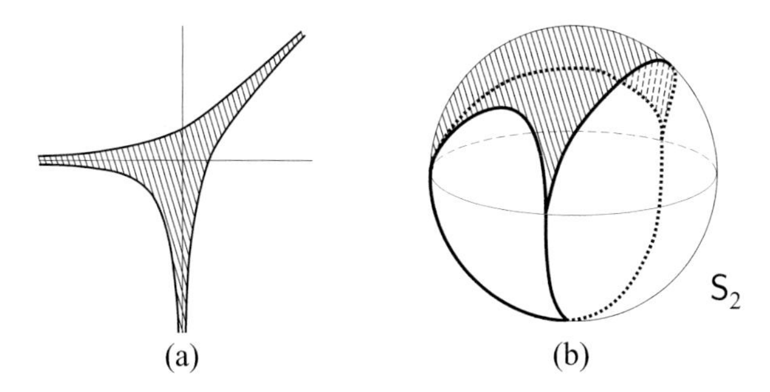

Fig. 11.3 The 'space helmet' (on the right) and its projection onto the plane (on the left), taken from [79] (©Cambridge University Press 2001; reproduced from *Ergodic Theory and Dynamical Systems*, Volume 21(6), 1695–1729 with permission)

Fig. 11.4 The authors of the expansive subdynamics paper: Doug Lind, Manfred Einsiedler, Richard Miles, and me (from right to left) at Warwick in 2011 (With permission of Tania Barnett)

of its time for the domestic user working over a low speed dial-up connection. It allowed Manfred and I to sit in our small house in Norwich and work with Doug in Seattle while sharing a whiteboard in real time. The project grew to include a real clarification of entropy rank (a measure of the 'size' of an action), and a complete description of the expansive half-spaces in terms of the prime ideas associated to the system [79] (Fig. 11.4). When Manfred went on to work with Doug in Seattle they developed these ideas further as part of a more systematic study of adelic amoebas [94].

Alf

Graham and I used another EPSRC grant to arrange for a visit from Alf van der Poorten during the academic year 2001–2002.[9] Alf was accompanied by his wife Joy, and we enjoyed several wonderful evenings with them both. One of those dinners ended up—as many did at the time—with Alf and me in

[9]Alf was a distinguished Dutch–Australian number theorist. He was born into a Jewish family in Amsterdam in 1942, and survived the war by hiding under a different name with a gentile family. His extraordinary story is sketched on Wikipedia [356] and in his obituary [154].

the garden holding glasses of wine, and with me cadging cigarettes from Alf. Many ex-smokers have a long period like this, in which they only smoke other people's cigarettes and thereby maintain the illusion that they no longer smoke themselves. On this occasion our nine year old daughter Adele saw us from her bedroom window, and her stern grilling the next morning helped ensure that it was the last cigarette I ever smoked.

Mathematically we were looking at the general problem thrown up by Yash Puri's work. In particular, we tried to characterise the realizable linear recurrence sequences, making only partial progress [111]. The central question was clarified much later by Gregory Minton, who completed the characterization of the arithmetic problem [217]. The other facet of the problem concerns sign questions related to linear recurrence sequences, a notoriously difficult problem.

We were also in the early stages of finding a good solution to a potentially awkward situation. Graham and I had started to write a book on recurrence sequences, with a focus on their arithmetic properties. On the other side of the world Alf and Igor Shparlinski had started to write something similar, but with much more of a Diophantine emphasis and bringing in Igor's formidable work applying estimates for trigonometric sums to linear recurrence sequences. We agreed to combine forces, and I started to pull the two manuscripts together into a coherent whole. With two small children and a busy day job, most of this was done on our home computer late at night, and for much of the time was carried out by email exchanges between Igor and me. Few things are quite as exhausting as a period of intense collaboration with someone in a significantly different time zone, and Igor was in Sydney. At one point Graham felt he was contributing too little and wanted to drop out, but I was keen to keep all four of us on board. I learned a great deal through this process, including much Mathematics from Igor and much about typesetting and editing from the mother of invention.

It was a complex project for several reasons, partly because the four authors were never in the same place at the same time (Fig. 11.5). One of the challenges was making sense of what in the end became a bibliography comprising 1382 entries, and the book has probably proved useful to others largely because of this. I had tried various ways to improve the handling of a database of references over the years, and this was my moment of conversion to BibTeX.[10] Ensuring that the eye was not offended by the *Journal of Number Theory* being

[10] BibTeX is reference management software that follows the LaTeX philosophy of separating the bibliographic information from how it is presented in the typeset document.

Fig. 11.5 Graham and me with Alf van der Poorten in 2001 (left), and with Igor Shparlinski in 2003 (right), in our house in Norwich (With permission of Tania Barnett)

abbreviated to *J. Number Theory* in one place and *J. Number Th.* in another was not manageable across such a huge list of publications without the assistance of a carefully collated and partially automated database. MathSciNet,[11] the web-based front end to the *Mathematical Reviews* database, had launched a few years earlier, and this was my first real exposure to the power of well-organised databases combined with effective software to exploit them.

We published the outcome of all this in the 'Mathematical Surveys and Monographs' series of the American Mathematical Society [110]. While the AMS were easy and pleasant to deal with, this had the slightly annoying consequence of small and more or less unusable royalty cheques in US dollars arriving for years afterwards.

Banff

By chance two workshops of interest for me took place in the Summer of 2003 in the same place, the beautiful Banff International Research Station in Alberta. The first of these was on the theme of 'The Many Aspects of Mahler's Measure' and took place at the end of April. Unfortunately on the day the

[11] *Mathematical Reviews* was launched in 1939, producing its first volume of reviews of published articles in the mathematical literature in January of 1940 under the guidance of the founding editor Otto Neugebauer. For many years a wall of volumes of *Mathematical Reviews* with covers in 'Princeton orange' was one of the signifiers of a mathematical library. The web-based access under the name MathSciNet started in 1996. A key feature of *Mathematical Reviews* is some of the underlying data structures, including a meticulous approach to identification of individual authors which remains robust in the face of name changes and the vagaries of translations between multiple languages. A brief account of the interesting history of the database is given in an article by Richert [255].

workshop was to start 60 cm of snow fell on Calgary International Airport. I was on a flight from London when this happened, and in the end my flight was one of very few that did manage to land there. Most of the participants from the USA particularly were delayed, but in the end almost everyone managed to get there and it was a highly successful event. In particular, it facilitated an exchange of ideas on the extraordinary connections between the values of Mahler measures, Deligne periods, and motivic cohomology.

The second was on 'Joint dynamics' and took place at the end of June. There were several exciting developments in the ergodic theory of group actions around this time. Bryna Kra presented on her recent work with Bernard Host on an important non-conventional ergodic theorem [153], and Dan Rudolph gave an account of a remarkable new method he and Benjy Weiss had used to prove that actions of amenable groups with the K property are mixing of all orders [263]. There were also several presentations on progress in rigidity theory.

The dramatic setting and beautiful places to walk make these two events stand out in memory from the large number of workshops and conferences I have attended.

Patrick and Gerry

Two senior figures working for British Telecom contacted the department out of the blue, looking to enrol for part-time PhDs in number theory. They had been doing some modules with the Open University which had rekindled an interest in Mathematics. After a few years Gerry Mclaren decided not to continue, but he had already been involved in a successful project on uniform appearance of primitive divisors across some families of elliptic curves [103]. Patrick Moss proved to be a quite exceptional mathematician. He quickly started producing interesting results, largely in the area of realizable sequences. He developed the notion of local realizability (for the part of a realizable sequence divisible by a fixed prime number) and related it to subtle questions about arithmetic conditions on a sequence that guarantee realizability by a group automorphism. He also proved that several sequences arising in number theory or combinatorics are realizable, including the Euler numbers and the Bernoulli numbers. Patrick also made real progress on 'realizability preserving' maps, showing that if a realizable sequence is sampled along the squares, or the cubes, and so on, then it becomes another realizable sequence. He also made the remarkable observation that the Fibonacci sequence $(1, 1, 2, 3, 5, \dots)$ when sampled along the squares and multiplied by 5 becomes a realizable sequence, a result that led to several later developments. In fact his thesis [224] was full of elegant ideas which led to other interesting work.[12] He also completed his thesis remarkably quickly, and we had to obtain special permission for him to complete his part-time studies several years ahead of schedule.

Sergiĭ and the Max Planck Institute

At various conferences I had conversations about Mathematics with Sergiĭ Kolyada, a mathematician at the National Academy of Sciences in Ukraine. We also discussed some vague ideas for conferences. These crystalised into an application to the Max Planck Institute for Mathematics to run a special program on 'Algebraic and Topological Dynamics' from May to July in 2004 and a European Science Foundation Exploratory Workshop 'Dynamical Systems: From Algebraic to Topological Dynamics' July 4–9, 2004. This was my first substantial role in organising conferences, aided enormously by Pieter Moree and the impressive administration of the Max Planck Institute in Bonn. Even before Brexit this was a natural choice, as it was a well-funded and superbly run research centre located within the Schengen zone.

It was with some trepidation that I told the assembled audience that Sergiĭ and I planned a conference proceedings. This is a daunting business, because it amounts to asking researchers to submit articles long before there is any guarantee the proceedings will be published at all. It also involves effectively soliciting articles that may later be rejected—and it certainly involves a substantial logistical and administrative burden. Cambridge University Press (CUP) published a long-standing London Mathematical Society *Lecture Note Series* many of whose volumes comprised conference proceedings, the American Mathematical Society had a series called *Contemporary Mathematics*, and Springer's famous Lecture Notes in Mathematics had in the past often comprised conference proceedings. It is the Springer Lecture Notes (of which there are several thousand) that often explain an entire wall of yellow spines in a Mathematics library. I contacted CUP and the AMS with an outline of our plans to publish a proceedings from the special program and workshop. CUP was surprisingly off-putting in response, being concerned about the longevity of the papers that would come forward, while the AMS was extraordinarily helpful. So we settled on the *Contemporary Mathematics* option. In the end

[12]Patrick graduated in 2003, with external examiner Prof. Shaun Bullet from Queen Mary University of London [224].

a good number of excellent papers came in, and the edited proceedings 'Algebraic and topological dynamics' appeared the next year [170] edited by Sergiĭ Kolyada, Yuri Manin,[13] and me.

At the time this felt like nothing more than an interesting episode. The program and workshop had gone well, and the edited proceedings proved to contain some highly cited and influential papers. The editorial side of the business and working closely with Sergiĭ were enjoyable and satisfying—but immensely time-consuming. In fact far from being a single episode this was to be the start of a long and happy collaboration.

Together with Martin Möller from Frankfurt and Pieter Moree from MPI Sergiĭ and I put together another application, which resulted in a Special Program from May to July 2009 and an International Conference that ran from the 20th to the 24th of July 2009 at the Max Planck under the title 'Dynamical numbers—interplay between dynamical systems and number theory.' This again resulted in a successful edited proceedings the next year [169], edited by Sergiĭ, Yuri, Martin, Pieter, and me.

More or less the same group went on to run a Special Program from June to July 2014 and an International Conference July 21–25, 2014 at the Max Planck once again, this time under the title 'Dynamics and numbers'. This also resulted in an edited proceedings with a good array of interesting papers [168] edited by Sergiĭ, Martin, Pieter and me.

It was only natural that Sergiĭ and I put together another application in the same spirit, this time with Anke Pohl from Bremen and L'ubomír Snoha from Banská Bystrica. We were successful again, and started making detailed plans for 'Dynamics: topology and numbers', comprising a conference in the first week of July 2018 at the Max Planck. Tragically Sergiĭ passed away suddenly on the 16th of May 2018, and the planned conference was a sad affair despite its scientific successes. Pieter, Anke, L'ubomír, and I edited a conference proceedings, which was dedicated to Sergiĭ's memory and contained articles about his life and Mathematics [219].[14] In the editor's preface to the final volume we used some words of Mike Shub in a memorial volume for Rufus Bowen many years earlier: Don't forget to say that we all liked him.

---

[13]Yuri Manin was a Russian mathematician who made significant research contributions in Diophantine and algebraic geometry particularly. He spent many years as one of the Directors of the Max Planck Institute, and passed away in 2023.

[14]Sergiĭ played a major role in reviving the Kyiv Mathematical Society and was its President from 2006 to 2014. He was awarded the State Prize of Ukraine in 2010. A short mathematical obituary appeared in Ukrainian [264] in addition to the articles in the volume [219].

The first three edited proceedings [168–170] and the events behind them brought Sergiĭ great pleasure, and all four of them reflected several strands of interesting interaction between dynamical systems and number theory.

Vicky and Helen

Graham Everest and I took on two jointly supervised PhD students at more or less the same time, Helen King[15] and Victoria Stangoe.[16] Helen worked on the arithmetic of elliptic divisibility sequences, and together with Graham published results on the appearance of prime powers in such sequences [119]. The main focus of Helen's work was in number theory, and while there was no formal arrangement Graham was her main supervisor.

Vicky began with some ideas contained in the earlier work of Chothi [50, 51], looking at the 'dynamical zeta function' of some of the systems studied there. In the simple case she looked at first, the dynamical system dual to multiplication by 2 on $\mathbb{Z}[\frac{1}{6}]$, there was an unexpected development. Vicky saw that the dynamical zeta function itself satisfied a functional equation that forced it to have singularities at a dense set of points on its circle of convergence. I was attracted to this argument for many reasons, one of them being that the key insight inside the calculation was something utterly elementary: All the divisors of an odd number are odd. This appearance of a 'natural boundary' in a dynamical zeta function shifted my perspective certainly on what is 'normal' or 'typical' for dynamical zeta functions. I presented this work at the Max Planck Institute for Mathematics in Bonn, and included a more speculative graph aimed at illustrating the wild behaviour in an associated 'prime orbit counting' expression. Christopher Deninger was in the audience, and had a different and much more sophisticated point of view within which the appearance of a natural boundary was not really expected, so our simple example attracted some interest (Fig. 11.6). We published this argument in a short paper in the

[15] Helen graduated in 2005, with external examiner Prof. Nelson Stephens from Goldsmiths College, University of London [166].

[16] Vicky graduated in 2005, with external examiner Prof. Richard Sharp from the University of Manchester [284].

Fig. 11.6 Discussing the natural boundary phenomenon from the thesis of Vicky Stangoe after my talk at MPI Bonn with Chistopher Deninger, Manfred Einsiedler, and Doug Lind, 1st July 2004 (Photo taken by Sergiĭ Kolyada and reproduced with permission from Irina Kolyada)

conference proceedings, including the speculative (and, as it turned out later, slightly deceptive) numerical data on the prime orbit counting function at the suggestion of the referee [108].

12

Two New Roles in Norwich

During the 1990s the Physics part of the School of Mathematics and Physics was closed, primarily because of falling student intakes. The remaining School of Mathematics was not without its troubles, and by the late 1990s was in an extremely difficult financial position. I was largely inattentive to all this, concentrating on my own teaching and research roles. This only changed when there was a discussion about who should take over as Head of School. The process was managed by the Dean of Science, who would take soundings and if necessary hold a vote.

There are periods in the life of an academic department that might easily appear in a C. P. Snow novel, and this was one of them. In our case there was a tied vote at one stage, for example. There were, however, many differences between the imagined Cambridge college where Lewis Elliot worked in 1937 (or indeed Christ's College where Snow was a Fellow when he wrote 'The Masters' in the late 1940s) and the School of Mathematics at UEA in the 1990s. The most significant of these was that the security of the financial underpinnings of the fictional college depicted in his novel had long since departed, and the threats of closures, forced mergers, and redundancies were and remain a real part of academic life in the UK outside a handful of institutions. In the midst of the process of choosing our new Head of School (at the time, for obscure historical reasons, the actual title was, confusingly, Dean of the School of Mathematics) one of my colleagues knocked on my door and said that in his opinion either I put myself forward or the emergent obvious candidate would become head—and would be unlikely to change the feeling of steady decline.

T. Ward, *People, Places, and Mathematics*, Springer Biographies,
https://doi.org/10.1007/978-3-031-39074-6_12

This conversation forced me for the first to engage with a 'threshold concept'.[1] I was one of the many people who spoke breezily and at times grumpily about 'the University' and even 'the Department' without consciously understanding that neither of those things really exist beyond being the aggregation of the capacities, efforts, attitudes, and concerns of their staff and students. This brief conversation made me start to think about my own role—beyond doing some research and teaching—in the future of the department. My initial response—on reflection, a pompous one—was to say I would stand but would not do so if anyone else did. After some to and fro and further rounds of consultations by the Dean, this led me to a somewhat delicate conversation with the obvious candidate, and resulted in my taking over as Head of School in 2002.

I was a Reader at the time, and one of the many difficulties I encountered was that part of the job of Head of School involved facing down some members of the Professoriate. With all parties still happily alive, on the question of what these differences were about I must be discreet.

Before I took over the University had arranged for a review of the school, led by two external mathematicians who wrote a report with recommendations about what to do next, with a separate confidential annex for the Dean of Science, the Vice-Chancellor, and the incoming Head of School. This external review was triggered by the unsustainable financial situation and an undoubted negative perception of the school in the wider university. The report and the review process itself were painful but useful, and within a few years the department had become more at ease with itself and had better relations with the wider university.

The latter was clear to me as a high priority. The spreadsheets with budgets and student number plans and the institutional planning cycles matter greatly, but in the end decisions about departments—including existential ones—are made by groups of people steered by feelings and instinct just as much as they are by logic. When faced with difficult decisions in Universities, we tend to follow Edgar, son of the Earl of Gloucester, just as much as the logical consequences of the financial forecasts. If that group, whoever they may be, thought the School of Mathematics was realistic about its position and full of dedicated colleagues willing to engage with the wider university and be good

[1] The notion of threshold concepts was brought into education through the work of Meyer and Land [201], who described it as 'passing through a portal, from which a new perspective opens up, allowing things formerly not perceived to come into view. This permits a new and previously inaccessible way of thinking about something.' The identification of such concepts is a real difficulty in using these ideas, but Mathematics is certainly filled with them.

citizens, then the massive deficit looked like something that could shrink if the School was adequately supported. If that group thought the School of Mathematics was full of arrogant colleagues who made no effort to engage with the wider university nor be good citizens, then it would be equally self-evident that the deficit would grow and that the School had no viable future.

> The weight of this sad time we must obey,
> Speak what we feel,
> not what we ought to say
> (Edgar, in Shakespeare's King Lear)

Being Head of School was among the more difficult roles I have held. At one distinct moment before I started, I said to someone outside the University that at least the current head had someone who checked up on how they were doing each morning—and then stopped short, because that person was me. There were some evenings where I placed a high stool at our kitchen counter in Hellesdon, arranged a telephone, a bottle of whisky, a jug of water, and a glass in a semi-circle around my notebook, and steeled myself for a few hours of being shouted at.

Much has been written about the role of Head of Department in a University context, and it is unlikely I can add anything useful to that. Any model is inadequate. An autocratic approach is not compatible with the complex mix of motivations and individualistic spirits of academics. On the other hand an entirely democratic approach simply replicates the culture of a school, and changing that culture may be the most pressing priority. The role is temporary, but has to be approached without worrying about one's own future. Arguably the candidate should be chosen from the reluctant willing rather than the enthusiastic eager. Above all, during the period of transition initiated in the 1980s, the role was predicated on a history of stable funding and tight control of student numbers but confronted a political environment of ideological dedication to market forces and a perception of higher education as an essentially private asset.

My initial term of office was three years, during which I had to make difficult decisions about my colleagues and friends in a small department, before handing over the reins to someone else to do the same. In the end I was asked to do a further two years, in part because of a turbulent reorganisation of the University from schools of study as the primary budget holding organisational unit to Faculties. The challenge of being asked to be manager of a small group of friends one year and not the next remains a central issue for middle management in Universities. The system made perfect sense in the era of block grants, where the role of University management was to apportion a secure income stream from government. As the system of funding higher education shifted, the demands placed on heads of department grew, and we all struggle with it.

The most startling thing I experienced was a dramatic shifting of my own perception of some colleagues. Like turning a kaleidoscope and watching the image become jumbled before settling into a new configuration, the people I knew best and thought I could rely on most rearranged themselves, and people I did not know well emerged as wonderful colleagues and drivers of change. I learned a great deal about myself and my colleagues through the process, both positive and negative. A painful lesson was that aspiring to get every single person behind a shared vision is too high a threshold for success. The larger and more satisfying lesson was that it can be just as exciting to facilitate what other people do as it is to do things oneself. From my point of view several things improved over the years, and the department became a happier place—but this was through the efforts of many people.

This period also reinforced and amplified a lesson that I had started to learn from interacting with Manfred Einsiedler. The sense that my contribution in education comprised the lectures, seminars, and tutorials I gave and my contribution in research was the theorems I proved and published shifted markedly. It became clear that enabling and supporting more talented people in both these activities would have far greater impact. It is of course possible that this lesson is nothing more than a sophisticated rationale for a long period in which my direct contribution with my own hands to both education and research is so slight.

My own research during those years was largely maintained by the efforts of Graham and Manfred, though there were some other visits and activities to keep ideas ticking over. One of these was a London Mathematical Society small grant which allowed Siddhartha Bhattacharya to visit from the Tata Institute in Mumbai. He stayed with us for a week, and both the children were very impressed with his soccer skills in the back garden. Siddhartha had made interesting contributions in studying the 'rigidity' of measurable conjugacies between algebraic dynamical systems, but for some reason we wanted to look at the question of rigidity for topological conjugacies. This is a delicate issue, but we ended up with a clean understanding of the situation in our setting, with topological rigidity in the context we considered being essentially equivalent to finite topological entropy [23]. This carried on a strand of thinking from the thesis work of Gary Morris concerning the implications of 'size' on dynamical properties. The context was superficially entirely different, but the interplay between having space in a dynamical system to allow certain behaviours certainly found an echo.

I visited Manfred in the summer of 2005, and on that trip spoke about the topological rigidity result both at Princeton and at Pennsylvania State. Anatole Katok,[2] a hugely influential and sometimes intimidating figure in ergodic theory and dynamical systems, was in the audience for the latter of these seminars. He had a reputation for being unable to hold back his enthusiasm, opinions, objections, or suggestions during seminars, at times more or less taking over entirely, so I was a little concerned when he became rather agitated during my talk. I readied myself for anything ranging from an explanation that our work was uninteresting to an error being pointed out—or to a fascinating and enlightening comment, and invited him to say what was on his mind. He proceeded to turn back to face the audience and explain that this form of rigidity (that is, topological rigidity that could arise for single maps rather than measurable rigidity brought about by the presence of several commuting maps) was a very delicate business which he liked a lot, and that the audience should pay close attention to understand this key difference (from measurable rigidity).

Graham ensured that there was always some research idea being passed back and forth with him, and was a constant source of encouragement. Manfred and I kept in touch—during the years I was Head of School he worked at the University of Washington in Seattle, Pennsylvania State University in State College, Princeton University, and Ohio State University—through visits, email, and NetMeeting. We worked on some ideas in entropy geometry and rigidity for commuting automorphisms both in a totally disconnected setting [83] and in certain connected group settings [84]. The latter result was later enormously generalised by Bhattacharya [24]. I also made several short visits to see Manfred and Doug in Seattle, and to see Manfred in State College and Princeton.

Sadly I have not kept the paper placemat from a coffee house in Seattle where Manfred and I first sketched a plan for a book. The original conception was that we would write a book that carefully articulates some required background and then proceeds in a methodical and motivated way to an account of the work done by Manfred Einsiedler, Anatole Katok, and Elon Lindenstrauss on the Littlewood conjecture in Diophantine approximation [95]. This 'book

[2]Anatole Katok was one of the major figures in the development of dynamics in the second half of the twentieth Century. Several obituaries detailing something of his contribution and life have appeared, including one by his former student Hasselblatt [146] in *Ergodic Theory and Dynamical Systems* (one of the journals he was involved in founding) and a short note in the *Moscow Mathematics Journal* [4].

project' grew in scope over the years, remains active, and is now a more or less permanent part of our lives.

The state of the project as I write this looks as follows. Our initial draft of the foundational ergodic theory involved became more ambitious and somewhat unwieldy. At a crucial moment Manfred saw that it could split rather neatly into two pieces. The first, which develops the machinery of ergodic theory, conditional measures, dynamics on homogeneous spaces of small rank, and recurrence, appeared in 2011 in Springer's Graduate Texts in Mathematics series [85]. The second concerned entropy and its manifold applications in number theory and homogeneous dynamics. For this entropy volume we joined forces with Elon Lindenstrauss, and the project has been in that liminal state familiar to all writers of being 95% completed and almost ready to go for some years [80]. The way in which this conceptual and technical framework is used in homogeneous dynamics is the topic of a third volume, which is at an earlier stage of development [90]. With a well honed habit of writing together, Manfred and I have also written other books along the way—a graduate text on Functional Analysis and Spectral Theory linked to modules he taught in Zürich [88] which also appeared in the Graduate Texts in Mathematics series, a still active and lengthy project on unitary representation theory also based on modules he taught [91], and a more elementary text on number theory and group theory linked to some classes aimed at bridging high school and university Mathematics for some talented and engaged Swiss students, together with Menny Aka which appeared in the Springer Undergraduate Mathematics Series [8].

These ongoing projects have been a wonderful thing for me for three reasons (Fig. 12.1). They mean I am still learning some Mathematics, albeit at a snail's pace. It continues the collaboration with Manfred, which has already reached a quarter Century. Finally, it maintains a link—albeit an extremely tenuous one in my case—with real students attending high-level modules somewhere.

The London Mathematical Society

The year 2002 was a busy time for several reasons. Susan Hezlet had moved to lead the publications at the London Mathematical Society (LMS), and she asked if I would be willing to join the Editorial Advisory Board. The system involved a quite large panel of subject specialists (in my case, ergodic theory and dynamical systems) who would triage submissions, arrange referees, and ultimately make a recommendation to the managing editors of the three journals (the *Bulletin*, *Journal*, and *Proceedings*, more or less separated by

Fig. 12.1 With Manfred and two of our books at the Springer bookstand during a Clay mathematics conference in Oxford in September 2017

length of article). It also meant dealing with some frustrated authors: Even then there were far more submissions deemed 'publishable' than the available page allocation permitted. The approach—most of the time—was for the managing editors of each journal to keep the backlog under control, which more or less translated into a page limit each month. This began a long and enjoyable association with the LMS, and I later joined their Publications Committee and carried out other editorial roles. It also began an interest in the mechanics, finances, and logistics of academic publishing more generally. Susan was a patient teacher, and I learned a great deal from the various gatherings of the editorial board and other committees or working groups.

It was great to be working with Susan again, and the regular meetings in Russell Square brought together a big group of dedicated mathematicians from all areas (Fig. 12.2). I had benefited many times from small LMS grants for various research visits, and had used their journals as both author and reader many times, so it was particularly satisfying to be able to give something back in this way.

Fig. 12.2 Shaun Stevens and me at a meeting of the London Mathematical Society editorial board in de Morgan House on the 9th of September 2011 (With permission from Susan Hezlet)

Ice and Other Perils

Sir David Eastwood took over as Vice-Chancellor of UEA in 2002, so I began my tenure as Head of Department with a huge financial deficit and unknown new leadership of the institution. A meeting was scheduled for each Head of Department with the new Vice-Chancellor, and I prepared for mine with some anxiety. Spreadsheets, proposals, research plans, all prepared with the aim of answering the question in the air of whether the department had a stable future at all. I entered the large and daunting Vice-Chancellor's office, which was later to become all too familiar, and he walked over to me, shook my hand in a friendly manner and asked "Your PhD was on something called ergodic theory—what is that about?" At the end of that first meeting I rather nervously asked what I should call him and the response was simple: "David".

It was a much friendlier beginning than I had expected, and his unequivocal backing for the department helped us greatly. He remained a weighty and supportive figure long afterwards, always happy to act as a referee decades later. My first meeting with the new Vice-Chancellor was a vivid illustration of the power of the obvious corollary to the frequently invoked maxim that a

University's only real asset are its staff and its students. If that is true, then very little is more important than how its leaders treat people.

His predecessor, with whom I had interacted a little on Senate and on Council, as incumbent Head of Department, and in the setting up of the review of the department, never gave the slightest indication he knew my name.

I was quickly pulled onto various bodies and committees whose titles, under the action of suitable permutations of the words, are universal in higher education: Information Strategy and Services Committee, Vice-Chancellor's committee on restructuring the University, ICT framework project board, Human Resources Committee, Academic appeals and disciplinary tribunal, Staffing Structure Panel (Finance), Working Group on Recognising and Rewarding Good Teaching, Working party on Governance, Staffing Structure Panel (ISD), and Chair of High Performance Computing group. While onerous, and difficult to combine with a normal teaching load, this brought me in contact with many different people and viewpoints. It was an unplanned but effective introduction to many aspects of the inner workings of a University.

The department was in a perilous position financially. At the time UEA ran an intricate 'Resource Allocation Model' labelled by academic years, and 'RM34' showed that four departments had 'unbridgeable deficits', one of them being Mathematics. This led to a series of difficult meetings with a group from the University Planning and Resources Committee to agree detailed plans to reduce expenditure and negotiate a loan to see the department through to a sustainable position. There was real solidarity between the heads most involved, but this was a stressful and difficult period. For various reasons, including external changes in the funding model, the finances stabilised and we started to be able to think about the future more positively. I was asked for several confidential briefings to help the relevant members of the University executive understand our thinking. These were mostly about projected numbers of students, research grants, staffing plans and non-staff costs, but they also reflected my concerns about the way the department felt about itself and the way the wider community saw us. My note on the 8th of February 2003 includes the line: 'As incoming Dean I have been working very hard to try and shift the School's *Weltanschauung* from beleaguered and—occasionally—almost paranoid, to a more constructive and optimistic view.'

Despite some of the difficulties involved in the role, I enjoyed being head of department and getting to know the wider University. My own career position was however somewhat awkward. In the normal course of events it would be the head of department who initiated the opaque process of putting someone forward for promotion to the School's Promotion Committee, which I chaired *ex officio*. This did not particularly concern me, and I was not aware that behind the scenes Graham Everest, whom we had managed to promote to a chair, and the kindly Dean of Science, Andrew Thompson,[3] initiated a discussion with the Vice-Chancellor and some external referees about this. These discussions eventually led to a meeting of the School's committee chaired by the Dean of Science. This is always a lengthy process, and after some rounds of letters and consideration within the university, a final interview panel for my promotion was arranged for January of 2005.

The department's usual Christmas meal was held in December 2004, a convivial and informal dinner at a Mexican restaurant in Norwich—too convivial in my case. The party—a mixture of colleagues and PhD students—decided to go on to the temporary Christmas ice skating rink in the centre of Norwich. Several things came together to disastrous effect that evening. I had, undeniably, had a considerable amount to drink. Many years earlier Tania and I had attended an ice-skating course in Birmingham, so I had some confidence in ice skating and may or may not have had some competence. The ice itself was deeply rutted and in poor condition. So—I was told later, no memory of that day remains—when our PhD student Helen King fell over, I chivalrously and briskly skated over to check that she was not hurt. The account of others present is that when I turned the skates sideways to brake I hit a deep groove and fell heavily on my side. From my point of view I returned to partial consciousness while being rushed through a long corridor in the Norfolk and Norwich University Hospital. It took some time to work out what I was seeing, which seemed to be fluorescent strip lights falling, fast and silently, about two metres in front of me. This perplexing scene eventually resolved itself with the realization that it was the bed I was on that was moving, down a long corridor illuminated with strip lights and very rapidly.

The real misfortune was that, understandably, so much attention was paid to my broken skull that we all missed the fact that the far more serious injury was in my back. Recovery from the head injury was slow but reassuring, but the medication I was on masked what had happened to my spine.

I attended the interview for my chair promotion still on extremely heavy pain killing medication, and the experience was all a bit blurred. The external assessor in person was Mark Pollicott, who asked about the 'abstract' or 'combinatorial' work on periodic orbits that Graham and I were working on. The panel were sympathetic to my medical predicament and we managed a more or less conventional interview, though I had to be delivered to the

[3] Andrew was a Chemist who worked in the development of anti-cancer drugs, cisplatin in particular [139].

door and collected from it by Tania as walking was extremely difficult. I met the Vice-Chancellor David Eastwood a few days later, and he said they had decided to promote me, increase my salary beyond the basic professorial level, and backdate this to the start of the academic year. He also said something that gave me pause then and gives me pause now: “In four or five years, you will need to decide whether your future lies in Mathematics or in leadership roles—it’s just a question of which you find more exciting”.

Life more or less returned to normal, and within a few months I was cycling to work again. Unfortunately the pain in my back and my legs grew worse and worse, and by Easter of 2005 I was in real trouble, somehow managing to cycle to work but having to hold meetings lying flat on the floor. Our General Practitioner and the osteopath he referred me to failed to diagnose what the problem was, and the osteopath certainly did more damage. I was progressively becoming twisted up and less able to walk, and the pain was extraordinary.

It was a physiotherapist called Julia Brudenell—someone recommended by a friend—who saved me. She took a careful history before doing anything and, as a result, worked out exactly what had happened in our very first session: The fall on the ice had broken facet joints, damaging and crushing the nerves connected to my right leg. Somehow the diagnosis alone was a huge step forward, and it marked the start of a road to recovery that took many years. Indeed the legacy is still with me—if I forget to do the exercises she showed me for a few days, or sit in the wrong way for a single evening, the referred pain starts to come back. There are other stranger consequences from the damaged nerves. For many years I had a sensation that I can only describe as a very cold squash ball cut in half and placed under my right foot, along with some other neuropathies.

An ancient tradition at many universities is for new professors to give an inaugural lecture. The practice ebbs and flows in many institutions, and David Eastwood had revived it at UEA. He was unable to attend mine, which took place on the 29th of November 2005. In his place the Dean of Science, Andrew Thompson hosted the event, and said some very kind words about me. The hand of Graham Everest in some of the stories and observations he made was clear. My talk was entitled ‘fourteenth Century planetary orbits to twenty-first Century number theory: What is ergodic theory?’, and I tried to trace the journey from Nicolas d’Oresme’s fourteenth Century remarks about the possibility of incommensurable planetary orbits[4]

---

[4]Nicolas d’Oresme was a fourteenth Century philosopher who made contributions in multiple areas. His (tenuous) connection with ergodic theory arises because he wrote about the idea of a model of the solar

to continued fractions and then optimistically venture into the connection between continued fractions and the modular surface $\mathrm{PSL}_2(\mathbb{Z})\backslash\mathrm{PSL}_2(\mathbb{R})$, and briefly mention the then recent work of Einsielder et al. [95] on the relationship between invariant measures on $\mathrm{SL}_3(\mathbb{Z})\backslash\mathrm{SL}_3(\mathbb{R})$ and Littlewood's problem in Diophantine number theory. I illustrated the idea of continued fractions using coloured sticks of different lengths, with the challenge of accurately working out the ratio between their lengths using a pencil but no ruler. Edward Acton, the Pro-Vice Chancellor for Education, was in the audience, and I suggested these sophisticated visual aids were made in his honour. Our parents made the trip from Dorset to attend the inaugural lecture, and it was wonderful to introduce them to some of our friends in Norwich and colleagues in the School of Mathematics.

At the start of the 2005–2006 academic year I became the Senate representative on the University's Planning and Resources Committee, and I remained on this body for six years in various guises. This exposed me to the key financial and strategic drivers that any University has to grapple with.

My term as Head of School ended in the Autumn of 2007, and I embarked on a year of research leave to try and recover my mathematical trajectory. The Mathematics department—with most of the work being done by a younger colleague, Mark Blyth—organised a wonderful meal on the 22nd of September 2007 at The Last Wine Bar in Norwich to thank me for the five years, which Tania and I greatly enjoyed.

Richard Again

After completing his PhD, Richard spent a few years working in software development before returning to UEA in a Mathematics learning support role for a few years. Graham Everest, Shaun Stevens, and I won another EPSRC award for a two year post-doctoral researcher to work on directional zeta functions. This allowed Richard to spend a few years as a full-time researcher, leading to a period of intense research activity.

The four of us worked on orbit growth problems for the S-integer dynamical systems that had been studied by Chothi and Stangoe [51, 284]. We were able to make real progress on the two extreme cases of S (a set of primes) being finite and having finite complement.

---

system in which the periods of the orbits of different planets could be incommensurable, which with an optimistic eye presages the concept of unique ergodicity.

In the finite case we were able to describe the prime orbit counting function and the limit points of some associated growth rates, explaining some of the things observed in Vicky's thesis. We also studied Mertens' theorem in this setting, and showed that the resulting asymptotic had the main term $k_T \log N$ in which the constant k_T depended on the set of primes S and the underlying map T in an extraordinarily intricate way. In the simplest case considered in [108] the constant turned out to be $\frac{5}{8}$, and in the next simplest case of the map dual to multiplication by 2 on the group $\mathbb{Z}[\frac{1}{15}]$ the constant was $\frac{55}{96}$. In this work we also demonstrated the natural boundary phenomenon for the dynamical zeta function in a much more general setting. The arguments and proofs brought in diverse tools from Diophantine analysis and number theory, and the final publication was dedicated to the memory of Parry [104].

The co-finite case (in which S contains all but finitely many primes) brought in entirely different ideas. Here the growth in the number of periodic points is extremely slow—bounded by a polynomial. This was new territory both technically and psychologically, because the natural paradigm in dynamical systems often involves exponential growth as a yardstick. Exponential growth arises naturally when the simplest maps are iterated. Our examples were on the one hand mixing simply because they were ergodic group automorphisms, but on the other hand looked a bit more like rotations in that they had no exponential growth in the number of periodic points. This meant that the dynamical zeta function was irrelevant as an analytical tool for studying orbit growth. Instead we developed an 'orbit Dirichlet series', a tool adapted well to study sequences that have polynomial bounds. The first main result was that for these examples the orbit Dirichlet series $\mathsf{d}_\alpha(z)$ was a rational function of the variables c^{-z} for certain whole numbers c in a finite set associated to the map α. This 'Dirichlet rational' property was unexpected in light of the strange behaviour of the dynamical zeta function in the case of S finite.[5] Moreover, the orbit counting function for α could then be related to the analytic properties of the Dirichlet series d_α, its abscissa of convergence, and the order of vanishing at the pole defining it. We also were able to find examples for which the orbit counting function had a precise asymptotic, and other examples for which it oscillated between two asymptotic expressions. This all involved some intricate calculations and proofs, including some messy

[5]This contrast between extremely bad behaviour at one end of the figure repeated on page 258 and (relatively) regular behaviour at the other raises the question of whether there is a different point of view that might put them on the same footing. This would be a notion of dynamical zeta function defined on the adeles, where the singularities that are too close together on the complex plane are teased apart and Tauberian relationships between analytic behaviour and growth rates are recovered.

counting ideas involving so-called inclusion-exclusion arguments, and it was some years before it finally appeared in print [105].

In January 2006 Richard and I attended a conference on Mahler measure at the University of South Alabama in Mobile. This was an enjoyable and interesting event, one of several conferences I attended over the years that reflected the extraordinary diversity of areas in which Mahler measures arose naturally. For some reason—I think we had finally understood a key step in an argument we had been struggling with—we were in a celebratory mood one lunchtime. This triggered a distinctly American experience. We had lunch at a restaurant in warm sunshine, sat outside under huge plastic mushrooms to provide shade, at the end of which I ordered two Irish coffees. After a few minutes the waitress returned to the table and said that the chef was not sure how to make these. In many places this would have been something we would call a 'spam's off' moment, quite reasonably. However, the attitude was very clearly that if a customer wants something they would do all they could to provide it. I told her it was fine if it was too much trouble, but an Irish coffee was essentially coffee, cream, and whisky—and, for some but not me, sugar. We sat there, enjoying the sunshine and warmth, and a long pause ensued. Then the waitress came out with a large tray and the message that the chef didn't want to get this wrong, so...The tray had two substantial cafetières of very good coffee, a large jug of cream, a bowl of sugar, two mugs—and a jug containing about half a litre of American Bourbon. It became a lengthy lunch, and a wonderful memory of a small restaurant in Alabama absolutely determined to serve their customers—no matter how unreasonable the order might be.

Alongside this, Richard and I had several other things we were thinking about. We looked at algebraic actions of $\mathbb{Q}^d$ (d copies of the additive rational numbers) and $\mathbb{Q}^\times$ (the multiplicative group of the non-zero rational numbers). Our motivation was to see what the algebraic machinery developed by Kitchens and Schmidt [167], which in principle worked for any countable acting group, looked like in these two extreme settings.[6] We were looking for

[6]They are extreme because the approach of Kitchens and Schmidt in part uses character theory to relate an algebraic dynamical system defined by a countable group Γ acting by automorphisms of a compact abelian group X to the structure of a $\mathbb{Z}[\Gamma]$ module for the dual group of X. Here $\mathbb{Z}[\Gamma]$ denotes the 'integral group ring' of the acting group Γ. For actions of $\mathbb{Z}^d$, this group ring is a ring of reasonable 'size': A notion of size called the 'Krull dimension' is finite. The same holds for the integral group ring of the additive group $\mathbb{Q}^d$, meaning that in this context algebraic $\mathbb{Q}^d$ actions were a modest next step in understanding. However, it is a real step because $\mathbb{Z}[\mathbb{Z}^d]$ is Noetherian (another 'smallness' property a ring may have) while $\mathbb{Z}[\mathbb{Q}^d]$ is not. In contrast $\mathbb{Q}^\times$ is an enormous group, and its integral group ring has infinite Krull dimension.

dynamical properties that behaved differently for actions of $\mathbb{Z}^d$, $\mathbb{Q}^d$, and $\mathbb{Q}^\times$, the latter two being 'actions of the rationals' in two different ways. The 'mixing of all orders' result with Klaus Schmidt [268] essentially amounted to noticing that the theorem we had sought was equivalent to an 'S-unit theorem' in number theory recently proved by Schlickewei [267]. It showed that a mixing algebraic action of $\mathbb{Z}^d$ by automorphisms of a compact connected group must be mixing of all orders, so our starting point was to try and prove the same for our 'actions of the rationals'. The deep Diophantine results that Klaus and I exploited could not cope with the problem, as the relevant equations could somehow slip away in the extra space available in $\mathbb{Q}^d$. However, the theory behind Schlickewei's work had continued to develop, and we found more recent and more 'uniform' results due to Evertse et al. [122]. This gave us what we needed to show the same 'mixing implies mixing of all orders' result for actions of $\mathbb{Q}^d$, and easy examples showed that the same could not hold for actions of $\mathbb{Q}^\times$. When we submitted the short paper that later appeared as [205] it generated one of my favourite referee reports, which likened the result of [122] to a 'big gun that the authors use to blaze away at the problem'.

We also continued the theme of directional dynamics, studying how various properties changed when viewed along different directions in the acting group of algebraic $\mathbb{Z}^d$ actions. In [206] we showed that the expansive subdynamics we had studied with Einsiedler and Lind [79] could be detected in some cases from the fine structure of directional periodic point growth. In [207] we looked at almost the opposite problem, looking for uniform or global statements instead of studying the fine structure of growth rates as the direction varies. Here we used Diophantine results to show that there were exponentially many periodic points in all directions, mostly for obvious reasons but in critical non-expansive directions for more subtle Diophantine reasons. In later work we showed a related form of directional uniformity for the mixing property [210, 211]. Richard went on to work at the Kungliga Tekniska högskolan (KTH) in Stockholm, and I visited him there.

A Brief Encounter with Finite Groups

Studying group actions as a whole in this setting, rather than looking at how dynamical attributes of directional sub-systems vary with the direction, brings in other questions. There were some early attempts to define a dynamical zeta function for $\mathbb{Z}^d$ actions in my thesis [320] and in the thesis of Mathiszik [192, 193] (I learned about this work because his supervisor Prof. Horst Michel gave a seminar on it at Warwick in October 1987). A convenient testing ground for some of these ideas was given by a variant of Ledrappier's example [175] in which the 'alphabet' group $\mathbb{Z}/2\mathbb{Z}$ is replaced by a finite group G. The specific question I became interested in was this: If two such systems are built using two finite abelian (that is, commutative) groups G_1 and G_2, and they have the same periodic point data, what can be said about the two groups? The system here is an action of $\mathbb{Z}^2$, so a 'periodic point' now means a point invariant under the action of a two-dimensional lattice in $\mathbb{Z}^2$. For a single transformation the possible periods are given by the counting numbers $1, 2, 3, \ldots$ which are rather easy to place in correspondence with the terms $z^1, z^2, z^3, \ldots$ appearing in a power series in a variable z. Beyond this convenient coincidence—that the subgroups of finite index in the integers are in one-to-one correspondence with the powers of the variable in a power series—there is a hidden and more powerful accident. The partially ordered set describing the lattice of subgroups of $\mathbb{Z}$ has an associated Möbius function (which is simply the classical Möbius function in number theory). This takes on the values 0 and ± 1 only, and in particular is bounded. For the lattice of finite-index subgroups of other groups the Möbius function is generally unbounded, and potentially plays a role whenever an inclusion-exclusion argument is used inside a growth rate calculation.

An easy argument for two actions of $\mathbb{Z}^2$ associated to finite groups in this way shows that if the periodic point counts match up, then the two groups G_1 and G_2 must have the same number of elements. By choosing various lattices in $\mathbb{Z}^2$ more and more information about the internal structure of the two groups could be elucidated. The natural conjecture was that the periodic data determined the group completely, and I was able to show this under the ugly assumption that the number of elements in the group was not divisible by the square of a Wieferich prime[7] larger than 4×10^{12} nor by 2^{10}, and published this in the *Journal of Number Theory* [331]. Some years later Christian Röttger, who had been a PhD student under Graham Everest, used Teichmüller systems of representatives for polynomials over finite fields in an

---

[7] My difficulty was related to the following kind of problem. Counting the number of solutions to an equation like $(2^n - 1)g = g$ in a finite abelian group G for all $n \geqslant 1$ cannot distinguish between a factor of G that looks like $\mathbb{Z}/p^2\mathbb{Z}$ or a factor that looks like $\mathbb{Z}/p\mathbb{Z} \oplus \mathbb{Z}/p\mathbb{Z}$ for a prime p if p has the strange property that whenever p divides $2^n - 1$ we also have p^2 dividing $2^n - 1$. It is known that the primes 1093 and 3511 have this property, and no others smaller than 4×10^{12} were known at the time (the search has since been extended to 6.7×10^{15}). A standard sort of heuristic argument, based on the optimistic belief that certain arithmetically defined quantities behave more or less randomly unless there is an obvious reason for them not to do so, suggests there should be infinitely many such primes. They are called Wieferich primes as a result of his proof that if x, y, z are integers with $x^p + y^p + z^p = 0$ with the product xyz not divisible by p, then p must be a Wieferich prime [352].

ingenious way to deal with all primes, giving a proof of the conjecture without my ugly hypotheses [259].

This all concerned the case of abelian groups used as the alphabet. The same question makes sense for any finite group as the alphabet, and deciding whether this could be done even in principle threw up a strange question in finite group theory. Roughly speaking, this concerned how much information could be discerned about the structure of a finite group from knowledge of the number of solutions to all simultaneous equations with variables in the group.[8] For abelian finite groups the answer is that the entire structure is revealed by such information, and this is an easy exercise which can be done in many quite elementary ways. I also needed something slightly stronger, getting closer to specifying which equations tell you which part of the structure. Even in the abelian case this was not obvious to me. In the end Les Reid from Southwest Missouri State University helped me sort out this abelian part of the argument.[9] In the general case, in which the group might not be abelian, it is all less clear, and I asked for help via the Usenet group `sci.math.research`. A helpful group theorist, David Sibley from Pennsylvania State, quickly responded with an elegant proof that the answer is also yes: The numerical data available from counting the solutions to all such equations determines the structure of the group. In fact there were two responses to my question. The first was a neat proof by David Sibley, and the second illustrated how difficult it can be to convey mathematical arguments to non-experts. With the best of intentions, William Rowan responded "I think the original question is answered by the theorem that two finite simple groups (or, more generally, groups with a single nonabelian characteristic subgroup making up the socle) which satisfy the same identities must be isomorphic. This is a consequence of the Jonsson–Hrushovski theorem which states that if the finite, subdirectly irreducible group A with nonabelian monolith is contained in the variety of groups generated by the group B, then A is isomorphic to a homomorphic image of a subgroup of B." Doubtless this is true, but it involved knowledge

[8]An 'equation' in a finite group means something like $x^2y^3x^{-1}z = e$, where e denotes the identity. The 'solution' is to identify the triples (x, y, z) with x, y, z in the group that satisfy the equation. In an abelian group we would conventionally write this in additive rather than multiplicative notation, so it would become $2x + 3y - x + z = 0$, which simplifies a little to $x + 3y - z = 0$. Several different equations can be studied at the same time, giving 'simultaneous' equations. The question thrown up was this: If you are told the number of solutions to all such simultaneous equations, can you work out the structure of the group?

[9]The specific statement I needed was this: If G is a finite abelian group and p is a prime, then the number of solutions to the equations $pg = e$, $p^2g = e, \dots, p^ng = e$ determine the p-primary component of G if and only if the number of elements in G is not divisible by p^{2n+2}. The 'if and only if' means that this bound—the $(2n + 2)$—is the exact answer, and one cannot do better.

I lacked then, lack still, and am most unlikely to ever acquire. Despite all this friendly assistance, the original question about periodic points seems hopeless because the simultaneous equations whose solutions count periodic points are a small and difficult to understand subset of all such equations.

Doug Lind made progress on a meaningful way to define a dynamical zeta function for actions of $\mathbb{Z}^d$, grouping together lattices according to their index [180]. He also showed how readily natural boundaries appeared in this setting, and started to explore the relationship between the counting of lattices in the acting group and the counting of periodic points in the space being acted on. Richard and I found much of interest in this paper, and were able to find some periodic orbit growth asymptotics for nilpotent acting groups [208] under the assumption that the dynamical system was rather simple (a full shift), and looked at some of the strange things that can arise for dynamics that is less uniform for certain algebraic $\mathbb{Z}^2$ actions [209]. Some years later we returned to one of the questions raised by Doug Lind, proved the existence of a natural boundary for the dynamical zeta function of certain algebraic actions on zero-dimensional compact groups, and gave further evidence in support of the idea that a single compact group automorphism would obey a Pólya–Carlson dichotomy [212].

Yuki

Richard took up another position a little before the end of the grant, leaving us with three months or so of post-doctoral funding. I mentioned this on a dynamics Usenet group, and it came to light that a PhD student of Karl Petersen at Chapel Hill in North Carolina was about to complete and had a small gap to fill before starting a post-doctoral position in Chile. This all took a little sorting out, but funding, timing, and visas all came together in the right way and so Yuki Yayama spent a few months in Norwich (Fig. 12.3). We worked on a different idea in directional dynamics, namely stability of the geometry of generating Markov partitions in expansive cones [347]. We made partial progress, but in some cases unwittingly made use of an argument from an old construction of generating partitions for automorphism of certain solenoids due to Wilson that turned out later to have an error in it [363]. This was all eventually sorted out by Nigel Burke and Ian Putnam [39].

Fig. 12.3 With Yuki Yayama and a feline assistant in our house in Norwich trying to understand Markov partitions on solenoids (With permission of Tania Barnett)

Tony, Sawian, and Apisit

Three PhD students bridged the period that began when I was head of school and ended when I was on the University executive.

Anthony (Tony) Flatters[10] was jointly supervised with Graham Everest. Tony worked on questions in arithmetic related to the presence of primitive divisors in families of recurrence sequences (a prime divisor of a term in a sequence of integers is called a primitive divisor of a term if it does not divide any earlier term; a 'Zsigmondy theorem' for a sequence says that every term beyond a certain point has a primitive divisor). Alongside his thesis [129] he used delicate estimates for values of cyclotomic polynomials to understand primitive divisors in certain Lehmer–Pierce sequences [128]. The earliest results in this area concerned sequences whose nth term is $a^n - b^n$, for which Zsigmondy, Bang, and many others had shown strong bounds on the point beyond which a primitive divisor is guaranteed. Tony and I looked at the same problem in positive characteristic, replacing the integers with polynomials over

[10]Tony graduated in 2010, with external examiner Prof. Samir Siksek from the University of Warwick [129].

a finite field. In this setting we were quickly able to establish a small effective bound for a theorem of Zsigmondy type [127].

Sawian Jaidee[11] and Apisit Pakapongpun[12] applied to do PhDs with scholarships from the Government of Thailand, and by then Graham's health meant he could not take on more supervisory duties.

I supervised Apisit on my own, and he started working on 'functorial' properties of closed periodic orbits for dynamical systems. The word functorial here may be thought of as follows. There is one world of dynamical systems (in this context, this just means maps on some space) and another world of the integer sequences that arise as their counts of periodic points. There are operations that can be applied on both sides: For example, the product or disjoint union of two dynamical systems may be formed; two sequences may be added together or multiplied term by term. And so on—in fact there are a great variety of such operations. The functorial problem is to understand how these operations look in the other world in each case. This work went in various directions, and we ended up publishing some of the results during his thesis [237] and continuing one aspect of the work after he had completed [238].

Sawian worked with Shaun Stevens and me as supervisors. We started by looking again at the simplest systems that we had studied with Chothi [50] and Stangoe [108]. A feature of the count of periodic points for compact connected group automorphisms that much of my own work grappled with was at the time a partially understood 'dichotomy'. The resulting sequences seemed to either be extremely well-behaved, with examples like the so-called Mersenne numbers $(1, 3, 7, 15, 31, 63, \dots) = (2^n - 1)_{n \geqslant 1}$ or they seemed to be wild in particularly difficult ways to deal with. The first example from Vicky's work [108] has a sequence of periodic point counts much like the Mersenne numbers, with the small but far-reaching change that the '3-part' is removed. That is, each term is divided by the prime 3 as many times as is possible, resulting in the sequence $(1, 1, 7, 5, 31, 7, \dots)$. More complicated examples repeat this for some set S of prime numbers. If the set S is either finite or co-finite (that is, contains all but finitely many primes) then it is relatively straightforward to predict things like how rapidly the resulting sequence grows, though the behaviour is erratic. If S is allowed to be infinite and co-infinite, the 'typical' case, then things are considerably more complicated. Sawian started

[11] Sawian graduated in 2010, with external examiner Prof. Oliver Jenkinson, Queen Mary University of London [160].

[12] Apisit graduated in 2010, with external examiner Prof. Franco Vivaldi, Queen Mary University of London [239].

looking at a smoothed way of studying how many closed orbits the systems have, leading to an analogue of something called Mertens' theorem in number theory. This smoothing or additional averaging allowed some progress to be made in her thesis [160].

For the usual party to celebrate the completion of their theses, Sawian and Apisit cooked a wonderful Thai meal in our kitchen in Hellesdon.

Shaun, Sawian, and I continued to work together after she had graduated and returned to Thailand, initially in a slightly different direction by looking at Mertens' theorem for a specific class of toral automorphisms, the 'quasi-hyperbolic' ones. This brought in a kind of resonance phenomenon and the resulting publication [159] included an interesting sequence. The result in part associates a certain parameter to a given quasi-hyperbolic toral automorphism, and we were interested in examples that would show how much this parameter could deviate from its typical value. This gave rise to a sequence of examples for which the parameter would be asymptotically much smaller than the generic case. The parameter was expressed as a complicated integral, and a colleague with expertise in the right sort of analysis, Paul Hammerton, helped us out by computing a few examples. This gave rise to a sequence of parameter values $(2, 4, 6, 10, 12, 20, 24, 34, 44, 64, 78, \dots)$ which seemed to be 'new' in the sense that it did not appear in the Online Encyclopedia of Integer Sequences [232], so I added it there. I also raised the question on the newsgroup `sci.math.research` in Usenet (Fig. 12.4).

I have highlighted this genuinely obscure problem because it illustrates the power of the connections afforded by the internet. Our little problem attracted the attention of Mark Daniel Ward, a professor of Statistics at Purdue with expertise in the right sort of asymptotic analysis. The problem turned out to be considerably more complicated than we had ever suspected, and Mark and his co-authors related it to something called the Israel–French conjecture and, via an intricate argument, found the exact asymptotics we had sought [131]. I have yet to meet Mark, but much enjoyed our email exchanges about the problem and its motivation—and his beautiful and intricate argument resolving our problem.

The three of us also continued to think about some questions that arose in Sawian's thesis work. The idea was that by careful constructions of infinite sets of primes we might be able to exhibit arbitrary growth parameters in Mertens' theorem, and even construct examples with such slow growth in the number of periodic points that the dominant term in Mertens' theorem would take an entirely different form. The real questions were number-theoretical, and rather quickly were beyond me. We were however able to exhibit one continuum of parameter values. Shaun, with a sound number-theoretic instinct, suspected

In trying to find an example of extreme behaviour in the dynamical analogue of Mertens' theorem for toral automorphisms, a seemingly rather subtle integral has arisen, and I'd appreciate any information about it. Apologies if this turns out to be well-known and easy!

Let J_k = \int_0^1 (1-cos(2\pi x)) (1-cos(4\pi x)) ... (1-cos(2k\pi x)) dx.

The two questions I would particularly value help with are:

1) It looks like J_k -> 0 as k -> \infty, but I don't see how to prove this. The function being integrated vanishes on the kth set of Farey numbers, and they are uniformly distributed, so this seems plausible.

2) 2^kJ_k is integral for all k\ge1 (this is easy to see if you write x=e^{2\pi i x} and think of it as a complex integral, for example). However, that sequence of integers (as k varies) is strange:

2,4,6,10,12,20,24,34,44,64,78,116,148,208,286,410,556,...

which is not (yet) in Sloane's online encyclopedia.

Ideas welcomed!

Best wishes Tom

Fig. 12.4 My question to a Usenet group on January 16, 2008

that there were other continua at different growth scales, and formulated the sort of statement about prime numbers that we would need to prove this. After some efforts he contacted the number theorist Stephan Baier, who quickly saw how to construct the sort of sets of primes we needed. There were still some technicalities to deal with, but this all came together with the construction of several continua of different growth rates for compact group automorphisms [13], which the four of us published together. By the time this paper finally appeared, I had moved to Durham.

Michael and Rama

I think I first met Michael Baake from the University of Bielefeld at one of the Max Planck workshops in Bonn, and we quickly found some interests in common. I ended up visiting his lively and welcoming research group five times, and always enjoyed the atmosphere that he created in his group, which involved students at all stages, post-doctoral researchers, visitors, professors all eating lunch together and talking about research questions across a huge range of interests. Everything from statistical physics to spectral properties of arithmetic dynamical systems to aperiodic order to crystallography to mathematical biology and more interested Michael, and he brought the same inclusive energy to all of it.

I also loved the city of Bielefeld and the interesting dynamics of a thirteenth Century city with a role in the Hanseatic League alongside a modern university founded in 1969. The main building of the University also intrigued me. It was enormous, containing shops, places to eat, and many other services along with academic departments. It also had repeated design elements on a huge scale, which brought to mind the huge hollow cylinder Rama that grew from the imagination of Arthur C. Clarke [55].[13]

Michael and I did some work together on diffraction spectra of the simplest algebraic dynamical systems, essentially showing that from the point of view of diffraction spectra configurations arising from $\mathbb{Z}^2$ actions that are mixing but not mixing of all orders could be indistinguishable from those arising in independent identically distributed processes [10]. The open and enquiring atmosphere of Michael's research group also lead to work that had links to things I was interested in; two particularly productive examples were work of Natascha Neumärker on 'relatively realizable' sequences [11, 227] and work of Angela Carnevale and Christopher Voll on Euler factorizations of orbit Dirichlet series [45]. One of the regrets I have about the way in which executive roles pulled me away from Mathematics is that it led me to stop interacting with Michael's wonderful research group.

A Note on the Office Door

On Tuesday the 4th of November 2008 I returned to my office in the early evening to find a note sellotaped to the door asking me to contact Edward Acton urgently. Inside I found several emails and phone messages saying the same thing. My assumption was that this was related to the inaugural lecture for a new colleague that I was organising, as I was lining up a suitable senior host, booking wine and bowls of crisps, and finding a room. Important certainly, but it seemed excessive to have such a sense of urgency from the Pro Vice-Chancellor for Education. I rang home to say I'd be running late and went over to see him in the Registry building.

I knew Edward a little through various committees we both attended in one role or another. Indeed I had fairly recently attended the University Learning

[13]It is of course quite possible that I imagined these pleasant visits as part of the *Bielefeld-Verschwörung*.

and Teaching Committee that he chaired as a supplicant for tragic reasons, to discuss the possibility of awarding a posthumous degree to one of our students who had died in his first year, allowing the family to witness the ceremony along with his own cohort. This was not a simple decision, and I was struck by the care and humanity with which it was discussed.

The reason for the urgency of the note on the door became clear at once. The Vice-Chancellor was unwell, and Edward Acton had been trying to carry out both roles, a clear impossibility. Ahead of an emergency meeting of Council that evening he wanted to know if I'd be willing to take over his current role following that meeting. I cycled home in a bit of a daze, and discussed this with Tania. My view has always been that part of the relationship with the University that had given me my first permanent job was that I would do my best to carry out anything asked of me, unless it was unethical or manifestly impossible. The latter was a real issue. I had spent the academic year 2007–2008 on research leave after five years as Head of School, and on my return had deliberately taken on a great deal of teaching and several doctoral students. The logistics alone looked more or less impossible. Nonetheless it was clear that I should say yes, and sort out the details later. In the normal course of events a typical appointment of this sort might involve preliminary meetings and interviews one Autumn, a formal appointment process the next Spring, with an eye to starting the next Autumn. Being asked to take on the role with almost no notice was daunting and turbulent.

There were also some technicalities related to governance questions like maintaining a majority of lay members on some key committees, so a new role of 'Associate' Pro-Vice Chancellor was created for me, and I attended my first meeting of the University executive at 9 a.m. on the morning of the 6th of November.

Colleagues in the School of Mathematics were wonderful over this strange process. Initially the whole matter was rather confidential, so I had to ask for volunteers to take over my personal tutees and much of my teaching at almost no notice, and with no explanation. I also had to migrate from my google calendar and `pine` email client on Linux onto the corporate Outlook dairy and email system, the first of many small adjustments to my life. The staff in 'VCO' (the Vice-Chancellor's Office) were welcoming and helpful, and my personal assistant Jacqui Churchill brought great kindness and huge knowledge to my assistance. Life quickly settled down into a new routine, with an extremely full diary and a new office to which my books and papers were eventually moved. For the first few months it still contained a few shelves of Russian History books relevant to Edward Acton's research field.

This transition really marked the end of the life that had started at Waterford Kamhlaba in the 1970s in which my primary identity was 'mathematician'.

> We all have lives that go on without us. I've a cricket me, who didn't stop like that was that. (From 'England' by Zaffar Kunial)

The feeling of being pulled in two directions by the responsibilities of being head of department and my own mathematical life—and the positive joy in seeing others accomplish things in education and in research—grew stronger. One foot remained on each horse, but almost all the weight gradually shifted to the executive role. In many ways the mathematician me vanished permanently: Retirement will allow me to get back to some projects, but the intensity and energy of attention I could have given to mathematics during the last 15 years of my career are now beyond my capabilities.

Life as PVC

I knew most of the academics on the University Executive fairly well, but the rest of the Vice-Chancellor's office and the professional services leadership were new to me. It was a friendly group, and I quickly found my feet, though nothing quite prepares you for an overnight transition into a life in which many days comprise back to back meetings without opportunities for eating or even comfort breaks. I also inherited a large portfolio of 'task and finish' groups and work-streams set in motion but not completed. It is all too easy to set things in motion in Universities, and rather more difficult to complete them or extract the tangible benefits imagined when they were started.

Several strands of work and challenges we were facing came together conceptually as evidence of an amiable collusion in parts of the University. There are many temptations in the life of a University, and UEA was in danger of succumbing to one of them. It is possible to slightly drift towards a form of unspoken contract in which nobody demands too much of anyone else. Questions about contact time and workload for students all too often become politicised, clustering around numbers that may grab headlines but have little meaning in terms of learning. Contact time is particularly susceptible to this, and is a frequent stick with which the sector is beaten. However, every University should be ambitious about the challenge that their educational offer provides. UEA was at the time in danger of becoming a place where all the subjective student satisfaction scores were high, while the tougher and more objective metrics like the proportion of students going on to graduate level employment were modest. Every University aspires to have high metrics on all measures, but there was a distinct feeling that some of the extremely high satisfaction scores were related to a lack of gritty challenge in the curriculum and the regulations surrounding academic progress. A habit of 'deemed credit'—permitted failure in a module that was a prerequisite for something in the next year, for example—had become entrenched. For my first few years I had a strong and determined Director of Taught Programmes, Geoff Moore, and his clear thinking developed into the articulation of a 'new academic model' which embedded more challenge into our regulations. Any change is controversial, and of course this created problems as well as solutions, but I hope contributed something to the evolution of education at UEA.

Something else I inherited was an Associate Dean in each of the Faculties. These roles are called different things in different institutions, but every University has a position of education lead in each Faculty or its equivalent group of more or less cognate discipline areas. Whether the role is called Dean, Associate Dean, Associate PVC, or some more inventive permutation of similar words, it has certain characteristics in most institutions. The area of responsibility is large, often involving the education of many thousands of students and many different disciplines. The formal authority and direct managerial reporting lines are limited and often non-existent. The workload can be daunting, and arises on multiple scales. Doing the job well often involves sacrificing much of one's own research or scholarship plans. These often thankless roles attract exceptionally dedicated colleagues, and it has been one of the great pleasures of the second half of my working life that I have spent so much time with colleagues who have chosen this particular form of service in higher education in four different institutions.

One of the institutional projects already underway when I took over was a branch campus in London. This was run as a joint venture with IUP (Into University Partnerships), and proved to be extremely complex partly as a result of the contractual configuration, and partly because Universities can struggle to work with organisations with a different and more business-like ethos. UEA withdrew from the venture a year or two after I moved to Durham in 2012, with a sense of relief in some quarters and frustration that this happened just as the difficult teething stages were behind them in others.

Once the dust had settled on a turbulent period for the whole institution, Edward Acton was appointed as Vice-Chancellor. At his suggestion I was enrolled on the marvellously titled 'Top Management Programme' run by the Leadership Foundation for Higher Education. This is an expensive business both in time and in money, and comprises a sequence of residential events with the same cohort of people in similar roles at other Universities. The mix of seminars, leadership courses, and psychometric profiling was not entirely to my liking, but I enjoyed it and the people I met while doing it. Particularly significant for me was the chance to work with people from Universities with entirely different missions, as it is easy to only ever interact with colleagues in similar types of institution.

At the time many of those 'similar types' had banded together as the '94 Group' [354] of smaller research-intensive Universities, partly as a defensive response to the growing lobbying power of the Russell Group of Universities founded in 1994. At the time the 94 Group comprised the Universities of Bath, Birkbeck University of London, Durham, East Anglia, Essex, Exeter, Goldsmiths University of London, Lancaster, Leicester, London School of Economics, Loughborough, Manchester Institute of Science and Technology (UMIST), Queen Mary University of London, Reading, Royal Holloway University of London, St Andrews, SOAS University of London, Surrey, Sussex, Warwick, and York. UMIST left in 2004, the London School of Economics left in 2006, Warwick in 2008; the departure of Bath, Durham, Exeter, Queen Mary University of London, St Andrews, Surrey and York in 2012 and of Reading in 2013 marked the beginning of the end for the group.

Most but not all of the education leads for the 94 Group met regularly at Woburn House on the North side of Tavistock Square in London. This formed a loose and supportive network, though even from the first few meetings it was clear that there were few issues on which the members saw things entirely the same way, and attendance was patchy.

At one of the 94 Group gatherings there was a presentation on one of the more ambitious Massive Open Online Course (MOOC) ventures, showing an impressive return on investment on one of the slides. These grew from correspondence and radio courses that became popular towards the end of the nineteenth Century. They grew to embrace technological advances like television and video recorders. A MOOC is an online course with the potential for unlimited participant numbers and a focus on open access. Many MOOCs allow for interactive community support alongside traditional course materials. The terminology was introduced in 2008 and there are now multiple large-scale platforms offering MOOCs. The older, broader, conception of Open Educational Resources (OER) and the ability to more efficiently re-use educational resources is probably a better way to frame the potential of the enormous shifts made possible by technology for widening access to educational opportunities. The acronym MOOC itself is problematic, as the central questions of balancing the cost of providing educational experiences of high quality against the possible income from doing so is changed but not removed by advances in technology.

The presentation intrigued me, and after it ended I asked if we could go back to the return on investment slide. On closer inspection the accounting legerdemain involved was an attribution of value to raising brand recognition worth several million pounds each year. My question about how that was measured was not universally welcomed, but seemed to me of central importance. Ever since then I have been insistent that we look at real costs openly and, where appropriate, decide to invest resources—that ultimately come from students or the public purse—for some wider gain rather than simply attaching a convenient amount of monetary value to an intangible asset. My positive view of MOOCs is largely about institutional learning—whatever the future of higher education looks like, an institution where everyone is comfortable with using the full range of remote, in person, synchronous, and asynchronous technological aids is going to be better placed. This certainly became significant during the rapid pivot to teaching and assessing online when the COVID pandemic hit the sector: The institutions with a track record of online education for whatever reason were far better able to cope.

The year 2012 was dubbed 'The Year of the MOOC' in a New York Times article and for many years enthusiasts argued that this type of technological innovation would fundamentally change the nature of Universities.

> In 50 years […] there will be only 10 institutions in the world delivering higher education and Udacity has a shot at being one of them. (Sebastian Thrun, Stanford professor and head of Google X, who founded KnowLabs which became Udacity, speaking in 2012.)

More measured reflections are probably nearer the mark. The early hype was overstated, but so was the sceptical reaction to the absence of a quick revolutionary change. Innovations made possible by technology and the thinking behind open reusable educational resources in higher education are genuinely revolutionary and will change the sector and society in profound ways, but it will be a long-term shift that builds and expands and learns along the way how to fit in with wider shifts in the world.

> As the MOOC snowball—it was never an avalanche, for all the feverous hope of former Blairite aides—crumbles and melts across the stinging and red-raw face of UK Higher Education, what should be our reaction? The icy water of the postulated disruption trickles down the collective institutional neck as we wonder if we should be jeering young upstarts in their game, or maintain our distance and dignity.

> Or are we too late? The painful brilliant ice thrown back in 2012 has become a greying slush, a dampening rather than a disruption. As the shining crystals decompose, should we be preparing for a brilliant new season of online education? (David Kernohan writing in 2015 [164])

One of the most stable features of higher education is the certainty among some commentators and professionals that technology is on the point of utterly transforming the sector. Every year some argue that next year the entire business model of full time higher education on a campus will come to an end. The logic is impeccable but shallow: Technology, and the extraordinary possibilities afforded by the internet, are certainly revolutionary, but I always saw them as adding to rather than replacing the more familiar educational processes with deeper roots. My view was that technology could enable us to do things better and differently—but that the thing we do was to use real resources to provide life-changing educational experiences and opportunities to students. Where the resource comes from is the substance of a legitimate and important debate, and how the opportunities can be extended to students from disadvantaged backgrounds and how technology can allow collaborative learning across continents are sensible questions. But the idea that the internet means this can all be sustainably done without the resources seems unlikely. Next year may yet prove me wrong, but the demand for on campus in person higher education has been stable for decades, and the imminent transformational disintermediation of the sector proclaimed at every conference on digital education for decades has not yet quite transpired. What has happened, and will doubtless accelerate, is constant change in how we operate and what technology can enable. A hundred years ago University-level examinations in Mathematics, for example, might include questions rendered trivial by software that could now be run on a mobile phone. How the subject is taught and assessed at this level has evolved to make sense of this. Some current examination questions might be rendered meaningless by access to an artificial intelligence engine, and we need to go through a similar evolution in understanding of how to make assessment meaningful again.

The 94 group fell apart shortly after Durham University, the University of Exeter, Queen Mary University of London, and the University of York left to join the Russell Group in 2012, and at the time of its dissolution it only had eleven remaining members.

Desdemona's Handkerchief

Much of the life of a University runs to pre-determined cycles of activity. National school leaving examinations, application deadlines through the single undergraduate system provided by the Universities and Colleges Admissions Service (UCAS), National Student Surveys, Teaching Excellence Framework submissions, Knowledge Exchange Framework submissions, Access and Participation Plan submissions, annual Assurance of Academic Quality and Standards reports, annual Degree Outcome Statements, Higher Education Statistics Agency returns, and national Research exercises—all run to schedules that are outside the gift of any one institution. Nonetheless it is the unpredictable event that often comes to dominate a period of time, more so for the executive and those immediately involved than for the wider University.

A striking example of this arose in November of 2009, when it came to light that an email server in the Climatic Research Unit (CRU) at the University of East Anglia had been hacked. This unit was a leading centre for research into anthropogenic climate change, and as a result was already a highly visible target of both well-organised, well-funded, climate denialist attacks and multiple *ad hoc* attacks by individuals.

What followed was, from a strictly logical point of view, a non-event: The enormous volumes of emails and data revealed that some of the scientists involved could be irascible in response to questions, communicated informally at times, and were doing the imperfect best they could to understand one of the most complex and multiply coupled physical systems ever studied. From a political and perceptual point of view it became a significant story, unhappily dubbed 'Climategate' in the popular press. Others closer to the centre of the story have written extensively and carefully about it, but some impressions stand out for me.

First, the science itself, while complex in detail, was relatively simple on a global scale. The increase in the retention of thermal energy from incident solar energy caused by increases in atmospheric Carbon dioxide is well understood and long established. Where the additional thermal energy goes in a given year, whether the consequences in phenomena from atmospheric water vapour to tree growth to melting permafrost mitigate or amplify and by how much, what the implications are in any one location—all of those are massively complex and have significant unpredictability when projected into the future. However the starting point of additional heat energy retention on a global scale is the simple thing that thousand of pages of hysterical newspaper coverage and hundred of hours of television never expressed with clarity and never really grappled with. Thus, for example, headlines may focus on local atmospheric temperatures and related phenomena while paying little attention to the vastly more significant thermal capacity of the deep oceans. The atmosphere matters greatly to those of us who live in it, but understanding what is happening to the world as a whole requires both ocean and atmosphere to be part of the modelling. One of the most absurd moments that illustrated the distinction between local and global, weather and climate, small thermal capacity phenomena and large thermal capacity phenomena took place on the 26th of February 2015. Senator James Inhofe of Oklahoma, Chair of the Senate Environment and Public Works Committee, brought a snowball into the Senate chamber and claimed in effect that this cold handful demonstrated that there was no global warming. It was indeed cold in the eastern United States that month, but NASA records show that 2015 was at the time the warmest year globally since the pre-industrial era, a record which was broken in 2016.

Second, once words like '...gate' are on front pages, it is impossible to convey any nuanced argument. A good example is the phrase 'hide the decline' inside one of the thousands of emails. Explaining the context takes ten minutes and a little thought, and involves how multiple different proxy measures for temperatures are combined and validated. It is an explanation of something completely innocuous—but proved far too nuanced for the febrile din.

Third, we have all benefited from the long march away from unthinking deference to experts. It never was beneficial to society that a simple assertion from the landowner, or the squire, or the priest, or someone in a white coat, was accepted by all simply because of a power imbalance. However the project feels only half completed: Deference to authority has been successfully diminished in many areas of life, but has not always been replaced by anything of substance. The huge diversity of our mother's reading included G. K. Chesterton, and she was fond of quoting from him. One of these quotations—as it turns out, a slightly misattributed one—came to mind often during this time.

> When men chose not to believe in God, they do not thereafter believe in nothing. They then become capable of believing in anything. (From Émile Cammaerts [44]; often attributed to G. K. Chesterton)

I have no wish to doff my cap to someone just because they are (for example) a scientist in a grand laboratory who has been working on the role of cholesterol in cardiovascular disease mechanisms for thirty years. However I would want to listen to their views on the efficacy of statins with a little more respect than I listen to the views of a random commentator, or to the wisdom of social media. A lack of common sense judgement of this sort became manifest particularly in the increasingly preposterous nature of much of the coverage of the story by the media. From ‘Question Time’ on the BBC to panels hosted by the Royal Society, the views of a serious researcher in a relevant field would be placed on one side, and the views of a newspaper columnist or economist with no scientific insight whatsoever on the other, with equal weight.

As we now contemplate the implications of the failure to address the consequences for the climate of the relentless growth in atmospheric Carbon dioxide levels during the crucial decade or two in which there were manageable pathways to keeping the rise in global temperatures from pre-industrial levels to below 1.5° C, a heavy burden of responsibility rests on how this question of ‘balance’ was handled by the media globally.

Finally, there was some solace to be found in the most unexpected places during what felt like a time of mayhem and considerable stress. One of these was an anonymous comment on the *Guardian* newspaper’s ‘Comment is free’ website.

> Like Desdemona’s handkerchief, Climategate offered absolute proof to those maddened by paranoia, but to the rest of us it remained just a handkerchief. (Posting to *Guardian* website by “thesnufkin”, 7 Jul 2010, 1:25PM)

There was of course a more parochial side to the chain of events. This included enormous stress for the staff in CRU and a huge diversion of effort for the University executive. Various inquiries, floods of Freedom of Information requests, responding to media enquiries—all took up huge amounts of time for many already busy people and a great deal of money.

Beyond the University itself, there were eight weighty inquiries. The House of Commons Science and Technology Committee conducted an inquiry into the affair, examining the implications of the incident for the integrity of scientific research, reviewing an independent review that UEA had asked Sir Muir Russell to lead,[14] and reviewing the independence of international climate data sets.[15] Other inquiries included an ‘Independent Climate Change Review’ chaired by Lord Oxburgh, an ‘International Science Assessment Panel’, an inquiry into the work of Michael Mann by Pennsylvania State University, an inquiry by the United States Environmental Protection Agency, another by the Department of Commerce, and another by the National Science Foundation.

[14] https://www.uea.ac.uk/library/the-independent-climate-change-email-review.

[15] https://publications.parliament.uk/pa/cm200910/cmselect/cmsctech/387/387i.pdf.

Much else followed, including an independent replication of the underlying models.

Assessing quite what it is that some of the stakeholders actually did believe is a more subtle exercise than simply listening to their public pronouncements. It is now abundantly clear that some of the corporate backers of the denialist campaigns had a full understanding of the science involved forty years earlier. For example, the oil company ExxonMobil had well developed internal models in the 1970s that projected trajectories of anthropogenic global warming exactly consistent with the forecasts from independent academic models and the later IPCC reports [291]. It is also a puzzling fact that nobody involved seemed at all eager to take up the massive arbitrage opportunity provided by their purported beliefs. If the sceptics believed what they were saying, the profits to be made in (for example) selling flood insurance based on models that took no account of the impact of the global warming they claimed was not happening were astronomical—but the opportunity was never taken up.

Vast effort was expended by huge numbers of people and three things emerge with clarity in hindsight. First, Desdemona's virtue was and remained unimpeachable. Second, nobody seemed to change the view they held about anthropogenic climate change before this all started, myself included. Third, between the date of the publication of the hacked email server in November 2009 and the COP 27 Sharm el-Sheikh Climate Change Conference in November 2022, NASA's direct measurement of underlying atmospheric Carbon dioxide levels showed an increase from 388 parts per million to 420 parts per million, and their best estimate of global temperature increase over the same period is on the order of 0.5 °C. Everything else is immaterial in comparison, and the time and money spent in inquiry after inquiry feels ridiculous.

Writing in the Summer of 2023 it seems that some of the long-predicted 'tipping point' phenomena are manifestly upon us. Nonetheless, the denialist campaign seems as strong as ever. Toby Young, who played an interesting brief cameo part in higher education as an erstwhile member of the Board of the Office for Students, is still active in denying the significance of anthropogenic climate change, and I follow his thoughts on social media with some interest. Meanwhile every molecule of Carbon dioxide we pump into the atmosphere inexorably absorbs some of the photons with wavelengths between 2000 and 15,000 nanometers, radiating some of that energy back onto the earth. The resulting additional capture of thermal energy is having the expected consequences (Figs. 12.5 and 12.6).

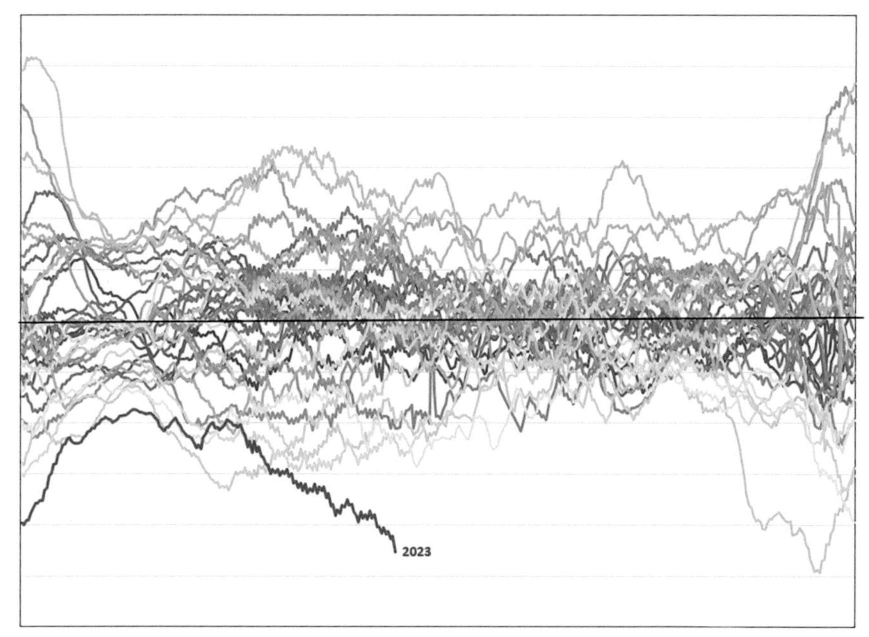

Fig. 12.5 The 'anomaly' (departure from the mean) of Antarctic Sea Ice extent (data from the National Oceanic and Atmospheric Administration (NOAA)) (Visualisation with permission from Prof. Eliot Jacobson). The sharp deviation on the right-hand side (in the Souther Hemisphere Summer) is the last time there was a significant El Nińo event; the red line is the plot of measurements for the Spring and early Summer of 2023

FOIA

The 'Climategate' episode certainly provided an involuntary crash course of a rather intense and not particularly enjoyable sort on the intricacies of the Freedom of Information Act 2000 (FOIA).[16] This piece of legislation was the fulfilment of a manifesto commitment by the government of Tony Blair.

> Unnecessary secrecy in government leads to arrogance in government and defective policy decisions. The Scott Report on arms to Iraq revealed Conservative abuses of power. We are pledged to a Freedom of Information Act, leading to

[16]https://www.legislation.gov.uk/ukpga/2000/36/contents.

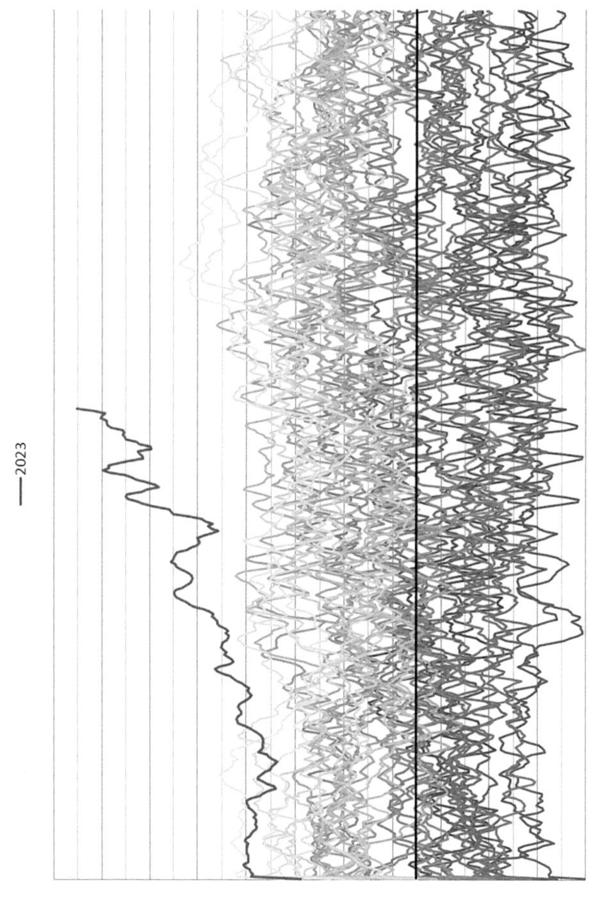

Fig. 12.6 The 2023 anomaly for surface temperatures in the North Atlantic (Visualisation with permission from Prof. Eliot Jacobson)

more open government, and an independent National Statistical Service. (From the 1997 Labour Party election manifesto)

While deeply mired in dealing with specific cases, or with the consequences of how poorly and ambiguously drafted the act was, it is easy to forget that the principles behind FOIA were clearly laudable and, if anything, have not gone far enough. The sweeping exemptions that parts of government use and the exemption for the private sector, mean the link to an informed and empowered citizen is a little optimistic. For most of us dealing with, for example, its widespread use in seeking out the exact contractual terms of a University's photocopier leasing arrangements it is often little more than another administrative burden. For a modern day Iago it is a tool for feeding their obsessions—but for many it is a genuinely useful way to hold powerful bodies to account and to uncover things that need to be brought into the open.

Perhaps the only person who has an un-nuanced view (remarkably, has two entirely contradictory un-nuanced views) of FOIA is the one person most directly responsible for bringing it into being. Tony Blair's enthusiasm was clear when he was an honoured guest at the Campaign for Freedom of Information's Awards in 1996.

> It is not some isolated constitutional reform that we are proposing with a Freedom of Information Act. It is a change that is absolutely fundamental to how we see politics developing in this country over the next few years [...] information is power and any government's attitude about sharing information with the people actually says a great deal about how it views power itself and how it views the relationship between itself and the people who elected it [...] People often say to me today: everyone says this before they get into power, then, after they get into power you start to read the words of the government on the screen and they don't seem so silly after all. (Tony Blair's speech to the Campaign for Freedom of Information Awards, 25 March 1996[17])

With the benefit of hindsight and some experience of the practical consequences of what became a poorly written law, his view had evolved somewhat and his enthusiasm had diminished a little.

> Freedom of Information. Three harmless words. I look at those words as I write them, and feel like shaking my head till it drops off my shoulders. You idiot. You naive, foolish, irresponsible nincompoop. There is really no description of

[17] Transcript from https://www.cfoi.org.uk/.

> stupidity, no matter how vivid, that is adequate. I quake at the imbecility of it. [...] Once I appreciated the full enormity of the blunder, I used to say—more than a little unfairly—to any civil servant who would listen: Where was Sir Humphrey when I needed him? We had legislated in the first throes of power. How could you, knowing what you know have allowed us to do such a thing so utterly undermining of sensible government? (From Tony Blair's memoir [26])

One of the consequences of 'Climategate' was that the Information Commissioner's Office (ICO) created a Higher Education Sector Panel for a time, and I ended up being on this. This—along with multiple other contacts, including direct meetings between the Commissioner and Edward Acton—eventually helped with the development of publication schemes and advice to the sector of how to handle floods of FOIA requests, and how to balance the interests of the researchers in research data prior to publication with the obligation to release data.

Graham

In 2008 Graham Everest told Shaun and me that he had stage 4 prostate cancer. This began a period of great turbulence for Graham and his family as the cruel disease advanced inexorably. Tania and I kept in close touch with both Graham and his wife Sue, and he wanted to keep thinking about Mathematics. During this time Graham and I had several long walks, talking a little about Mathematics and much more about our shared Christian faith, our families, and some of his many interests. Some of the things we had always enjoyed like the wine, cheese, and poetry evenings the Everests sometimes hosted gradually came to an end as the illness progressed.

Anish Ghosh, Shaun Stevens, Sanju Velani (from the University of York), and I put together a plan for a retirement conference for Graham. This was funded by the London Mathematical Society and the Number Theory Foundation and was a successful but poignant event. It took place at UEA from the 14th to the 16th of December 2009, under the title 'Diverse faces of Arithmetic'. The speakers reflected different aspects of Graham's mathematical interests. Manfred Einsiedler (Eidgenössische Technische Hochschule (ETH) Zürich) and Franco Vivaldi (Queen Mary University of London) spoke on homogeneous and arithmetic dynamics respectively; Kirsten Eisenträger (Pennsylvania State University) and Alexandra Shlapentokh (East Carolina University) spoke on his recent interest in model theory; Kálmán Győry (Lajos Kossuth University of Debrecen), Andy Hone (University of Kent), Valéry

Mahé (Université de Franche-Comté), Nelson Stephens (Royal Holloway, University of London), Chris Smyth (University of Edinburgh) and Igor Shparlinski (Macquarie University) spoke on various topics in number theory.

Our last project together was an expository paper on Hall's conjecture in Diophantine number theory, attempting to trace a 'repulsion' phenomena through some of the rich history of Diophantine equations.[18] Some of Graham's hand-written notes from our conversations about this trailed off under the influence of the morphine he was on, but we managed to see the project through to completion.

On the 30th of July 2010 Tania and I were on the A303, driving down to visit family in Dorset, when my mobile rang. As soon as I saw the call was from my PA Jacqui Churchill, I knew that Graham had passed away.

> The first author [Graham Everest] dedicates this paper to the staff of Mulbarton Ward and the Weybourne Day Unit in Norwich. The second author dedicates this paper to the memory of his friend and colleague Graham Everest (1957–2010). (Dedication at the end of our paper [117])

Graham passed away before learning that our last paper together had won the Lester Ford prize for research exposition, which he would have enjoyed very much. Alongside his research interests he believed in the importance of education and exposition and he had won a UEA Teaching Excellence award in 2005.

He was ordained into the Anglican priesthood in 2006 and by the time of his death had been defrocked and reinstated as a result of some of his own struggles. His funeral took place at St Peter's church in Cringleford on the 10th of August 2010. Tania, Raphael, and I were supposed to be on holiday in Paleokastritsa that week, so the two of them went ahead and I moved my flights at the last minute to be able to attend. The church was full, with people reflecting the many different aspects of Graham's life in attendance. He had started at UEA immediately after completing his doctorate and the University community was well represented. I was honoured to be asked to lead the final

[18]The repulsion concept here is related to a phenomenon that is widely observed but only understood in rare instances. The simplest example is Catalan's conjecture that the only consecutive powers are 8 and 9. More formally the conjecture is that the only integer solutions to the equation $x^a - y^b = 1$ with $a, b > 1$ and $x, y > 0$ are $x = 3, a = 2, y = 2, b = 3$ corresponding to $3^2 - 2^3 = 1$. This was proved by Mihăilescu [203] in 2004. More generally, Pillai conjectured in 1945 that for any $k \geqslant 1$ the equation $x^a - y^b = k$ has only finitely many integer solutions. Equivalently, powers 'repel' each other in that different powers of integers cannot be close together infinitely open. This problem remains open, with active research on more constrained versions.

prayers and began them with a passage from a much loved book by Thornton Wilder concerning the mysterious continuity of the links to those we love before and after their death.

> Soon we shall die and all memory of those we have known will have left the earth, and we ourselves shall be loved for a while and forgotten. But the love will have been enough; all those impulses of love return to the love that made them. Even memory is not necessary for love. There is a land of the living and a land of the dead and the bridge is love, the only survival, the only meaning. (From 'The Bridge of San Luis Rey' by Thornton Wilder)

The School of Mathematics at UEA created the Graham Everest Memorial Fund to remember something of Graham's deep commitment to the life-changing power of education.

The Netherlands

Through a series of chance events and personal contacts, four other countries had particular relevance to my mathematical life. Early on it was the USA of course, where I attended conferences and workshops frequently. From about 2008 on, it was Germany, Poland, and the Netherlands that dominated my research visits. During 2010 Gunther Cornelissen came up with the idea of arranging for a visiting position at Utrecht, and this led to me becoming the F. C. Donders visiting professor for a semester in 2011. This was difficult to combine with my PVC role at UEA, but was a wonderful time. I taught a short module on ergodic theory, and more generally was able to enjoy being a small part of the vibrant and highly connected network of people in the Netherlands interested in number theory and dynamical systems. The ease of moving around in the country meant that joint meetings or informal gatherings were easy to arrange and brought together a diverse mix of mathematicians.

The logistics were helped by the proximity between Norwich and Utrecht, but nonetheless it was a tiring business. For twelve weeks I took the 06.15 flight from Norwich to Schiphol on a Thursday, and the 21.20 flight back on a Friday. The airport in Norwich was not far from home, so I cycled to the airport each time, and the 'D' gates of Schiphol airport became a familiar home from home. I spent each Thursday night at a friendly basement bed and breakfast in Utrecht close to Gunther's house. Gunther and his family were extremely hospitable, and sometimes Tania would come over as well. Utrecht

Fig. 12.7 With Manfred Einsiedler, Klaus Schmidt, and Doug Lind for a thesis defence in Leiden in June 2015

(and Delft and Leiden, where we had made several visits also) holds a place in our hearts forever as a result of this semester.

Among many enjoyable visits to the Netherlands, I took part in several doctoral thesis defences. These are formal and public event, with full academic regalia—either of the host University or of the examiner's own University (Fig. 12.7).

We celebrated one such event with a boat trip which produced an irresistible moment for any admirer of the writings of Jerome K. Jerome (Fig. 12.8).

Music

Music at UEA had a long and influential history, and by 2010 had a strong presence in certain areas including electroacoustic music. Sadly the financial position did not match the quality of the research and teaching, largely because of the usual mix of student numbers, decisions about regulated fees taken externally, and research income. After some agonising discussions and scrutiny of financial models by the executive team we took a proposal to close the school

Fig. 12.8 Three ergodic theorists in a boat: with Doug Lind and Manfred Einsiedler, in Leiden in June 2015

to Senate, starting with a programme of teaching out from 2011. These are among the most painful decisions any leadership team ever makes, and the entire process was understandably contentious. This all came to a head with the key decision at Senate—technically a decision to make this recommendation to the governing Council, but in everyone's mind the real decision point—on the 9th of November 2011. There was a noisy demonstration outside the building where Senate was to be held, and someone there managed to grab hold of my tie and try and pull me over backwards. It was a slightly frightening incident, and I ended up requiring some medical attention.

It came as a huge relief when our own security staff viewed their own footage and confirmed that the culprit was not a UEA student, but a local political activist who was pursuing an agenda of his own. Our own students were vocal in opposition to the proposed closure, expressed great loyalty to the department and its staff, and generated substantial external campaigns and petitions—but their campaign never crossed into anything inappropriate.

I stood in for the Vice-Chancellor at the poignant graduation ceremony for Music in July 2012, but by then attention had turned towards making the teach-out work well, and finding ways to continue the special place of music at UEA in other ways.

Robert, Matthew, and Stefanie

Once again three PhD students were studying with me across a significant change in my professional life: Robert Royals[19] jointly supervised with Anish Ghosh, Matthew Staines,[20] and Stefanie Zegowitz[21] jointly supervised with Shaun Stevens. I will say more about Robert and Stefanie later, as most of our work together happened while I was at Durham.

Matthew started working on the reverse Mahler measure problem in the context of algebraic dynamics. The forward problem might be thought of as computing the Mahler measure of a polynomial, and understanding the possible values arising. The reverse problem asks for a description of the set of polynomials with the same Mahler measure. He quickly made some interesting progress, and went on to look at some other reverse problems in the setting of automorphisms of compact groups, with shared Mahler measure replaced by other natural notions of equivalence. This began for me a way of viewing the space of automorphisms of compact groups as a single space (something that might be called a 'moduli space' in other contexts), onto which various notions of equivalence might be imposed. This point of view dominated my thinking for some years (Fig. 12.9).

> Let $\mathcal{G}$ denote the collection of all pairs (G,T) with G a compact metric abelian group and T a continuous automorphism.
>
> My mathematical autobiography: Try to say something non-trivial about the space $\mathcal{G}$, or about $\mathcal{G}$ /meaningful equivalence relation.
>
> Sample question: As you move around in $\mathcal{G}$, which dynamical properties are 'rigid' (granular) and which are 'flexible' (smoothly varying)?
>
> If you fix G then $\mathrm{Aut}(G)$ is totally disconnected in its natural topology (Iwasawa 1949).
>
> The point is to vary G as well as T.

Fig. 12.9 A slide from a seminar I gave at the University of Sheffield on the 14th of November 2016

[19] Robert graduated in 2015, with external examiner Prof. Sanju Velani, University of York [261].

[20] Matthew graduated in 2013, with external examiner Prof. Chris Smyth, University of Edinburgh [283].

[21] Stefanie graduated in 2015, with external examiner Prof. Oliver Jenkinson, Queen Mary University of London [369].

13

From Sillery to the Office for Students

On the 14th of March 2012 I was contacted by a search consultant called Mike Dixon about the role of PVC Education at Durham University.

Two days earlier Durham, along with Exeter, Queen Mary University of London, and York, had announced that they would be joining the so-called Russell Group of Universities in August 2012. When the news of this significant rearrangement of the deck chairs among some of the larger research Universities came through, I was attending a residential event as part of the so-called 'Top Management Programme'. A member of the Executive of Durham University was also in the group, and this came as news to him, showing just how secretive the negotiations had been.

The realignment was in part a response to the beginning of a huge shift in public policy, which can only be explained by a detour into the history of numbers and fees in UK higher education. My brief, partial, and amateur account of part of this history not only will betray to any expert my own limited understanding, but merely scratches the surface of a complex legacy. Shortly after the passage of the Higher Education and Research Act 2017 David Kernohan produced a useful article entitled 'Major reviews of higher education through history'[1] bringing together 22 Acts of Parliament, starting with the Universities (Wine Licenses) Act of 1743[2] and ending with the Higher Education and Research Act 2017; 20 major reviews, starting with the Haldane report of 1904 which led to the Haldane principle that allocation of research

[1]https://wonkhe.com/blogs/major-reviews-in-history/.

[2]http://www.legislation.gov.uk/apgb/Geo2/16/40/contents.

T. Ward, *People, Places, and Mathematics*, Springer Biographies,
https://doi.org/10.1007/978-3-031-39074-6_13

funds should be determined by researchers and recommending the creation of the University Grants Committee, and ending with the Browne review of 2010; and seven White Papers starting with the 'Educational Reconstruction' paper of 1943 and ending with 'Success as a Knowledge Economy' in 2015. Sadly the 1743 'Act to prevent the retailing of Wine within either of the Universities, in that Part of Great Britain called England, without Licence' was repealed by the Statute Law Revision Act 1887.[3]

The forces that shaped my career most began with decisions our parents took at the start of their careers in the early 1950s, but were also a by-product of the various white papers and policy changes that brought universities like Warwick and East Anglia into existence in the 1960s.[4]

The English Genius

Despite great differences in branding, campuses, rankings, wealth, demography of the student body, and history, one of the startling features of Universities across the UK is how similar they are or aspire to be. This is largely the outcome of the relentless pressure of absurd rankings and league tables, many of which attempt to impose a simple linear ordering on experiences of immense and barely comparable complexity. It is also an inevitable consequence of the way Universities in the UK have been funded. This has remained the case in the face of repeated efforts to encourage diversity and innovation by governments of all types.

The ancient Universities, four in Scotland, two in England, and one in Ireland, had medieval and early modern origins, and existed prior to 1600 CE. Nine so-called 'red brick' Universities were added in the main industrial cities of England in the nineteenth Century. One of the most significant of these—in a context created by the fact that for much of their histories the two ancient Universities in England had excluded Jews, Catholics, and dissenters—was the founding of what is now University College London in 1825, with no clergy on its governing body and no religious requirements for staff or students.[5]

[3] https://www.legislation.gov.uk/ukpga/Vict/50-51/59/enacted.

[4] Repeating that step back in time of sixty or seventy years also encompasses the founding ideas of the research area I entered, built on the foundational ideas of recurrence discussed by Henri Poincaré in 1890, the ergodic theorems of George Birkhoff and John von Neumann in the early 1930s, and the insights into flows on surfaces due to Gustav Hedlund and Eberhard Hopf in the late 1930s.

[5] The name of this institution is not a simple matter. As far as I can tell, it was London University from 1825 to 1836, University College, London from 1836 to 1907, University of London, University College from 1907 to 1976, University College London from 1976 to 2005, and since 2005 it has retained

University College London also expanded access to higher education to the lower classes. As with any other expansion of access to higher education, this was not universally welcomed and provoked some negative reactions.

> Ye Dons and ye Doctors, ye Provosts and Proctors,
> Who are paid to monopolize knowledge,
> Come make opposition by voice and petition
> To the radical infidel College
> [...]
> But let them not babble of Greek to the rabble,
> Nor teach the mechanics their letters;
> The labouring classes were born to be asses,
> And not to be aping their betters.
> 'Tis a terrible crisis for Cam and for Isis!
> Fat butchers are learning dissection;
> And looking-glass-makers become sabbath-breakers
> To study the rules of reflection;
> "Sin:ϕ and sin:θ"—what sins can be sweeter?
> Are taught to the poor of both sexes,
> And weavers and spinners jump up from their dinners
> To flirt with their Y's and their X's.
> (From 'A Discourse Delivered by a College Tutor at a Supper Party'; Winthop Mackworth Praed, The Morning Chronicle, July 19th, 1825)

However it was the second half of the twentieth Century that saw a rapid growth in numbers and the creation of the sector as most of its participants now experience it.

A white paper [40] in 1956 authored by the then Lord Privy Seal R. A. Butler, Secretary of State for Scotland James Stuart, and Minister for Education David Eccles brought into being a new type of higher education institution called a 'College of Advanced Technology' (CAT). The initial suggestion was that 24 existing technical colleges would become Colleges of Advanced Technology, but in the end only ten came into being over the next few years: Battersea, Birmingham, Bristol, Brunel, Chelsea, Loughborough, Northampton, Salford, and Welsh under the CAT title, and Bradford Institute of Technology. These all came under the control of the Local Education Authority (LEA) and the salaries of lecturers were linked to those of teachers,

University College London as its legal name but has branded itself as UCL. Many other Universities have equally convoluted histories concerning their name, but this one is particularly striking as it is one of the largest institutions in the UK.

contributing to difficulties in filling posts through the late 1950s. In 1961 the Ministry of Education reached an agreement with the LEAs that brought in a new arrangement under which the Colleges of Advanced Technology would be directly funded by government grants, creating for a few years a higher education sector outside the conventional University model.

Over the next few decades all of these Colleges of Advanced Technology either became Universities or were subsumed into Universities: Battersea CAT became the University of Surrey, Birmingham CAT became Aston University, Bristol CAT became the University of Bath, Brunel CAT became Brunel University, Chelsea CAT was later absorbed by King's College London, Loughborough CAT became Loughborough University, Northampton CAT became City, University of London, Salford CAT became the University of Salford, the Welsh CAT merged into Cardiff University, and Bradford Institute of Technology became the University of Bradford. Some of these Universities retain distinctive features of their Advanced Technology origins, but they do not now stand out from many other institutions in this way.

The University Grants Committee approved the founding of seven 'plate glass' Universities in the late 1950s and early 1960s. The name was coined by the barrister Michael Beloff: 'I have chosen to call them the Plateglass Universities. It is architecturally evocative; but more important, it is metaphorically accurate' (from [20]). These were the University of Sussex founded in 1961, York and East Anglia in 1963, Essex and Lancaster in 1964, Kent at Canterbury and Warwick in 1965. What became known as the Robbins Report [256] after its Chair Lord Robbins was commissioned by the government in 1961 and published in 1963. This major report recommended the rapid expansion of the number of Universities, and that the Colleges of Advanced Technology should all become Universities. Many of the new institutions founded as a result are also sometimes referred to as 'plate glass' Universities. The Robbins report articulated some enlightened principles.

> Throughout our Report we have assumed as an axiom that courses of higher education should be available for all those who are qualified by ability and attainment to pursue them and who wish to do so. (From the Robbins Report [256, para. 31])

It also tried to express four natural objectives of a 'properly balanced' higher education.

1. Instruction in skills suitable to play a part in the general division of labour;
2. taught in such a way as to promote the general powers of the mind [...] to produce not mere specialists but rather cultivated men and women;

> 3. the search for truth is an essential function of institutions of higher education and the process of education is itself most vital when it partakes of the nature of discovery;
> 4. the transmission of a common culture and common standards of citizenship. By this we do not mean the forcing of all individuality into a common mould: that would be the negation of higher education as we conceive it. But we believe that it is a proper function of higher education, as of education in schools, to provide in partnership with the family that background of culture and social habit upon which a healthy society depends. (From the Robbins Report [256, para. 25–28])

The fourth objective in particular placed these institutions in the Humboldtian ideal, as a community of scholars and students with the union of teaching and research central to the work of the individual scholar. Many of the institutions from this wave of development also had ambitious plans to define a 'new map of learning.'

> The creation of the new universities offered a chance for syllabus reform that could not have arisen during expansion of the old. The opportunity to participate lured to the campuses innovators who would otherwise have been chary of sacrificing years of research to the demands of organisation. It was a co-operative endeavour in which the youngest assistant lecturer, the doctoral down fresh on his academic cheek, could advise as much as the senior professor—indeed he might even be listened to. The founding fathers of Plateglass had more power to shape an academic structure than the most despotic heads in an older university. (From 'The Plateglass University' by Beloff [20, Ch. 3])

The ambition for a radical break with tradition was real, even if the author did not quite manage to break free of language predicated on the belief that the typical academic involved would be male. Beloff goes on to write of 'the redrawing of boundaries between old subjects, and in many cases their abolition' and that the 'bias is towards the liberal rather than the vocational'. As with the Colleges of Advanced Technology, while traces of this spirit remain in some of the plate glass Universities, they do not authentically stand out in these regards any longer.

There have been several other innovations in the provision of higher education. Some local authorities established Colleges of Higher Education. These were largely intended to serve local populations and potential students who might not otherwise think of going into higher education. Here too there has been a steady drift away from the distinctive and towards the generic. Edge

Hill University in Ormskirk and Bournemouth University, for example, no longer stand out especially strongly from many other Universities.

Perhaps the largest homogenisation of the sector took place in 1992 with the end of the 'binary divide'. Following the election of a Labour government under Harold Wilson in 1964, the Secretary of State for Education and Science Anthony Crosland articulated a plan for a 'binary system' for higher education, with autonomous Universities funded through the University Grants Committee and a differently funded and controlled public sector comprising Technical and Further Education Colleges. He was remarkably prescient about what the consequences of naive competition and rankings might be.

> The Government accepts this dual system as being fundamentally the right one, with each sector making its own distinctive contribution to the whole. We infinitely prefer it to the alternative concept of a unitary system, hierarchically arranged on the "ladder" principle, with the Universities at the top and the other institutions down below. Such a system would be characterized by a continuous rat-race to reach the First or University Division, a constant pressure on those below to ape the Universities above, and a certain inevitable failure to achieve the diversity in higher education which contemporary society needs. (From Anthony Crosland's speech at Woolwich Polytechnic on the 27th of April 1965[6])

The intention was to address the skills requirements of industry by upgrading the status of technical education to equal the status of the University sector—that they would be 'separate but equal'. This thinking led to a white paper in 1966, 'A plan for Polytechnics and other Colleges' [61], which recommended that colleges with the most potential should become regional Polytechnics. These would become 'large and comprehensive' providers of technical and vocational higher education through full-time, part-time, and sandwich courses offering degrees validated by the Council for National Academic Awards. The trajectories of these institutions is an interesting story in itself, well described in a book by Pratt [248]. At one stage there were more than 30 Polytechnics.

The Further and Higher Education Act 1992[7] shaped the destiny of the sector for the next 25 years. This closed down the Universities Funding Council and the Polytechnics and Colleges Funding Council, and brought all higher education institutions, including three that were directly controlled

---

[6]These and some other difficult to find speeches and documents have been helpfully collated on the website of the Higher Education Policy Institute (HEPI).

[7]https://www.legislation.gov.uk/ukpga/1992/13/contents.

by the Department for Education[8] in England together under the control of the newly created Higher Education Funding Council for England (HEFCE). HEFCE was formally launched on the 2nd of June 1992, and took over its funding role from the 1st of April 1993. At the time the sector in England comprised 81 Universities and 50 Colleges. The funding model had four main components: Teaching ('T funding') was allocated based on contracted numbers of students, with different costs of subjects reflected in a set of bands; Research ('R funding') allocated based on the various iterations of the national Research Assessment Exercise (RAE); 'Non-formula funding' for needs that fell into neither T nor R; and a regular round of capital funding to support strategic developments. The act also required HEFCE to set up the Higher Education Statistics Agency (HESA), to collect enormous volumes of data about institutions.

The 1992 act also closed the 'binary divide', with all the Polytechnics becoming Universities.

While some differences particularly in the terms of employment remain in places, the broad picture is once again of homogenisation. Across most measures of performance, spread of subjects, mix of students, and contribution to research most of the former Polytechnics do not stand out from the so-called 'pre-92' Universities. Indeed the phrases 'pre-92' and 'post-92' are not widely used outside the sector, and would have no significance for current applicants or students. The outcome of the most recent national Research Excellence Framework gave Northumbria University, which had been Newcastle Polytechnic until 1992, the highest 'research power' ranking of any University outside the Russell Group, a telling example of how rapidly things have changed.

The frequently reiterated aim of promoting greater diversity and innovation in the sector is constantly hampered by league tables, uniform key performance indicators, and regulatory pressures. From newspapers to politicians to regulators, the instinct to define one institution as 'worse' or 'better' than another prevents the flourishing of an environment in which one institution is simply 'different' to another for good reasons. The sociologist Sir Howard Newby, former Vice-Chancellor of the Universities of the West of England, Southampton, and Liverpool and at one stage the Chief Executive of HEFCE, pithily described these kind of processes as the 'English genius for turning diversity into hierarchy' [228].

[8] The Open University, founded by the Labour government of Prime Minister Harold Wilson and given its royal charter in 1969, Cranfield Institute of Technology, and the Royal College of Arts were, for different reasons, directly overseen by the Department for Education.

University Fees

The first zephyrs of the hurricane to come had already started under the Labour government that came into office in 1997 led by Tony Blair, with the introduction of the principle of financial burden sharing between the student and the taxpayer in 1998. These changes were controversial, and led to widespread demonstrations, including several at UEA, but the policy did at least have two meaningful starting propositions. Firstly, that both the country and the individual benefit from higher education, and so both should contribute. Secondly, that some higher education institutions were facing existential threat due to a funding crisis.

The superficial tension was about the principle of students paying fees, a bitterly contested Manichaean question. The real debate behind the scenes was about the combination of numbers and fees. HEFCE introduced Student Number Controls in 2009–2010 following a significant overspend as a result of the 2008 cohort entering higher education being larger than planned, though in fact there had been some controls on numbers since 2006. For the years that Student Number Controls existed, much of my role involved playing a small part in the delicate game of ensuring that the university admitted the number of students in each subject that our HEFCE contract specified, and within the University trying to use the predictable block grant from HEFCE to best effect. The tight controls on numbers certainly protected universities, but it also constrained them. It was closely associated with the fact that students from disadvantaged backgrounds were far less likely to experience higher education: With *de facto* rationing of University places, the understandable pressure from wealthier households to win those places was difficult to resist. The simple uncomfortable truth was this: The refuse collector's taxes helped to pay for the stockbroker's child to go to Oxford, while the refuse collector's child was unlikely to benefit from higher education at all.

Retaining number controls allowed no solution to this politically difficult fact, nor to the growing demand for widening access to higher education. Releasing the controls on numbers entailed either a blank cheque commitment from taxpayers, or a reduction in the 'unit of resource' (the amount of funding a University would have for each student) with each additional student. This was the real Manichaean issue, and it was frustrating to watch the sometimes intemperate political posturing on all sides fail to confront the underlying issues in an open way.

Prior to the Education Act 1962[9] Universities charged fees of students, and this was a significant part of the income of institutions. For the student this meant that higher education required family wealth or winning various competitive scholarships. Our parents, for example, were able to study at the University of Birmingham in the 1940s because they won competitive scholarships, our father's scholarship being worth £186 per annum. The resulting atmosphere is beautifully captured in Anthony Powell's depiction of a scene in an unspecified university (based on Oxford) in the 1920s. The setting is a tea party hosted by the academic Sillery, and the guests include a group of students who had become friends while at an unspecified public school (based on Eton College).

> During one of these pauses, Sillery, pottering about the room with the plate of rock-buns, remarked: 'There is a freshman named Quiggin who said he would take a dish of tea with me this afternoon. He comes from a modest home, and is, I think, a little sensitive about it, so I hope you will all be specially understanding with him. He is at one of the smaller colleges—I cannot for the moment remember which—and he has collected unto himself sundry scholarships and exhibitions, which is—I think you will all agree—much to his credit.' (From 'A question of upbringing' by Powell [247])

It was indeed a question of upbringing.

A Golden Age?

The Education Act 1962 required local education authorities to pay the tuition fees of students embarking on a full-time first degree, and to provide them with a means-tested maintenance grant. To a large extent this was catching up with changes in society following the Second World War: Most local education authorities were already paying student tuition fees and providing maintenance grants. This contributed to the rapid growth in both the number and the size of universities discussed earlier, and enormously expanded the opportunities open to a talented student 'from a modest home'. In some sense it repeated for higher education the progressive reforms to school education brought in by the Education Act 1944[10] (dubbed the 'Butler Act' after the President of the Board of Education from 1941 to 1944).

[9] www.legislation.gov.uk/ukpga/Eliz2/10-11/12/enacted.

[10] www.legislation.gov.uk/ukpga/Geo6/7-8/31/contents/enacted.

The 1962 Act was formally repealed by the Teaching and Higher Education Act 1998,[11] which brought in tuition fees and student loans. This also repealed the Education (Student Loans) Act 1998,[12] and parts of the Education Act 1996.[13] These changes were triggered by the Dearing Report [67], the product of an inquiry commissioned by Gillian Shephard, the Conservative Secretary of State for Education and Employment. The report made clear the scale of investment needed to meet the growing demand for continued expansion in higher education, and engaged with the controversial possibility of a shared contribution model.

> We are particularly concerned about planned further reductions in the unit of funding for higher education. If these are carried forward, it would have been halved in 25 years. We believe that this would damage both the quality and effectiveness of higher education. We are also concerned about some other immediate needs, especially in relation to research. [...] We recognise the need for new sources of finance for higher education to respond to these problems and to provide for growth. We therefore recommend that students enter into an obligation to make contributions to the cost of their higher education once they are in work. Inescapably these contributions lie in the future. [...] We do not underestimate the strength of feeling on the issue of seeking a contribution towards tuition costs: nor do we dispute the logic of the arguments put forward. A detailed assessment of the issues has, however, convinced us that the arguments in favour of a contribution to tuition costs from graduates in work are strong, if not widely appreciated. They relate to equity between social groups, broadening participation, equity with part-time students in higher education and in further education, strengthening the student role in higher education, and identifying a new source of income that can be ring-fenced for higher education. (From The Dearing Report [67])

It would not be the only time that an inquiry of this sort, destined by its remit to generate controversy, bridged two different governments. The landslide election [60] brought in a Labour government led by Tony Blair, with a huge 179 seat majority. A means-tested student contribution of £1000 as an up-front contribution was introduced in 1998, which crossed the Rubicon.

This fee rose in line with inflation, reaching £1225 in 2007. The Higher Education Act 2004[14] made the fundamental shift to a loan with deferred

[11] www.legislation.gov.uk/ukpga/1998/30.

[12] www.legislation.gov.uk/ukpga/1998/1/contents/enacted.

[13] www.legislation.gov.uk/ukpga/1996/56/contents.

[14] www.legislation.gov.uk/ukpga/2004/8/contents/enacted.

income contingent repayments, and the fee was increased to £3000 from 2006. The language used was of 'top-up' fees, supplementing the direct grant to universities from government via HEFCE, which on each occasion was reduced by a similar amount. As a Head of Department, this rescued us from an impossible financial situation caused in part by how the internal cost weightings of different subjects was handled at UEA. The relief was short-lived, despite the indexing of the fee, which raised it to £3290 by 2010.

Markets

The Labour government commissioned the Independent Review of Higher Education Funding and Student Finance in 2009, in response to the growing pressures on university finances. Lord Peter Mandelson, the Secretary of State for Business, Innovation and Skills, invited it to consider the 'balance of contributions to universities by taxpayers, students, graduates and employers'. The membership of the review panel was itself contested, and in particular Lord Mandelson turned down demands from the National Union of Students to have a representative on the panel.

The review was chaired by Lord Browne of Madingley, former chief executive of BP. As with the Dearing report, it was commissioned by one government and received by another. The final report, 'Securing a Sustainable Future for Higher Education' appeared in October of 2010.[15] Just as the Robbins report articulated some principles and aspirations for higher education, the Browne report defined some principles. These reflected changes in society, and positioned the student as an agent actively exercising power as a consumer. The report also addressed the fact that at the time there were at least 20,000 suitably qualified applicants unable to find University places each year.

Principle 1: There should be more investment in higher education—but institutions will have to convince students of the benefits of investing more
Principle 2: Student choice should increase
Principle 3: Everyone who has the potential should have the opportunity to benefit from higher education
Principle 4: No student should have to pay towards the costs of learning until they are working
Principle 5: When payments are made they should be affordable

[15] https://www.gov.uk/government/publications/the-browne-report-higher-education-funding-and-student-finance.

Principle 6: There should be better support for part time students
(From the Browne report)

The recommendations in the report tried to achieve multiple objectives: To enable growth in access to higher education from disadvantaged students, to fund universities adequately and sustainably, and to create a dynamic market in which students would act as economic agents making decisions about value for money. To this end the report proposed no cap at all on tuition fees, but that fees above £6000 would attract a marginal 'tax' that grew to reach 75% at £12,000. The tax would fund a national scholarship scheme to enable students facing financial hardship to access higher education, and outside the perverse behaviours arising from education being seen as a Veblen good, meant that no university would charge more than £15,000 for a home undergraduate. The university would receive the fee direct from government, and the student would repay them through an income dependent loan. It was more or less a graduate tax on future graduates, but with the individual student choice element designed to promote market behaviours, and with a wide range of fee levels built into the system.

The Browne report also addressed the question of how the sector should be organised as a whole. At the time higher education in England involved four different agencies. Funding for both education and research and student number controls was run through HEFCE, quality and standards were controlled by the Quality Assurance Agency (QAA), monitoring of performance in widening participation for students from less advantaged backgrounds was overseen by the Office for Fair Access (OFFA), and a body set up to make rulings on student complaints after internal procedures have completed, the Office of the Independent Adjudicator (OIA). The proposal was to replace all four of these with a single body, the Higher Education Council. This Council was expected to take a targeted approach to regulation, giving more autonomy to Universities, and to make an annual report to Parliament explaining how it is using public money and protecting the investment students would be making in their own education. Sadly many of the ideas in the Browne report never really came to fruition, this included.

The general election of 6 May 2010 produced no overall majority. After some days of negotiations Gordon Brown resigned on 11 May 2010, and a coalition government was formed by the Conservatives and the Liberal Democrats. The new Prime Minister was the Conservative David Cameron, and Nick Clegg from the Liberal Democrats was Deputy. For the first four years the Minister of State for Universities and Science was David Willetts, a thoughtful and careful politician who later wrote a magisterial account of the history and role of higher education in the modern era [361]. The Secretary of State for Business, Innovation and Skills was the Liberal Democrat Vince Cable.

One of the most contentious issues facing the coalition government concerned tuition fees. The Liberal Democrats, perhaps reflecting a long history of grappling not with the ugly prose of the art of the possible in government but with the heady poetry of opposition, stood on a platform of abolishing tuition fees altogether. These proposals did not, for example, engage in any meaningful way with the complex difficulties of number controls that the policy entailed. Indeed, instead of grappling with this the party orchestrated dramatic public pledges signed by candidates against any increase in fees. In several parliamentary seats containing universities this issue undoubtedly influenced the result. The National Union of Students, with a clear sense of how the Browne review might go, produced a powerful campaign, cleverly using public commitments to this pledge.

There are, unsurprisingly, contested versions of how the negotiations that created the coalition handled this. David Cameron claimed later that the Conservatives were willing to compromise as this had been such a defining commitment for the Liberal Democrats, and was surprised that this was not demanded in the discussions. Whatever the reason, the coalition government responded to the Browne review by raising the fee cap to £9000 alongside a proposed 80% cut in direct funding to universities. This was fiercely contested in parliament, on the streets, and in the form of a judicial review. The review failed to halt the changes, and so the new fee came into effect for students entering university in 2012.

The proposals in the Browne review would have certainly created price differentiation, but the decision to cap the fee at £9000 more or less prevented this. The naive expectation of many commentators—and of the government—was that a range of fee levels would emerge naturally, but it seemed obvious that no university could afford to signal that it was going to deliberately spend less on education, nor that it lacked confidence in its own degrees. Essentially all universities immediately put their fees to the cap, and the opportunity for real price signalling that had been sought politically was lost. Whether it had been a sensible idea to insert such a simplistic market pressure into higher education in the first place is less clear. Indeed, one of the most perplexing aspects of the changes introduced since Dearing is this: It has been the governments most vocal in support of market dynamism and consumer pressure who have been most consistently wrong-footed by actual market behaviours. The last few decades have been a litany of attempts to prevent the outworking of market pressures arising from interventions designed to enable those market pressures to be expressed.

The ease with which the Liberal Democrats abandoned their solemn public pre-electoral pledge on tuition fees, and the perplexing way in which the nuanced proposals in the Browne review became the blunt instrument of a *de facto* single fee level are perhaps best explained by the complacent lack of seriousness with which the Secretary of State in the coalition government approached the whole matter.

> I'm quite good at delegating. I worked pretty well with the Tory ministers in agreeing they should have their own little empires and so on but obviously at the end of the day I had to understand the subjects that I was responsible for and it took a long time. For things like university finance, one of the reasons I think we got into so much trouble over tuition fees was the mind-boggling complexity of the subject. And it's a bit like the Schleswig–Holstein question: somebody was dead who understood it and David Willetts I think sort of understood it and I gradually understood it but by then we'd already made the key decisions which were probably wrong. So I think the complexity and massive nature of the material we were dealing with came as a surprise. (Vince Cable, in an Institute for Government interview by Jen Gold and Peter Riddell on the 7th of July 2015[16])

The ministers with their 'little empires' referred to with so little courtesy included David Willetts, arguably the outstanding higher education minister of the modern age.

Alongside the changes in how universities were funded were changes in how big they could be. With the price fixed, volume became the only real variable to determine the resources that could be generated from home students. In pursuit of this market dynamism, the government announced that from 2012 entry the cap on student numbers would be lifted for applicants with A-level grades of AAB or higher, or equivalent qualifications. This began a process of slowly removing student number controls entirely, with the exception of certain regulated programmes like Medicine. The lifting of caps for AAB+ in 2012 immediately changed the landscape, and made it logical for the Russell Group of universities to consider pulling in those institutions expected to be able to grow as a result. This was one of the reasons for Durham, Exeter, York, and Queen Mary to join the Russell Group, which increased the size of that powerful lobbying group to 24 institutions.

[16] www.instituteforgovernment.org.uk/ministers-reflect/person/vince-cable/.

The tensions between numbers and the ultimate cost to the Treasury of fee levels remains an unresolved issue for each successive government. The long term consequences of the Darwinian competition embedded into the system for the mix of subjects offered and the regional spread of access to higher education are hard to predict, but there is certainly a risk of inequalities being exacerbated as a result.

The question of higher education fees and how it had been handled by the coalition government became part of a wider set of issues that saw the Liberal Democrats lose all but eight of their 57 parliamentary seats in the 2015 general election [64].

Brexit

The 2015 election campaign also saw another instance of complacent hubris on a yet greater scale, with the Conservative manifesto promising what became the Brexit referendum.

> ...put these changes to the British people in a straight in-out referendum on our membership of the European Union by the end of 2017 [...and to deliver...] annual net migration in the tens of thousands, not the hundreds of thousands. (From the 2015 Conservative election manifesto)

The so-called Brexit referendum and the many unresolved problems that flowed from it are too complex to address here, but some of the characters involved in this monumental act of national self-harm are well captured by Cyril Connolly's phrase from 1938.

> Were I to deduce any system from my feelings on leaving Eton, it might be called The Theory of Permanent Adolescence. It is the theory that the experiences undergone by boys at the great public schools, their glories and disappointments, are so intense as to dominate their lives and to arrest their development. (Cyril Connoly, in 'Enemies of Promise' [57])

The Russell Group of Universities created a 'Brexit group' to work through some of the potential implications, and for some reason I was on this. At the very first meeting the chair, Sir Leszek Borysiewicz (Vice-Chancellor of the University of Cambridge at the time) surprised many of us by suggesting that the leave campaign might easily win. They proved to be prescient words. The meetings were interesting because we were often joined by quite senior figures from, or close to, government. The gulf in perception between the University representatives on the group and some of the pro-Brexit politicians and 'special advisers' we met was considerable. Many of the discussions went as follows. We would point out that things like the Erasmus programme, membership of the European research infrastructure, and so on, would be disrupted and needed to be renegotiated, and that if this did not happen then we risked huge damage to both the education and research missions of UK higher education. The response was a courteous and eloquent version of simply reiterating 'that's project fear' and nothing significant progressed.

Our biggest area of concern quickly moved beyond the parochial concerns of higher education and focused on the implications for the economy and society of voluntarily erecting trade barriers to our nearest neighbours, and the implications for society of becoming a more inward-looking and nationalistic country. The likely scale of economic damage if we ended up with what came to be called a 'hard Brexit' was both large and obvious, and our anxiety became focused on these central economic and societal questions. Not only does it give me no satisfaction, it fills me with despair that essentially all our concerns came to pass.

Looking back at the wreckage, one obvious lesson is the immense power of the perfectly framed but meaningless hendiatris: 'Take back control', 'that's project fear', 'get Brexit done', 'Brexit means Brexit', and 'oven ready deal'.

Growth and Social Mobility

Just as climate change needs both a macroscopic viewpoint concerned with global thermal capacity and planetary scale albedo and a microscopic one concerned with scattering of individual photons by individual Carbon dioxide molecules—and everything in between—higher education can only be understood on multiple scales. The figures quoted in this section come from multiple sources, including the Office for National Statistics (ONS)[17] and the The Chartered Institute of Personnel and Development (CIPD).[18] A particularly helpful summary of the changes in higher education over this period is contained in a Higher Education Policy Institute (HEPI) briefing note by Nick Hillman [150].

[17] https://www.ons.gov.uk/.

[18] https://www.cipd.co.uk/.

In the early 1950s there were 85,000 full-time students in the UK, and they were taught by about 8000 staff. There were 18 Universities in 1952, and the University Grants Committee had a staff of 16. Research was largely controlled and funded directly from Government departments. Roughly 60% of the working age population were in employment, made up of 96% of the male working age population and 46% of the female. Manufacturing accounted for 40% of total employment, and so-called knowledge workers accounted for less than 25%.

I have talked earlier about the rapid growth in the number of Universities and the numbers of people able to access higher education. These developments met some resistance, but were welcomed by many politicians who saw the power of increased access.

> The record of the expansion of higher education in the 1980s is a good one that stands to the credit of all who have made it possible. But there is scope over the next 25 years for even greater advance. Assuming an average annual economic growth rate of 2%, in 25 years time Britain will be more than half again as rich as it is today. That is to say, our per capita income will be more or less at the same level as it is in the United States currently. This more affluent society will be built on better education and will itself want to be more highly educated. [...] This is not just a British phenomenon. Increased participation is important for all of us in Western Europe. [...] One of the great trends of the next quarter of a century will be our increasing integration into Western Europe, and increasingly close links between our institutions of higher education.
> (From a speech by the Secretary of State for Education Kenneth Baker at Lancaster University in January 1989[19])

The UK in 2022 has around 150 Universities employing roughly 150,000 staff, with many other colleges and other organisations providing higher education. There are more than 2 million students in full-time higher education, and there has been a sustained sectoral shift in employment from manufacturing to the knowledge sector. Roughly 70% of working age women are in employment.

The 2005 study by Blanden et al. [27] contributed to social mobility moving out of sociological and economic study and into the political mainstream. This report was undoubtedly influential, and its conclusions have become accepted fact in all quarters. It is important to note however that even its main conclusions have been contested, and there are some trenchant and cogent

[19] From the website of the Higher Education Policy Institute (HEPI).

critiques of its methodology [137]. Partly as a result of this report, the belief that inter-generational mobility in income, in access to higher education, in home ownership, and in many other indicators seemed to be both low by international standards and slowing was embraced as a pressing concern by politicians of all parties.

There have been many studies trying to bring this work up to date, and on the 25th anniversary of the Sutton Trust a sobering report by Eyles et al. [123] was published.

> Today's research shows how the picture has developed in more recent times. While we have seen some progress in narrowing education gaps, especially in access to university, the report shows how far opportunities are still determined by family background. Most worryingly, it predicts a fall in mobility for poorer young people driven by the impact of the pandemic. This is exacerbated by a cost of living crisis, and increasing divides in who can afford to buy their own home and those who can't. The authors warn that this risks a 'step change' down in mobility prospects for today's low income young people compared to other nations. (From [123])

For many research universities success in recruiting large numbers of students may yet prove to be a Pyrrhic victory. Most Russell Group universities have attempted to navigate a path through the Scylla of the cost of internationally competitive research and the Charybdis of a 'unit of resource' (income for each home student) that has been more or less flat in cash terms ever since the 2012 implementation of the Browne review by pursuing the simple expedient of growing fast. Arguably this has diminished both the intensity and the quality of education they can provide, and the quality of research they can undertake.[20]

This is a vast and complex area of research, and there are serious researchers who take different views on the central question of whether social mobility has slowed in the country. Nonetheless, the overall numbers raise the following macroscopic issue for the contribution that higher education can make to society. Higher education as an engine of expanding opportunities and social mobility may have been somewhat successful during a period of rapid growth in numbers and in knowledge economy jobs. The real challenge is to maintain even the limited progress made thus far without the huge affordances of this growth. The simple reality is that the claimed objective of all governments—to

---

[20] This argument has been well made from an economist's point of view by Nigel Thrift, former Vice-Chancellor of Warwick University [296].

enable everyone to achieve their potential educationally if they wish to—plays out very differently if it is about competition for University places and professional jobs rather than being built on expansion in them. The political consensus around what is sometimes called 'widening participation' evolved through a period in which it had no real impact on the children of the wealthy and powerful. It is not at all clear how this will play out when it does start to have an impact.

I take a blunt view of these complex questions. If everyone is supported to explore all they might be capable of through education at all levels, if the gap in formal qualifications between the state school system in the round and the 'best' schools both private and state as measured by examination results narrows, and if barriers to higher education are lowered, then the children of the wealthy and powerful might very well grow up having to clean their own toilets and mow their own lawns. They may also grow up to face much stiffer competition for the most highly remunerated careers. The incoherence and obfuscation of public policy in this area suggests that politicians of all parties, while eager for the accolades accruing to those who make lofty speeches dedicated to expanding social mobility, are not universally eager for this to really happen.

Politics

Life on a university executive involves its share of politics internally, but much of the life of a university is driven by the external political world. When I first took on the PVC role the external world for higher education was dominated by two 'sector bodies' or quangos, and two politicians. The first of these bodies was the Higher Education Funding Council (HEFCE, a 'non-departmental public body'), which oversaw both the education and research mission of the sector viewed as a whole, running a system that funded Universities via block grants and contracts for teaching. The second was the Quality Assurance Agency (QAA, a 'non-profit organisation'), which oversaw academic standards in a careful and at times cumbersome way. The dominant political figures were the Secretary of State for Education, and someone we tended to call the 'Universities minister'; both the job titles changed frequently. Rather confusingly there was a Secretary of State for Innovation, Universities and Skills at the time, John Denham. The remit included responsibility for the 'development, funding and performance' of higher education. In practice HEFCE was a highly effective buffer between government and the sector, and the politicians involved tended to respect the role of HEFCE and the principle of University autonomy. The significant role HEFCE played in the development of the sector was reflected in the calibre of the people who served as its Chief Executive: Sir Graeme Davies from 1992 to 1995, Sir Brian Fender to 2001, Sir Howard Newby to 2006, Sir David Eastwood to 2009, Sir Alan Langlands to 2014, and Dame Madeleine Atkins to its closure in 2018.

Two ministers had particularly significant impact on the sector. The first of these was David Willetts, who was in charge when the most significant shift in funding from public to private occurred. The second was Jo Johnson, who brought in the Higher Education and Research Act 2017 (HERA).[21] This act closed HEFCE, split education from research, and brought into being a market regulator called the Office for Students (OfS). Philosophically this completed the journey from a shared perception of higher education as an entirely public good to a predominantly private one. Ideologically this was intended to unleash 'market dynamism' across the sector and lead to new entrants, mergers, forced closures, downward pressure on prices, upward pressure on quality—the whole panoply of things one learns about the impact of unfettered market forces in elementary economics. The eagerness with which some of the politicians involved talked about Universities going bankrupt was not shared by those of us concerned about the well-being of students, staff, or the parts of the country where the local University played a significant and at times counter-cyclical role as an economic anchor.

In September 2017 I—naively—sent in an application to join the board of the incoming Office for Students. I was concerned, as many were, that there was a real risk the new body would go too far in decoupling itself from the context, knowledge, and network created by HEFCE and thought someone familiar with the routine business of delivering higher education might have something to offer. I was called to London for an interview, and to my amazement found the panel did not have a student representative on it. I raised this during the conversation as an example of the sort of question where someone with experience of the *ancien régime* might be useful.

Once again I found myself in a significant interview somewhat medically hampered, though this time much less seriously than had been the case when I was interviewed for a chair at UEA. My arm was in a padded sling following a strange accident at home in Philoxenos—while cleaning one of its many rooms a long sliver of glass had impaled itself into the palm of my hand. This all seemed relatively minor—I pulled it out, cleaned the tiny wound, and rang my older sister Kristi the doctor. Her immediate question—"how do you

[21] https://www.legislation.gov.uk/ukpga/2017/29/enacted.

know it hasn't broken off inside your hand" reflected a lifetime of experience in medicine and proved to be prescient. Tania drove me to an NHS walk-in centre, and they quickly established using X-rays that a substantial amount of glass remained deeply embedded. This all become a bit of a drama because hands are both complex and important. The final—successful—step involved a rather gruesome shoulder block to anaesthetise that arm, a tourniquet, and a tricky operation to locate and remove the almost invisible glass fragments with tiny cross-sections. The hand surgeon and their team were impressive and ultimately successful, and it was only in hindsight that I realised quite how badly this could have ended up without their skill. The experience of spending a few years in the USA once again made me appreciate what lay behind the fact that the entire process that began with my walking into a walk-in centre and ended with a brilliant hand surgeon in a modern facility was entirely about medical care rather than how it would all be paid for.

The interview took place in early September of 2017 and the Office for Students was to come into being at the start of 2018. The timing seemed a little strange when a long period of silence was broken by an email on the 28th of December 2017, thanking me politely for applying and saying I would not be needed. Within days it became clear this was a bullet dodged, because on the 1st of January 2018 the six new appointments to the Board were announced. One of these was Toby Young, a figure sufficiently controversial that I would have been placed in an impossible position had I been appointed. His background was in journalism, and he had been involved in setting up 'free schools' (a model of state funded high schools that could be set up by interested groups). He was a friend of the Secretary of State for Foreign and Commonwealth Affairs Boris Johnson and his brother the Minister of State for Universities, Science, Research and Innovation, Jo Johnson. Both enthusiastically defended the appointment during the first week of January.

The appointment provoked such a strong reaction from the sector that Toby Young resigned within a few days, but the original decision could not be undone. The Office for Students, in its first moments, had appointed to its board someone with no real connection to higher education and a problematic legacy of highly offensive tweets, leaving the impression that this had more to do with the fact that he was a friend of senior politicians in government at the time. The need for the new regulator to be visibly independent of government and manifestly adherent to the 'Nolan Principles'[22] was undermined immedi-

[22]John Major as Prime Minister created The Committee on Standards in Public Life as an advisory non-departmental public to advise him on ethical standards. It led to the creation of a code of conduct dubbed

ately. It had, at the first possible moment, made clear that it was willing to be a creature of the political whim of the government of the moment, and would not resist political pressure. The consequences of this decision are with us still. Whenever the Office for Students steps into politically contested territory, as it often must, the ghost of Banquo is in the room.

Months later the official watchdog, the Commissioner for Public Appointments Peter Riddell, investigated the strange sequence of events, and ruled that the Department for Education had breached the Governance Code, had made 'avoidable mistakes' including failing to carry out adequate background checks, and that ministers had interfered in the process.[23] Judgements of this sort seem devoid of consequence, but perhaps operate as a gentle nudge on the tiller in a peculiarly English way.

I was invited to a 'Friends of HEFCE Reception' in the Cholmondeley Room of the House of Lords on the 27th of March 2018. This was a poignant event in the circumstances, but was something of a celebration as well. In its quarter century of existence, HEFCE had overseen enormous growth in the size, impact, and power of the sector.

> For 25 years HEFCE has been committed to improving the lives and life chances of successive generations of students and, alongside the Research Councils, it has played a vital part in promoting world-class research. It always had the trust of universities and colleges and it always retained the self-confidence to speak truth to power. It will be missed. (Alan Langlands, writing in the booklet produced for the 'Friends of HEFCE' event)

Along with the two particularly influential ministers David Willetts and Jo Johnson, a relatively obscure technical decision by the Office for National Statistics (ONS) in December 2018 cast a long shadow over future developments. Before this student loans were more or less paid out to Universities from the Treasury via the Student Loan Company (SLC) and recorded as a matching asset in the public finances, with the awkward question of non- or partial repayment postponed into the distant future. It was a national version of putting recurrent spending onto a credit card with no real plan for paying it off.

---

the 'Seven Principles of Public Life' or 'Nolan principles' after the first chairman of the committee, Lord Nolan. They comprise selflessness, integrity, objectivity, accountability, openness, honesty, and leadership.

[23]https://publicappointmentscommissioner.independent.gov.uk/wp-content/uploads/2019/04/Commissioner-for-Public-Appointments-Investigation-OFS-Final.pdf.

> the decision that Office for National Statistics (ONS) has reached on the recording of student loans in the national accounts and public sector finances and provides background on why we have been reviewing the treatment of student loans. We have decided that the best way to reflect student loans within these statistics is to treat part as financial assets (loans), since some portion will be repaid, and part as government expenditure (capital transfers), since some will not. We describe this as the partitioned loan-transfer approach.
> (From the ONS report 'New treatment of student loans in the public sector finances and national accounts'[24])

The underlying issues were already being widely discussed, and the Longitudinal Education Outcomes (LEO) data set was already becoming part of the landscape. LEO, whose name is a fascinating insight into how the purpose of higher education is viewed in some circles, is an extraordinarily complex data set that combines information at individual level concerning salaries, income tax, social security benefits, and student loans. Its origins lay in a research project approved by David Willetts as Universities Minister and carried out by the Institute for Fiscal Studies which used anonymised data from student loans in England alongside tax and benefits data to assess graduate outcomes from a financial point of view. Before this work was published the coalition Conservative and Liberal Democrat government passed the Small Business, Enterprise and Employment Act (2015)[25] which *inter alia* authorised the permanent linking of the data sets and commissioned the Department for Education to publish the data regularly. This produced longer term information about graduate salaries, and hence repayment of student loans to the Treasury, broken down by subject and institution. The Resource Accounting and Budgeting (RAB) charge, the estimate of the real cost to the Treasury of the student loan book each year, entered common parlance and the true estimated cost to the Treasury of the system entered political debate more explicitly.

When HERA was implemented the sector—or at least the small piece of it I was most aware of—experienced a sudden silence, followed by a blizzard of one-way communication. The silence was the end of frequent informal contacts, gatherings of advisory committees, and the quiet interventions out of the public view of a steady hand on the tiller of the sector. The blizzard became a pattern of lengthy and complex consultations, often with impossibly short turnaround times.

---

[24] https://www.ons.gov.uk/.

[25] https://www.legislation.gov.uk/ukpga/2015/26/contents/enacted.

> As the dust begins to settle on the 699 pages of Office for Students' (OfS) consultations and accompanying documents published on Thursday and providers across the sector begin to draft responses (deadline March 17th), it feels like there is a gaping chasm between the sector and its regulator. Language in the accompanying press release with references to 'crack downs', 'tough regulatory action' and 'protecting students from being let down', jars with a sector which has contributed so much throughout the pandemic.
> (Sally Burtonshaw of London Higher in a HEPI blog on 26 January 2022[26])

In fact the two big consultations in 2022 on student outcomes and 'low quality' courses comprised 151,689 words. This is approximately the length of the Torah and the Qur'an combined, but it is unlikely that the two consultations will have comparable influence, nor that they will be the subject of reverence or even study over a comparable time period. The Office for Students has not quite issued as many commands as the 613 commandments from the Torah, but it is well on the way.

> Yet, when faced with a swathe of proposals which, if implemented together, would change the shape and nature of higher education in England as we know it, it is important not to lose sight of the 'big picture' and the realities of what is at stake. For, what we are facing now is not a series of seemingly independent consultations concerned with the minutiae of regulation, but a multi-pronged and coordinated assault on the values our higher education sector holds dear.
> (Diana Beech of London Higher in a HEPI blog on 7 March 2022[27])

The duration in office of the Secretary of State and the Minister with responsibility for higher education shown in Table 13.1—and how this has changed over time—is a telling indicator of how little weight the sector has politically.

The political climate leading up to HERA in 2017 and the years since have changed higher education in the UK profoundly, but not entirely as intended. There is no real price mechanism at work for domestic students, though there is a shadow version via the widespread use of scholarships. The expected large number of new entrants to the sector has not really materialised, and is no longer much talked about. The failure to confront the real choices about student numbers, the unit of resource for home students, and the Resource Accounting and Budgeting (RAB) charge is driving the most selective

[26] www.hepi.ac.uk/2022/01/26/minimum-thresholds-are-we-really-protecting-students/.

[27] www.hepi.ac.uk/2022/03/07/a-test-of-spirit-for-english-higher-education/.

Table 13.1 Days in office of the relevant ministers since 2007

Minister	
David Lammy	1,048
David Willetts	1,525
Greg Clark	300
Jo Johnson	974
Sam Gyimah	325
Chris Skidmore	231
Jo Johnson	43
Chris Skidmore	156
Michelle Donelan	873
Andrea Jenkyns	110
Robert Halfon	Incumbent
Secretary of State	
John Denham	708
Michael Gove	1525
Nicky Morgan	729
Justine Greening	543
Damian Hinds	562
Gavin Williamson	784
Nadhim Zahawi	293
Michelle Donelan	2
James Cleverly	61
Kit Malthouse	49
Gillian Keegan	Incumbent

Universities to turn increasingly to international students for a sustainable future. New phrases have entered—and in some cases drifted away—from the education lexicon. We all learned to solemnly talk about 'sector capture'—the thesis that HEFCE, which oversaw massive expansion in the opportunities afforded to diverse students through education as well as an enormous growth in the research power of the country, had become captured by the sector and therefore could not be trusted to regulate it. We learned how to read press releases about 'boots on the ground' interventions from the Office for Students in a spirit of neither alarm nor mockery. We tried to avoid talking about wave after wave of headline-grabbing moral panics, three of the most prominent being 'grade inflation', 'free speech', and 'decolonisation'. All are interesting and important issues, but none have been well framed in the public debate.

Grade Inflation

'Grade inflation'—which for someone involved, as I have been, with trying to improve learning for much of my career, could also legitimately be called 'the excellent educational leaders I have worked with doing their job well for many years' or 'students rising to the challenge and opportunities of more diverse and authentic assessments'—is of course old. It is old even in its modern sense, and even as seen in a mass higher education system.

> Data from a 1974 national survey of 134 colleges was presented to verify that grade point averages had increased .404 points from 1965 to 1973. Approximately two-thirds of the increase occurred since 1968 and the 1968 to 1970 period showed the highest average annual increments. Essentially, the same pattern and magnitude of change was revealed for college subgroups classified on the basis of size, geographic area, curricular emphasis, degrees offered, and public–private. Possible actions to counter the trend were cited but rational initial steps proposed more research to identify the scope of the problem and reasons for the movement. Dialogue to develop an institutional or unit perspective was considered an essential phase. Grade inflation was considered to be but symptom of a broader problem: namely, of an increased concern for student views and subsequent instructional innovations to adapt to these views. (Abstract of a report by Arvo E. Juola [162])

My view on all this comes from some simple observations. As long as we use criteria based assessment—that is, assessment by demonstration of learning outcomes—then there is no logic to complaining about the outcomes without evidence coming from inside the process. The conclusion that 'inflation' is the product of more lenient marking or lower expectations rather than a mix of different things including evolution in educational practice and behaviours may or may not be true, but is impossible to justify on the evidence available. Equally, I am aware of no grounds to have greater confidence in the integrity of grades acquired long ago than in the grades acquired today, unless one is content to simply observe the outcomes and infer something. A simple example of this was the various shifts between the balance of continual assessment and formal examinations in both secondary and higher education under the influence of various ministers and the global pandemic. There is robust evidence that changes of this sort alter the overall performance of women and men differently, and if one simply looks at outcomes then this has the perplexing consequence of 'inflation' for one group but 'deflation' for another.

Equally absurd is the notion that in my leadership role for education I would be raising standards if I issued a dictat to redesign assessments with the aim of producing lower grades, and lowering standards if I did the reverse. In fact both would be a violation of the principle that subject experts have ownership of their programmes and how they are assessed. My job was to support them to do this carefully, with external input and transparently, rigorously assessing the outcomes for the programme—and then to defend the outcomes, whatever they may be.

Whose Free Speech?

'Free speech' has also generated a great deal of noise, and has been seen as sufficiently important to find time for legislation on the matter in the UK.

> We have gathered today in support of freedom of speech—the cornerstone of Western democracy. But supporting free speech is no longer enough. Free speech is now something that has to be defended. (From a speech by Minister of State for Higher and Further Education Michelle Donelan at a Policy Exchange event on the 26th of April 2022)

Well—not quite on the matter, and certainly not concerning 'free speech' as someone schooled in the First Amendment of the United States Constitution might understand it. Indeed, such a person might be puzzled by the idea of a freedom of speech principle that applies to one specific part of society only. It has a flavour of a Freedom of Speech (on Tuesdays) law. This bill,[28] the Higher Education (Freedom of Speech) Act 2023 has been taken through Parliament by a Government that has done more than most to restrict free speech and constrain the right to peacefully protest, but perhaps saw political advantage in creating a culture war narrative about free speech on University campuses and, above all, in Student Unions. As far as we can tell, the intent is that the freedom to give a speech in a University building must be rigorously protected—while the pavement outside the building becomes an arena in which the right to express a controversial view is more and more limited. Freedom of speech also seems to be defined in a creatively narrow way—that a speaker saying things the right sort of politician approves of must be given a platform, but the audience ought not express their views by, for example, walking out. It

[28]https://bills.parliament.uk/bills/2862.

is right to challenge any body—on a university campus or elsewhere—that seeks to 'no platform' a voice, but trying to in effect create an obligation to platform seems a cumbersome and odd response, which leaps over the middle ground of protecting actual freedom of expression for all parties, and instead risks valorising just one.

When the relevant Minister first discussed this on television, she was clear that the criterion would be blanket protection for all 'legal but offensive' speech. Within hours the Government was tweeting clarifications that this did not include (for example) Holocaust denial, which is probably one of the easiest example of legal but offensive speech to apprehend in the UK. For Universities this will lead to a muddle. We are required to adhere to Section 26(1) of the Counter-Terrorism and Security Act 2015[29] (usually called the 'Prevent Agenda', meaning an obligation to prevent radicalization, which is rather tricky to define—indeed I am possibly expressing radicalizing views right now), the complex guidance flowing from the Single Equality Act 2010,[30] and have come under intense pressure from the same branch of Government to sign up to the International Holocaust Remembrance Alliance (IHRA) definition of anti-semitism, whose entire purpose is to grapple with the difficult question of 'legal but offensive' speech in that specific context. Above all, the entire legislative and political effort of the last few decades has gone into positioning the student as a customer, who is being invited to view a lecture or seminar with unwelcome ideas much as they might view a worm in a salad they had purchased. Fortunately the vast majority of students are extremely sensible people who are more than open to having their ideas challenged—but who are often engaged with nuanced concerns about equality and diversity.

To this rich diet of obligations, market and consumer ideology, and explicit requirements that control, regulate, and prevent free speech has been added an obligation to protect it. Working in a University does not always entail believing impossible things, but it increasingly involves pursuing incompatible objectives.

No group can be entirely happy with the current mess, and the instances—which certainly exist—where voices have been stifled or suppressed on a University campus find their origins in more complex phenomena with deeper roots than those that the legislation will address. Indeed, there are academics

[29] https://www.legislation.gov.uk/ukpga/2015/6/contents/enacted.

[30] https://www.gov.uk/guidance/equality-act-2010-guidance.

who with good reason feel they have been hounded out of an institution because of their views, and the current clash of expectations is not sustainable.

Sadly most of the deeper questions about the issue are not widely discussed. One of these is to ask how the years of effort by successive governments to position higher education as a consumer-led transactional purchase have created a climate in which some student groups feel empowered to demand that nothing they dislike is said out loud on a campus. Another is to deal more honestly with the reality that a threat to free speech on a University campus tends to revolve around cancelled events, 'disinvited' speakers, and—sadly—some academics undoubtedly hounded out of an institution, while the largely ignored threats to free speech in wider society involve arrests, prison time, recording of 'non-crime hate incidents', the opacity of the operations of the Defence and Security Media Advisory Committee, and so on.

> Therefore aiding, abetting, counselling, conspiring etc those offences by posting videos of people crossing the channel which show that activity in a positive light could be an offence that is committed online and therefore falls within what is priority illegal content. The result of this amendment would therefore be that platforms would have to proactively remove that content. (Statement in the House of Commons by the Secretary of State for Digital, Culture, Media and Sport, Michelle Donelan on the 17th of January 2023)

> Rules introduced by the Cabinet Office in 2022 specify that the social media accounts of potential speakers [at a conference] must be vetted [...] to check whether these people have ever criticised government officials or government policy. The vetting process is impartial and purely evidence-based. The check on your social media has identified material that criticises government officials and policy. It is for this reason [...] that I am afraid we have no choice and must cancel your invitation. (From an email quoted in the article 'Rees-Mogg's blacklist is positively Soviet' by Edward Lucas in *The Times* newspaper, 24th of April 2023)

A government genuinely committed to free speech might profitably begin, and arguably end, by staying its own hand from passing laws like the Public Order Act 2023.[31] We also might all benefit from reducing the rate of creation of hideous neologisms like 'non-crime incident'.

---

[31] https://bills.parliament.uk/bills/3153.

And keep an eye on everyone, see what they do. Report back when, um, I don't know. When it makes sense. (Gardner Chubb in 'Burn after reading' [56])

There are free speech advocates on every campus ranging from absolutist to nuanced, and there are others who (for example) think it important to routinely temper an absolutist approach to free speech with concern for various groups of people and their feelings of safety or belonging on campus. Each of these represent serious ideas, with proponents of great moral and intellectual weight. My own thinking is closest to those seeking to reduce constraints on free speech to the absolute minimum as required by the law. What absolutely eludes me is the rationale for treating Universities or Student Unions differently to the rest of the country. I would like to see free speech being defended across the board, on a University campus and in the public square.

The highest values of universities are that they are places of intellectual inquiry and challenge, of unfettered and open investigation, of constant questioning and testing of ideas, of exposing and evaluating assumptions. They should be places of questioning and not deference, of analysis and not ideology, of intellectual openness and not comfortable exclusion, of teaching students to think and not follow, and they should be driven by the demand that arguments are accepted only when they stand the test of scrutiny. This openness to question and challenge should lie at the foundation of research, and should permeate student's education at all levels. (Robert Zimmer[32] speaking at the *Times Higher Education* international conference in 2010)

Confronting the Colonial Legacy

Decolonisation has become part of the coded language in which words like 'woke' are used to produce an entire musical scale of dog-whistles. I am sure that some foolish things are said and done on University campuses, as they always have been, but the challenge of constantly interrogating our unspoken and contextual assumptions should always be a welcome one. In some cases a process in curricular reform that would be labelled 'decolonisation' for one audience could legitimately be called 'modernising', 'preparing students better for a globalised world', or simply 'improving' for another.

[32] Robert Zimmer (1947–2023) was an American mathematician specializing in ergodic theory, differential geometry, and rigidity phenomena in Lie group actions. He served as the President of the University of Chicago from 2006 to 2021.

Within Mathematics no serious person can doubt that the sort of education I experienced told a story of development through classical period Greek developments, a mysterious gap, then the emergence of a great European tradition starting more or less with the publication of the *Liber abacci* by Leonardo of Pisa in 1202. There is nothing wrong with this picture in itself, and the colourful and interesting events, people, and incidents within it are probably broadly genuine—but it largely ignored the equally lengthy and compelling tale of Mathematics as it developed in the Arab world, in China, in Africa, or in Polynesia. Similarly, it valorises the heroic male figure with no interrogation of the class, political, and gender context that enabled their contributions and stifled the efforts of others.

In fact one of the books I inherited from our father is a particularly revealing example of this habit: Eric Temple Bell's 'Men of Mathematics' [18] is a colourful and entertaining account of the lives of between 34 and 39 men (depending on how many of the Bernoulli family are counted in the list) and one woman, placed in chronological order and celebrating their contributions to the discipline (Fig. 13.1). All are located in a canonical story of Greek Mathematics ending with the death of Archimedes in 212 BCE with the next character being Descartes, born in 1596 CE. It is an entertaining read and does much to bring the human side of the people involved to life—but it does highlight a narrow and distinctly eurocentric view of how mathematics developed, and emphasises the heroic male figure as the driver of events with little regard for the societal context.

It seems self-evident that the considered life of the critically engaged product of higher education in any discipline should include an openness to understand the world from differing perspectives, have some insight into how historical processes like colonial conquest or slavery cast shadows into modernity and into the unspoken assumptions within every discipline, and should be informed by something of the history of their own discipline viewed from other places.

The erasure of the history of Mathematics in Africa is a fascinating story in itself. A tiny example concerns the way in which counting was done in the Khoisan languages (needless to say, that grouping of languages is itself a product of work by a philologist based outside the region). There is strong evidence that much of the academic understanding of how this worked is based on a tiny study from 1866 led by someone who had a prior belief that the specific language studied was 'uniquely inarticulate because it uses click consonants' (see [301]). The careful tracing of this particular thread starts to give some insight into the larger tapestry telling the story of how much of Africa's role in the world was erased.

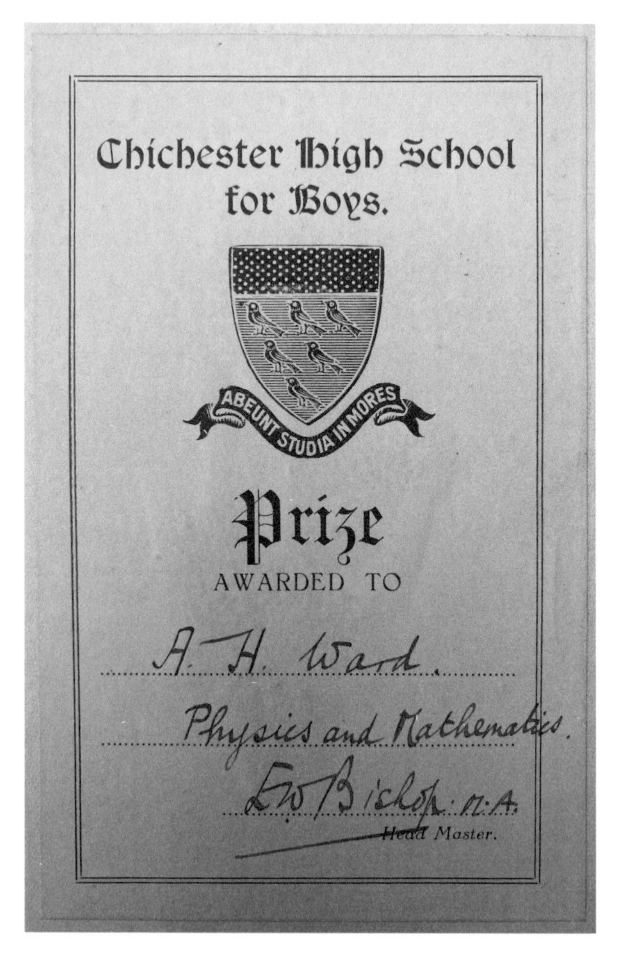

Fig. 13.1 The bookplate in the copy of Bell's book given to our father as a school leaving prize in 1943

But however common in the popular imagination, the commencement of modern history with these most famous of feats of discovery, presented as if to be viewed on a trapeze in the center ring of a three-ring circus, obscures the true beginnings of the story of how the globe became permanently stitched together and thus became "modern". It also so dramatically miscasts the role of Africa that it becomes a profound mis-telling. [...] The teaching of history about this era of iconic discoveries[33] is confoundingly silent not only on that decade [1488–

[33] Europe's fifteenth Century 'Age of Discovery' and the opening of trade routes between West and East through the voyages of Vasco de Gama, Ferdinand Magellan, and Christopher Columbus.

> 1498] but on the nearly three decades from the Portuguese arrival at Elmina[34] until their landing in India. It was this moment, when Europe and what is nowadays styled sub-Saharan Africa came into permanent deep contact, that laid the foundations of the modern age. (From 'Born in Blackness' by Howard Waring French [130])

A phrase similar to 'decolonisation' in being freighted with sometimes coded meaning, and in how it is used by multiple parties in various tones of voice, is 'white privilege'. The myopia induced by my own experiences and the historical and geographical contexts I have inhabited is unavoidable, but I have no doubt that many of the doors and borders that I passed through were easier to navigate because of the colour of my skin or the passport I held. However it is a single visceral moment that most vividly remains with me as a constant reminder of how my skin colour and how others see it has impacted on my life at certain critical moments. Shortly after we returned to Columbus in 1993, some small administrative slip meant I did not receive the routine annual reminder to renew the licence tag for our Honda Civic, which took the form of a sticker on the rear licence plate. I am normally diligent about things like this, and try to keep on top of the organisational details of modern life carefully. Very late one night a police car flashed its lights behind me and blipped its siren. I pulled into the next available parking lot and wound down my window. When the police officer said my licence tag was out of date I somehow was still in a UK frame of mind, so confidently announced "surely not", and jumped out of the car to go and look...only to find the clearly frightened and angry police officer several metres away from me, legs in a wide crouch, with a handgun pointed at me. This all was sorted out relatively quickly—my strange accent doubtless helped—and they let me drive on rather than being towed, but it is easy to imagine how differently this might have played out.

Gas Laws

Both research and education in UK Universities are largely publicly funded, even if that now comes hidden in the form of the Resource Accounting and Budgeting (RAB) charge. It is entirely reasonable that this entails government involvement in both the definition of what good outcomes mean, and in

[34] Elmina Castle was built on the South coast of what is now Ghana by the Portuguese in 1482 as a base for the trade in slaves, gold and ivory.

the measurement of those outcomes. Doing this meaningfully and without unintended consequences is more than difficult, and raises fundamental questions about the role of knowledge creation and education in society.

On the research side there is some understanding that, for example, climate change raises problems that are inherently multi-disciplinary, complex, and long-term. Both funders and Universities are finding ways to facilitate risk-taking inside large projects to rise to this sort of challenge. The situation is far from perfect, but (for example) the simple-minded use of citations over a short period of time as an outcome measure, and a narrow view of the impact of research have become more and more nuanced. Citation metrics over a relatively short period and journal rankings using them are often used to measure the importance of research, and this introduces an artificial pressure which at times may have almost directly the opposite effect. Two striking examples illustrate some of the dangers. A paper in algebraic geometry by Hain and Zucker [141] has become important in a certain part of number theory and is widely cited—but the first citation to it showing in *Mathematical Reviews* appeared ten years after its publication. Many of the fields associated to Business adhere rigidly to the Journal Impact Factor (introduced originally to help librarians make purchasing decisions, now often used to measure the quality of papers) as a proxy for quality of research outputs. The behaviours this has promoted do not seem to be well aligned with encouraging research with high societal or policy impact [142].

The picture in education is less benign, and the narrow focus on how the purpose of higher education is defined will inevitably drive unplanned changes. The central problem of how to deliver the 'axiom that courses of higher education should be available for all those who are qualified by ability and attainment to pursue them and who wish to do so' from the Robbins Report of 1963, maintain a unit of resource adequate for high-quality education, and prevent ballooning debt be it public or private, remains. For a mathematician the word 'axiom' chosen in the Robbins Report stands out—the axioms are the things we take to be self-evidently true rather than work out what they mean and how they can influence practice.

Like most people who devote their lives to higher education, I see delivering on the promise of that axiom as absolutely central to the positive role education can make to society. But I also—as a mathematician—cannot help wishing we would talk more openly about the interaction between numbers, adequacy of funding per student, and the risks of kicking the can down the road by slowly reducing how much is spent on the education of each student while expecting the experience to deliver more and more. Higher education has its own form of gas law: The unit of resource multiplied by the number of students is capped by the limits of how thinly stretched the system can become and the willingness of the country to absorb the Resource and Budgeting charge.

As these inherent contradictions slowly exert their pressure on the system, we will see more and more narrowing of the vision of what the point of it all is. It is understandable that the Treasury wants the 'quality' of a higher education to be assessed in terms of future earnings, but it is naive to imagine that the pressure this produces will not have consequences.

> Too much and for too long, we seemed to have surrendered personal excellence and community values in the mere accumulation of material things. Our Gross National Product, now, is over $800 billion dollars a year, but that Gross National Product—if we judge the United States of America by that—that Gross National Product counts air pollution and cigarette advertising, and ambulances to clear our highways of carnage. It counts special locks for our doors and the jails for the people who break them. It counts the destruction of the redwood and the loss of our natural wonder in chaotic sprawl. It counts napalm and counts nuclear warheads and armored cars for the police to fight the riots in our cities. It counts Whitman's rifle and Speck's knife, and the television programs which glorify violence in order to sell toys to our children.
> Yet the gross national product does not allow for the health of our children, the quality of their education or the joy of their play. It does not include the beauty of our poetry or the strength of our marriages, the intelligence of our public debate or the integrity of our public officials. It measures neither our wit nor our courage, neither our wisdom nor our learning, neither our compassion nor our devotion to our country, it measures everything in short, except that which makes life worthwhile. And it can tell us everything about America except why we are proud that we are Americans.
> (From a speech by Robert F. Kennedy at the University of Kansas on the 18th of March, 1968)

14

Durham

Our son Raphael would be doing his GCSE examinations in the Summer of 2012 and we thought, perhaps naively, that moving to a new school for the sixth form should not be too disruptive. Our daughter Adele had started a Physics degree at Durham in Autumn 2011, and some aspects of the distinctive approach to education at Durham was attractive. I was also interested in what it would be like to step into a senior role in a new University where I knew nobody. I had been at UEA for many years, and there were familiar faces everywhere I went on the campus.

My own notes from the initial telephone conversations with the search consultant were daunting: 'The hottest PVC role in the market this year; found themselves spoiled for choice—calibre phenomenal; 16 really interesting individuals'. Somewhat hyperbolic, but it certainly was an interesting role.

We had already arranged to visit David Pierce and his wife Ayşe Berkman in Istanbul for the last week of April, and our minds were quickly dazzled by the sense of contested religious history in that extraordinary city. On the 30th of April we were all in the Kariye Musuem, a beautiful Greek Orthodox Church building that later became a mosque, when my phone buzzed. I rushed out of the dark coolness of the beautiful interior onto a hot and dusty street to hear Mike Dixon say 'Game on'. It took me a moment to work out what was meant—that I had been shortlisted for the Durham role.

The formal part of the selection process took place on the 14th and 15th of May. I drove up to Durham from Norwich early on the 14th, and spent that afternoon in a series of informal meetings. The next morning began with a tour of the University estate and ended with a buffet lunch for candidates in the

T. Ward, *People, Places, and Mathematics*, Springer Biographies,
https://doi.org/10.1007/978-3-031-39074-6_14

rather grand Senate Chamber in Old Shire Hall in the centre of Durham. The efforts to keep the candidates separate on the first day were not particularly effective, and I had introduced myself to what turned out to be the other external candidate simply on the basis that he was wearing a suit and waiting outside a room I had been sent to. It turned out we had both been put up at the Durham Marriott Hotel on Old Elvet, in the centre of the town. We joined forces for supper, and decided to express our disapproval of a stern warning in the correspondence that alcohol would not be covered as an expense by ordering rather more wine than was strictly appropriate. Nobody seemed to mind, and in hindsight this was another indication that the recruitment process was not a particularly efficient machine.

The buffet lunch was memorable for several reasons. It entailed one of the few universal truths about university catering: Relentless un-evidenced optimism about what can legitimately be described as 'finger food'. In this case the primary source of structural integrity came from rapidly softening lettuce leaves, which were intended to support various rather liquid options like coronation chicken. Given the huge role food plays in the life of Durham University, my amused thought that perhaps managing this was a key part of the selection process was not entirely ridiculous. More difficult than the collapsing lettuce leaves was the gradual realisation that among the friendly people sharing this difficult task were some internal candidates, one of whom was voluble about their disapproval of the post being advertised externally at all.

The formal interview involved a familiar type of panel, three of whose members stand out in memory. The Chair of Council, Anne Galbraith, combined genuine friendliness with acute questions reflecting her legal background. The PVC Research, Tom McLeish, was full of enthusiasm and clearly liked the idea of a mathematician joining the executive. He was to become a great friend, and often stayed with us if he needed to be in Durham late in the evening. We left a key in a boot for him, and often only realised he had slept here when we met over breakfast. One of the more memorable evenings with Tom was in February 2016, as he happened to be staying with us the day that the Laser Interferometer Gravitational-Wave Observatory announced the direct observation of the gravity wave GW150914. Everyone in the house was treated to an excited but cogent explanation of what this all meant, and a hug to celebrate the event. Research leads, both in industry and higher education, are often exceptionally bright and interesting people, but even in that company Tom McLeish stood out [360]. In many ways he was an eighteenth Century polymath, and I never found a domain of thought in which he was not both interested and well versed. While holding down a demanding role on the executive of Durham University, at one point he held joint research grants in each of the four faculties. After stepping down as PVC, he took up his ideal job, becoming the inaugural Professor of Natural Philosophy at the University of York, with an affiliation to the Centre for Medieval Studies and the Humanities Research Centre. He became a lay preacher in the Church of England in 1993, and wrote extensively on the relationship between faith and science [195, 196]. He passed away on the 27th of February 2023, and his last posting on Facebook was a quotation from the Song of Songs: 'Love is as strong as death...many waters cannot quench Love'.

Of course the most influential panel member was the Vice-Chancellor Chris Higgins,[1] whose great love for the institution and strong personality came across clearly. When the interview ended he jumped up to shake hands, explained that he also had a daughter studying at Durham, and that it would be several days before they could reach a decision as the internal candidates would be interviewed on other days.

I spent an hour or two with Adele and some of her friends, and tried to help with some Mathematics problems they had. Then the long drive home began, and somewhere near Lincoln my phone rang. I pulled over at the Farm Cafe just past Holbeach and took the call, which was Chris Higgins offering me the job. This was an unexpected development both in substance and in rapidity, and I learned some time later that they had decided not to go ahead with the planned later interview. This left a more than disgruntled new colleague, and in hindsight was an indicator of a certain degree of impetuousness within the institution at the time. I rang Tania to let her know the news, which came as a shock to her: The abstract possibility of a move to Durham was a different proposition when rendered so concrete and immediate. I continued my drive across that part of Lincoln and into Norfolk with the sun setting behind me, enjoying the sensation of an enormous Norfolk sky, loud music playing in the car.

A few weeks later my phone rang just before 5 p.m. on a Friday. This is something of a witching hour in many walks of life, and all too often emails or telephone calls at this time signify something urgent—or something that another party wants off their mental desk before the weekend. In this case it was another indication of a degree of turbulence in the appointment process: A clearly harassed person in Human Resources at Durham asking if I could

[1]Chris Higgins was a distinguished molecular biologist, who did fundamental work on the so-called ABC transport mechanisms [149]. Prior to the position of Vice–Chancellor at Durham, he held several senior roles including Director of the Medical Research Council Clinical Sciences Centre and Head of Division in the Imperial College Faculty of Medicine.

urgently contact all my referees, because they had missed my request that I be notified before any of them were contacted. I was on a crowded local train out of London at the time, and it took some effort to sort it all out and make appropriate apologies.

Leaving UEA

I let the Vice-Chancellor at UEA, Edward Acton, know at once. The timing was not ideal from his point of view, and he tried hard to persuade me to delay the move as long as possible, but for many reasons it really needed to be September 2012. Leaving UEA after such a long time was painful. The executive team organised a lunch and some kind gifts, including a handsome bottle of 18 year old Glen Garioch single cask whisky. It was more than a decade before the right opportunity presented itself to open this handsome gift, and that was a poetry dinner we hosted in Durham in November 2022 attended by Alan Houston.

A particularly kind gesture from the Vice-Chancellor was to arrange a dinner for a small group of close friends at Wood Hall the 12th of July 2012, which marked a rather grand goodbye. The guests included Shaun Stevens and David Stevens from the School of Mathematics, the pastor of our church, and some other old friends. The School of Mathematics arranged a wonderful and less formal send-off on the 30th of August, which was great fun—and ended with some karaoke using white board markers as microphones. David Stevens managed to get a brief slide show with pictures gleaned from social media running on the display boards in the department, and it all felt like a proper farewell to the department that had so shaped my life.

It was poignant to leave many friends at UEA and in Norwich, but I was excited at the prospect of moving to a new part of the country and to a new University.

Alongside the appointment process this was already a busy period. In addition to the various demands of the PVC role I happened to be attending several conferences and making other visits. Soon after visiting David and Ayşe I attended a conference entitled 'Ergodic Theory and Dynamical Systems: Perspectives and Prospects' at the University of Warwick from April 16th to 20th. This was a large international conference, and as ever it was a pleasure to spend some time at Warwick. In May I went back to Vienna for a workshop at the Erwin Schrödinger Institute on 'Periodic Orbits in Dynamical Systems' (Fig. 14.1).

Fig. 14.1 Dinner in the apartment of Klaus and Annelise Schmidt in Vienna on the 22nd of May 2012, with Doug Lind, Klaus and Annelise Schmidt, and Manfred Einsiedler (Photo with permission of Evgeny Verbitskiy)

I ended up running with the Olympic torch through part of Norwich on the 4th of July. Two students from UEA also ran with the torch, because of extraordinary work they had done for charities. The reason I had this honour was far more mundane and much less impressive. Samsung had sponsored three slots for UEA and wanted a member of the University executive to run one of the legs. I had simply responded to the email. It was enjoyable—my leg happened to be near Norwich Central Baptist Church which we attended, and many friends gathered along the few hundred metres. Afterwards the family and a few friends stayed in the centre of Norwich and had a meal in town, with me still in the Olympic running suit and carrying the torch. We stopped often to let children take photos holding the torch, and the torch made various visits to places like the UEA nursery.

In June I went to another conference in Uppsala, and shortly after we drove down to Dorset for a large family gathering to celebrate our father's birthday at Coasters Cottage. A few days later we went back to the Grim's Dyke Hotel in Harrow for a grander event to celebrate my father-in-law's birthday. I attended my last UEA congregation in July, where I gave the oration for an honorary degree award to Prof. Sir Martin Taylor, a distinguished number theorist who had been a great friend to the School of Mathematics at UEA.

We squeezed in a short holiday at the Pleasaunce Hotel in Overstrand on the North Norfolk coast, and then Tania left for a few weeks with her cousins in Germany and I packed up the house. Adele, Raphael, and I had a poignant last day in Norwich, visiting favourite places and reminiscing about the many happy years we had all spent there.

Fig. 14.2 A monthly meeting as managing editors of the *Bulletin of the London Mathematical Society* with Shaun Stevens in the School of Mathematics at UEA

Shaun Stevens and I were the managing editors of the *Bulletin of the London Mathematical Society*, and this was one of the few mathematical things I kept going during my time as PVC. Our habit of carrying this out by monthly meetings in person (Fig. 14.2) made it essential to ask someone else to take on my part of it. David Evans in the School of Mathematics kindly agreed to take on my role at short notice, so from Issue 5 of Volume 44 of 2012, the managing editors became Shaun and David. Other people at other institutions helped out with similar editorial roles, including one at *Ergodic Theory and Dynamical Systems*.

Looking back at all this it is easy to understand why much of 2012 seemed a bit of a blur at the time.

Elsevier

The publisher Elsevier, which had become the *bête noire* of the mathematical community over the vexed question of journal pricing, contacted me on the 11th of June 2012 to ask about joining a small advisory council for their Mathematics group. My view was that they—along with several other publishers—were indeed extracting far too much from University libraries, and that the economic model of publicly funded research and time of authors, editors, and referees resulting in outlandish profits for publishers, with the literature unavailable to anyone outside the largest universities, was unsustainable and morally troubling. This opinion might take one towards boycotting them, which many thoughtful mathematicians did, or it might incline one to talk to them, which was my choice.

This question of engagement or isolation keeps cropping up in contexts of varying importance. I was personally involved in an extremely minor way in the Anti-Apartheid Movement's efforts to campaign for cultural, sporting, and financial isolation of South Africa at the time. The complexity of the issues thrown up by subsequent boycotts have tended to make me see too many sides to any given question—though I am certainly comfortable that the South African regime at the time had to be isolated. Calls to not interact with a given entity have often made me struggle with the way in which upbringing produces blind spots. Listing the countries which execute minors, or the countries with members of legislative bodies in those positions for religious reasons, should more routinely provoke us to ask difficult questions of ourselves and countries with whom the UK has been historically allied with the same intensity we ask them of others.

The advisory group met once or twice a year, and I hope at least helped Elsevier understand some of the perspective of working mathematicians. The meetings usually involved a short day of discussions and a rather grand dinner, and a fair amount of travel as the locations deliberately moved around the world. It was fascinating to hear the publisher's perspective and to see some of their plans for the future, but I am not sure if we had much of an impact.

The gathering in Paris produced a particularly entertaining evening. The dinner was to take place in a Michelin-starred restaurant, and we had been asked in advance for any dietary preferences from a short list of options. My office had let them know that I would prefer vegetarian food and fruit instead of dessert, a formula I always use to survive too many grand dinners in rapid succession. In the event I wish we had not been asked, because the restaurant clearly did not approve and was not at all shy about making this clear. It was a complex meal with multiple courses, and for each of these I was brought a dish with meat, politely asked if there might be a vegetarian alternative as ordered, experienced almost comical tut-tutting, and eventually had it replaced with a slice of cucumber and half a tomato or something similar. This happened again with the main course, and eventually a bowl of over-cooked scrambled eggs so heavily salted as to be almost inedible was placed in front of me. As one might expect the bread, the wine, the cheese later—and the company—were all delightful, but it all seemed very strange and avoidable had they simply made the perfectly reasonable stipulation in advance that they were not minded to cater for vegetarians.

Moving

The physical move to Durham ended up being the largest move of our lives in several ways. It meant packing up and leaving a house after 20 years of living there, and after bringing up two children in it. The house was not large and we had filled every corner, including the attic. Over the years I had squeezed bookshelves into every space available, and without ever deciding to do so we had accumulated a substantial library. The move also entailed packing up my office at UEA, with a large number of books and papers.

The distance was not so large as to involve shipping, which in some ways brings its own clarity and discipline—but it was an overnight trip for the movers, and two lorries were involved. Tania's health was not good at the time, and her initial positivity about the move had dissipated significantly as it came closer in time. We certainly did not have the capacity to do much in the way of slimming down the household in advance of the move. The final slight complication was that it involved the household moving from 23 Overbury Road in Norwich to a house we had rented from Durham University to find our feet in a new city, and my office moving from the campus of UEA to an office in the Mathematics Department in Durham. The office at this stage included about 25 shelf metres of books and four large steel filing cabinets of Mathematics papers. In the end this all entailed one large removal truck and one medium-sized one, and we opted to do the packing ourselves. It was an exhausting process for the movers and for us. Tania was away in Germany for the last week or two, and ended up making the journey by train. I drove to Durham with Adele, Raphael, one of Raphael's friends, and our cat Toby.

Raphael was only going to be with us in Durham for a few weeks. After much thought and discussion the three of us had decided he would stay on in Norwich, and some lovely friends we had met through St Paul's Church in Hellesdon years earlier had offered to have him live with them. This was a painful business for all of us, and in the end after spending a year with them he decided to move to Durham. As a result his last 2 years of school were far more disrupted than we had all hoped it would be. Fortunately the local high school in Durham he ended up moving to was enormously helpful, and not only admitted him into Year 13 at short notice, but supported the transition in detailed ways. He managed the move with aplomb, and ended up successfully entering the university he had hoped to.

Tania and I spent the first year in Durham living in Mountjoy House on the science site of the University. This was convenient in location and almost absurdly beautiful, with views over Durham and its Cathedral. It was also irredeemably cold and damp, being on the North slope of a hill in permanent shade, with large leaky single-glazed windows and a primitive heating system using fuel oil. By chance the winter of 2012–2013 was a long and cold one, and at times we took to wearing heavy warm clothing indoors. We decided a second winter there was not an attractive option, and started looking for a new house in earnest in the Spring of 2013. We knew what we would call it—Tania had long wanted to name a house Mishpocha—but we struggled to find a suitable house. The word Mishpocha is Yiddish but derived from a Hebrew word meaning family; it means extended family in the broadest sense.

We started attending King's Church Durham almost immediately, which met in a large room in the Student's Union building a short walk away from Mountjoy House. Every now and then the room was unavailable, and on one of these occasions we were to meet in the Calman Centre on the university campus. As we walked down the hill to this I remarked to Tania that meeting in a new space was likely to mean meeting new people, because nobody had a usual seat in which to sit. This came to pass, and after the service I was talking to a couple I had not met before when a mutual friend came up and said "I'm so pleased you've met—Edith wants to sell a house and Tom wants to buy one". We thought enough of this to follow up, and after some twists and turns we bought that house, and attached the nameplate Mishpocha to its door. We did the move across Durham ourselves, using the church van and with much assistance from Raphael and church friends.

Durham University

Life on the executive of Durham University was fascinating and at times stormy. The University is both ambitious and complex, with various aspects of modern administrative and managerial structures of Schools and Faculties overlaid on earlier structures of Colleges and a strong sense of history, at times uneasily. The relationship between the executive and the governing Council was also less than ideal at the time, and at times genuinely fraught. Above all it was one of the smallest institutions in the Russell Group, but had boundless ambition in research. Much of the research done in Universities is funded by Research Council grants, and these generally only cover 80% of the 'full Economic Cost' of the activity. An even larger cost for many institutions is the time allocated to faculty members for research, which usually is a larger commitment of resource than the so-called 'QR' (Quality-related Research) income can cover. The shortfall falls onto the income generated by education, and so a key driver of financial stability is the balance in scale between the two areas of activity.

> All research grant proposals and fellowship applications submitted will be costed on the basis of full economic costs (fEC). If a grant is awarded, research councils will provide funding at 80% of the fEC. The organisation must agree to find the balance of fEC for the project from other resources. Universities and other higher education organisations will use the Transparent Approach to Costing (TRAC) methodology to calculate full economic costs. This methodology has been validated by UKRI and is subject to ongoing review.[2]

Quality-related Research income is based on the outcomes of the national assessment of research outputs and their impact, currently called the Research Excellence Framework (REF). This replaced the Research Assessment Exercise which began with an exercise run in 1986 by the University Grants Committee and chaired by Sir Peter Swinnerton-Dyer. A similar exercise was conducted in 1989 by the Universities Funding Council under the title 'research selectivity exercise'. The first Research Assessment Exercise *per se* was run in 1992, and a legal challenge led to a judge warning that the process needed to become more transparent. Subsequent Research Assessment Exercises in 1996, 2001, 2008 saw changes in the details of the approach and a growing concentration on the assessment of a small number of research outputs to avoid creating a pressure to publish a high volume of lower quality work. The details of who was included changed in technical ways, but the broad pressure that encouraged the 'poaching' of research stars is unrelenting. Following several reviews, including the 'Review of Research Assessment' led by Sir Gareth Roberts[3] and a report by the House of Commons Science and Technology Committee in 2004[4] chaired by Ian Gibson, the system was replaced by the Research Excellence Framework (REF) which ran in 2014 and, following a review by Lord Nicholas Stern,[5] in 2021. The largest trades union involved, the Association of University Teachers (AUT, now incorporated into the

[2] www.ukri.org/councils/epsrc/guidance-for-applicants/costs-you-can-apply-for/principles-of-full-economic-costing-fec/.

[3] https://web.archive.org/web/20070720232304/http://www.rareview.ac.uk/reports/roberts.asp.

[4] https://publications.parliament.uk/pa/cm200304/cmselect/cmsctech/586/586.pdf.

[5] https://www.gov.uk/government/publications/research-excellence-framework-review.

University and College Union (UCU)), adopted a policy opposed to the exercise in 1996.

> The RAE has had a disastrous impact on the UK higher education system, leading to the closure of departments with strong research profiles and healthy student recruitment. It has been responsible for job losses, discriminatory practices, widespread demoralisation of staff, the narrowing of research opportunities through the over-concentration of funding and the undermining of the relationship between teaching and research. (From https://www.ucu.org.uk/rae2008)

For the first few weeks my office was in Old Shire Hall in the centre of Durham, with an impressive anteroom and magnificent but inconvenient surroundings. My office was large, with an outer office for a Personal Assistant. The internet in the building was unreliable and the wireless and mobile networks unusable, so I took to leaning out of, and at times escaping through, a window to get a signal.

After those few weeks the executive, along with many parts of student services, moved into the newly built Palatine Centre, also a magnificent building in its own way. The central part of the executive were in a shared open-plan office together with the executive team support. This was my first experience of working in a genuinely open-plan environment, and like everyone who does so I found it to have both advantages and disadvantages. The ease of snatching a few words with the Vice-Chancellor or the Treasurer was useful, while the noise and lack of privacy was not. In principle we were hot-desking, which worked to a limited extent. One—of many—stories about the unconventional open-plan executive office was that it was a messy compromise. The original plan (I was told) was for the Vice-Chancellor and Deputy to have conventional offices with windows looking across Durham to the Cathedral, and the other members of the executive would have offices without windows on the other side of the building, with the executive assistants in an open-plan space between the two. This plan provoked a revolt, and the compromise found was for all to join the open-plan space. University stories should all be treated with much scepticism, but those that end in a plan dissolving on contact with reality resulting in a messy compromise tend to be the more credible ones.

The biggest lessons taught by the first few months of life in the open-plan environment were about the balance between physical infrastructure, digital infrastructure, and human behaviour. Because of the huge capital sums involved, the tangible outcomes, and the glamour of events to open buildings, we readily slip into prioritising the physical over the digital and the human. Two slightly absurd things were particularly educational for me in this regard. Due to historical accident, the members of the executive were on several different Exchange servers, which did not talk to each other at all reliably. This meant that no confidence could be placed in the versions of diaries visible to each of the Personal Assistants, who had to resort to looking at each other's screens or passing a laptop around to try and schedule meetings. An even more absurd thing was triggered by a commitment I had made before starting. Durham had hired a people coach to try and change some problematic cultures on the university executive. She had contacted me as soon as my appointment was announced to suggest I should attend a 'Theory U' event shortly after arrival, along with two other members of the executive. When the day came to start the journey to the Presencing Foundations Program in Cape Cod on October the 26th 2012, we discovered that there were three of us standing outside the Palatine Centre with our luggage, waiting for three different taxis arranged by three different PAs to take us to the same flight from Newcastle Airport to Boston, where we were to stay in three different hotels.

I had dinner the next night in Boston with Paulina Lubacz, the University Treasurer, and as usual was entertained by some aspects of life in America. I noted down what I ate because it was so evocative of the culture: 'Sean's simple chicken' comprised a 'pounded breaded free-range corn-fed chicken breast cooked with tomatoes roasted with a drizzle of olive oil and capers served over garlic mash with baby spinach leaves and balsamic vinegar'. I wondered what might happen if Sean was asked to rustle up something complicated, but thought it best not to ask. Asking at the hotel for a nearby church produced another moment distinct to the country: They helpfully found an Episcopalian Church just round the corner, 'doing Rite 1 at 9 and Rite 2 at 11:15'. Both corporeally and spiritually I felt a little out of my depth.

The Presencing event was interesting for several reasons. It was led by Otto Scharmer, the author of the guide we were given [265]. It pushed us into various essentially Jungian workshops far outside my existing experience. I approached it all with an open mind, and found some of the workshops far more powerful and insightful than I would have imagined possible. For some of the exercises I was in a small group with people far more familiar with presencing, who viewed some quite extraordinary events with greater equanimity than I did. Despite my instinctive prior views, the exercises produced insights for various members of the group—myself included—that seemed meaningful and useful, informed by things unspoken.

That said, some of the associated texts were more impenetrable than anything I had encountered in Mathematics. Individually familiar words seemed to be concatenated so as to produce sentences of gnomic confusion, but perhaps I did indeed 'Go to the threshold and allow the inner knowing to emerge.' I did not, however, attempt to persuade any of my colleagues on various University executives to use interpretive slow-motion dance to access our unspoken collective wisdom.

It also placed us in luxurious accommodation on Cape Cod during Hurricane Sandy. We were at one edge of the worst impacts, but it was alarming enough to see the power of the wind and violent rain.

Stefanie and Robert Again

At the time of the move to Durham I was part of the supervisory team for two PhD students, Robert Royals and Stefanie Zegowitz. Robert decided to stay on in Norwich and did most of his thesis with his main supervisor Anish Ghosh, on a problem in Diophantine number theory [136]. Robert also had a chapter in his thesis concerning a problem of mine on 'adelic perturbation'. This grew from earlier work with Richard Miles, Vicky Stangoe, and Graham Everest and was my attempt to extract a general arithmetic principle of the following sort. For generating functions with integer coefficients given by a linear recurrence, an 'adelic perturbation' means changing the coefficients by factoring out the primes from a chosen and possibly infinite set of primes. The terminology might sound strange, in that a 'perturbation' should be a small, potentially an infinitesimal, change. In this context one way to think of this is to start with the empty set of primes, and then see a 'small' perturbation as including a single very large prime, or including a 'thin' or 'sparse' infinite set of prime numbers.

My hope was that the resulting family of functions obeyed something called a Pólya–Carlson dichotomy, meaning that they are either rational (extremely well behaved) or admit a natural boundary (as badly behaved as possible in some sense). Robert and I made progress on some special cases, and many years later this chapter of his thesis came to feature in an unexpected new collaboration. Jakub Byszewski in Kraków and Gunther Cornelissen in Utrecht were working on some orbit-counting problems in a different setting closer to algebraic geometry, and had stumbled into the same type of arithmetic problem. As I had at various times talked to Jakub about related matters he looked carefully through Robert's thesis and found the exact result they needed. This became an appendix to their paper [43], and has become part of a different setting in which this Pólya–Carlson dichotomy phenomenon seems to arise.

Stefanie's other supervisor was Shaun Stevens at UEA, and her project concerned topological dynamical systems with a group of symmetries, studying in particular what happens to the periodic orbits when the quotient space is formed. The special case of a single involution symmetry formed the simplest case, where the structure of the group plays no real role. In this simple setting forming the quotient space is 'halving' and going in the reverse direction to build a cover space sitting above the original system is 'doubling'. We wrote up the possible diversity of relationships that are possible between the arithmetic and growth properties of closed periodic orbits in a halved or doubled system [286], as the aspects of the problem that came purely from dynamical systems were so clear in that setting. Stefanie spend some of her time with me in Durham, but the main input needed for the general case in which the structure of the group really plays a role came from Shaun and Stefanie, leading to Stefanie's first publication on her own [370].

Richard and the IAS

Richard Miles and I won some funding for him to visit Durham as a Senior Research Fellow of the Institute for Advanced Study in 2013. He gave some talks both about Mathematics and about his experiences of working in some of Sweden's free schools. We continued to develop some of the work that Robert Royals and Matthew Staines had been doing. Specifically, we made more progress on the idea that a Pólya–Carlson dichotomy could be proved for the dynamical zeta functions of group automorphisms, inspired in part by a suggestion from Benjy Weiss that we might find useful ideas in a book by Segal [272] with more on natural boundaries in complex analysis than is readily found in the standard 'second course in complex analysis'. After the event I noticed that the review of this lengthy and interesting book in *Mathematical Reviews* ends by saying 'It is also invaluable as a reference for any mathematician interested in, but perhaps not actively working in, any of the fields covered by these nine chapters—and that represents a sizable fraction of the profession.' We were in that sizable fraction and found useful ideas in it, but still struggled to get much closer to the general case, which remains open. We did however make some progress, and contacted Jason Bell at the University of Waterloo for help with some of the arithmetic of sequences we needed. He joined us as a helpful co-author of the final version [19]. I have never met Jason, but this was an enjoyable collaboration.

We also began to sketch out some ideas for a sort of survey of the viewpoint that I had started to think through with Matthew Staines, viewing the space of all compact group automorphisms as a whole and trying to understand three different things. One was to fix some dynamical attributes—entropy and zeta function for example—and try to understand the 'fibre' of group automorphisms with those attributes. Another was to determine for a given dynamical invariant what possible values it could take on, in particular whether it could vary smoothly or was inherently discrete for some reason. This was part of the context for my somewhat obsessive pursuit of a Pólya–Carlson dichotomy for the dynamical zeta function of group automorphism: Did the 'quality' of the zeta function deteriorate continuously away from being rational or did it immediately leap to the extreme behaviour of having a natural boundary? Finally, for a given notion of equivalence between two such systems you might hope to describe the quotient space (that is, a space with one representative chosen from the collection of all group automorphisms equivalent to it in that sense). All generate interesting questions, most of which remain open. As with the short paper Richard and I wrote on mixing properties of actions of the rationals [205], our submission generated an entertaining referee report which included the line 'Figure 2 is a blast' (Fig. 14.3). This appeared in the proceedings 'Recent Trends in Ergodic Theory and Dynamical Systems' of the Vadodara conference in honour of Dani's 65th Birthday in December 2012. I managed to attend this conference, but diary constraints from work pulled my return earlier, and the reasonable suggestion from Tania that I should be at home on Christmas Day itself pushed my arrival later, so in the end I was only able to spend two nights in India.

Colleges

Durham is a collegiate University, and at the time there were 16 Colleges, two of them at a campus in Stockton. The word 'college' is used widely in UK universities with several different meanings. Like the colleges of Oxford and Cambridge, the Durham colleges are 'listed bodies' in The Education (Listed Bodies) (England) Order 2013[6] under the powers given to the Secretary of State by Sections 216(2) and 232(5) of the Education Reform Act 1988.[7] Unlike those of Oxford and Cambridge, the Durham colleges do not deliver teaching (with the exception of a historic college with religious foundations that has its own academic staff offering college-based programmes). The

[6]https://www.legislation.gov.uk/uksi/2013/2993.

[7]https://www.legislation.gov.uk/ukpga/1988/40/contents.

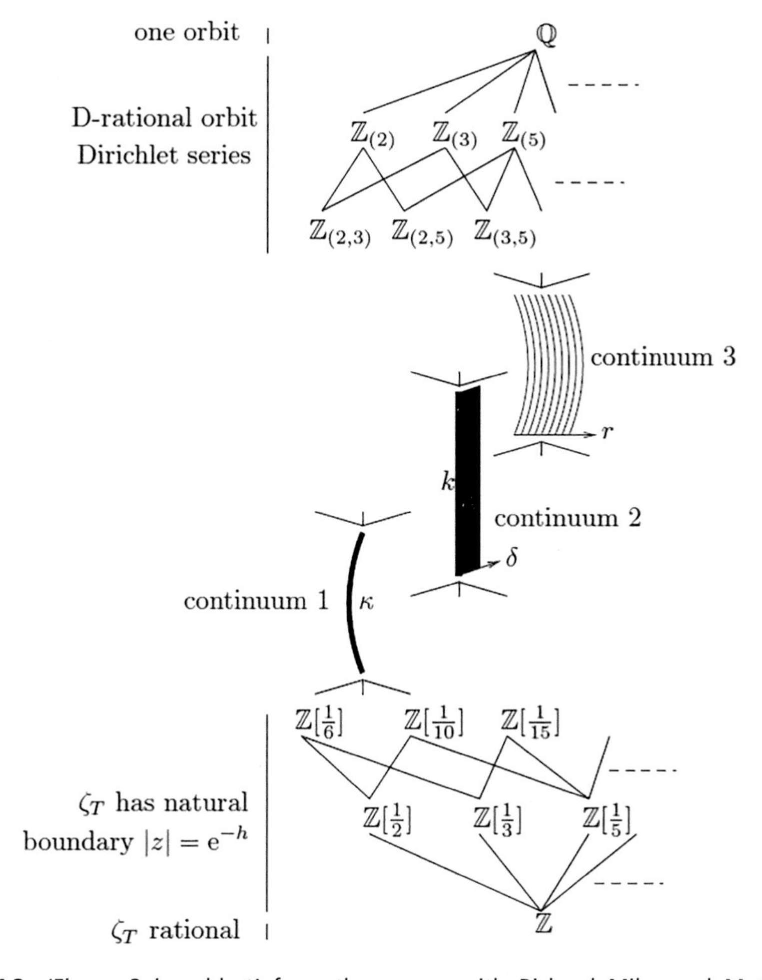

Fig. 14.3 'Figure 2 is a blast' from the survey with Richard Miles and Matthew Staines [204]. The diagram tries to visualize how certain growth invariants for compact group automorphisms vary among the automorphisms of one-dimensional solenoids arising as algebraic factors of a single automorphism (©American Mathematical Society; First published in *Contemporary Mathematics* Vol. 631 (2015), published by the American Mathematical Society and used with permission)

Durham colleges focus on the residential, pastoral, and social aspects of student life, and provide a strong sense of community and identity.

Tania and I were warmly welcomed by the network of colleges, largely in the form of invitations to attend various 'formals'. Partly out of environmental conviction and partly in order to survive the sheer weight of calories in this friendly onslaught, I was recorded in the various systems as being vegetarian and preferring fruit to dessert. This usually worked effortlessly, but one of the early grand dinners gave me an insight into the amount of effort going on behind the scenes. At a black-tie celebration event in University College ('Castle') soon after we arrived, a complicated dessert that looked a little like an artist's drawing exercise on a black slate was brought to me. It was something like a cube, a cylinder, and a sphere of various kinds of ices and the like. I asked if they might find me a piece of fruit instead, expecting they would find an apple and chop it up or something equally straightforward. I had forgotten all about it within a few minutes, and assumed there had been a mix-up when a black slate complete with the complicated geometric dessert was placed in front of me. It took a moment to work out that rather than a repeat delivery this was an exact replica, with each of the geometric shapes made of thin slices of several different pieces of fruit. I felt guilty about the amount of work involved in the kitchen during an already busy service, and made sure to thank the people involved afterwards.

The various dinners and receptions produced many memorable moments. At one of them—during the period of standing around with a drink in hand—I found myself in a circle with the Vice-Chancellor and six or seven other colleagues. Quite by chance he mentioned something about his time as a student at Durham, turned to his left and said 'you studied here too, didn't you?'. This went on round the group and it became clear that everyone there apart from me had been a Durham student at some point. When this had gone round the circle I turned to the Vice-Chancellor and said "well Chris, as you know I studied at Warwick". Without a pause, and certainly with the best of intentions, he looked at me and said "Well you've done very well for yourself". It was a wonderful and characteristically Durham moment.

At the time members of the university executive were encouraged to take on roles in college life, and we did this in two ways. The two colleges at the Stockton campus sometimes found it less easy to feel part of the University as a whole, so I became the chair of council for John Snow College. This was a wonderful community, with a mix of largely international business students and home pre-clinical medical students. Creating a sense of community in a college 35 kilometres away from the rest of the campus required great dedication from the Principal and their staff, and I was enormously impressed by how successful this was. We also both joined the SCR (Senior Common Room) of one of the colleges located on the Bailey in the centre of Durham, St Cuthbert's Society. This led to many interesting evening research presentations and discussions reflecting the wide range of interests of the students and the Principal.

The Teaching Excellence Framework

The long and sometimes frustrating history of attempts to improve education in Universities at national level changed in tenor in 2017 following an election commitment.

> We will ensure that universities deliver the best possible value for money to students: we will introduce a framework to recognise universities offering the highest teaching quality (From the 2015 Conservative election manifesto)

This began a series of consultations that eventually became the Teaching Excellence Framework (TEF). I was involved in some of the various groups looking at how this might be done, including a Russell Group body that tried to influence it. The initial idea from the government and its advisers was to simply develop a formulaic methodology based on various metrics associated to a course of study, and output a measure of overall 'Teaching Excellence'. This all became quite contested, and early on some Universities did not want to participate at all. My view was that this took us into difficult territory: The commitment was in a manifesto of a party that went on to win an election, so the question of whether we liked the idea seemed not particularly relevant. At Durham and at Leeds I pushed for us to be as involved as possible, engage with pilots, and so on.

The dividing lines between many Universities and government slowly emerged. Many of these involved reasonable arguments on both sides, but one was absurd.

The implicit link to 'value for money' in the election manifesto provoked much discussion. Much of the income from education for UK Universities is conditional on certainly regulatory requirements, but the political dream of using some sort of metric teaching assessment to drive through fee changes remained on the table but always seemed too difficult. From my point of view there were two big problems, both rather practical.

First, the robustness of the methodology involved would be severely tested and possibly found wanting. If a marginal judgement from a TEF panel resulted in the loss of tens of millions of pounds of recurrent income to any organisation, legal challenge would be almost automatic. This would have tied the process up in judicial review and delivered outcomes that served the interest of nobody involved.

Second, there is the paradoxical nature of the punitive ideology itself. The framers of much of the regulatory and legislative changes that were to eventually reach the unhappy staging post of the Higher Education and

Research Act 2017 imposed their own market-driven ideology onto the sector. Universities and their leadership come in all shapes and sizes, and I am not naive enough to imagine they are uniformly saintly in their actions at all—but the legal entity is in most cases an exempt educational charity.[8] They are not—with a small number of exceptions—entities that seek to, or indeed can, make a profit which is distributed to shareholders or other stakeholders. The logic of creating a system that identifies problems in the teaching at a University and then responds by in effect further depriving it of resources seems fundamentally opposed to the interests of the students involved. It is a sharp dividing line in how one thinks of the sector: One way to view the complex story is to think of a set of weighing scales, with an understanding of the public good arising from the creation of graduates in one pan and a belief that standards will be driven up if Universities are forced to compete in the other. If it is tilted in one direction, then there is if anything a compelling logic to saying that a poor TEF outcome should attract additional support and funding to protect the students involved. If it is tilted in another direction, then it makes sense to use the instruments of the market to financially 'punish' a University that delivers a poor TEF outcome. This distinction manifests itself again and again in different ways and the gulf between the two conceptions of the University grows ever wider. Roughly speaking, the state of the sector now reflects the outcome of a dogged pursuit of the market ideology, and the more processes are created and demands are imposed in the interest of 'student protection' the less protected the students really are.

We were also not at all keen on the terminology, because 'Teaching' suggests a particular didactic style of education, in which a 'Teacher' has the responsibility to convey a syllabus to the student. Our view was that the education we aspired to was much more of a shared project, in which the student is just as active an agent as the 'teacher'. Many of the things we saw as being distinctive about the educational experience we enabled—group projects, research involving undergraduates, industrial placements, projects—sat uneasily under the label 'Teaching'. On this we were pushed back, and it remained the 'Teaching Excellence Framework' or TEF.

The focus on metrics and the idea of a simple algorithmic approach disturbed us for two main reasons. One is that metric measurement of educational

[8]Charities in the UK are usually regulated by the Charity Commissioners. An 'exempt' charity is not unregulated—it is an entity created by an Act of Parliament or a Royal Charter that is regulated by another body. In the case of (most) Universities in the UK, that other body has been the Office for Students since 2018. The Office for Students has a Memorandum of Understanding with the Charity Commission clarifying the relationship. Universities are still obliged to comply with Charity Law.

experiences are notoriously difficult to make reliable and meaningful. Another was the prospect of huge effort being diverted into gaming the system by influencing the metrics. I had some sympathy but little liking for the other view, as metrics always feel like they are conveying important and objective information. This ended up as something like a score draw: The metrics would indeed be used, but only to formulate an 'initial hypothesis'. This would then help support a panel to come to a final judgement using a narrative submission as well as the metrics.

How the outcomes would be expressed remained an unresolved area of difference. The experience of higher education is complex, and the natural solution seemed obvious: A base level of adequacy accompanied by a suite of outcomes expressing different aspects of that experience. So one course (or University) might be particularly good at employer engagement, another at rigourous academic stretch. On this we simply hit a brick wall: Not only was the government determined to have a single outcome, they wanted it to be Gold, Silver, or Bronze. This difference—the idea that experiences of real complexity and diversity can all be placed in one linear ranking meaningfully—has pervaded the system and always will. I argued for the view that a better system would be closer to 'QA-plus': A basic 'Quality Assurance' threshold to be met, and then a portfolio of things that could reflect the different emphases of different Universities and the extent to which they were really delivered.

There were many other points of difference about the relative weightings of different metrics, but the most difficult remaining issues concerned statistics, where to some extent we collided with C. P. Snow's 'two cultures' problem. The first statistical issue went to the heart of the rather complex metric methodology built into the system. On a given measure, expressed as a percentage, a positive or negative flag for that measure required being 2% above or below the 'benchmark' (constructed from the measurement for a group of similar institutions). The approach to how the benchmarks were going to be made up was itself difficult—for example, considering how regional differences in salaries and employment rates be included in employment metrics—but the statistical stumbling block was simpler. The Russell Group contained some large institutions, and it seemed clear that a tiny number of big Universities would in effect define the benchmark they were to be measured against, making it impossible to exceed it by the required 2%. In some of the meetings it was difficult to retain much confidence that our interlocutors fully understood the issue.

The truly absurd statistical problem went to the heart of the original proposals, which were to do this vast and onerous exercise at the level of departments or even individual courses. When this is attempted, you quickly run into the problem of a lack of statistical validity for metrics derived from small cohorts. Here too I am not sure the issue was ever fully understood by every one involved, and there were several discussions asking us in effect to come up with a clever formula that would take as input numbers without statistical validity and magically output classifications with statistical validity.

The biggest debate concerned the purpose of the whole exercise.

Was it to add yet more to the enormous amount of data available to applicants, little of it used? There was a long history of attempts to improve the transparency of data available to applicants, and much effort went into devising things like the 'Key Information Sets' (KIS) bringing together educational information, accommodation costs, other financial information, student satisfaction scores from the National Student Survey (NSS), employment and graduate salary data, and information about the Student Union. The KIS was introduced as a result of a government white paper 'Students at the heart of the system' published by the coalition government in 2011.[9]

Was it aimed at 'enhancement'? That is, using the outcomes to support educational improvements in Universities? If this was the objective, then different design choices would be made throughout.

Was it really about the unshakeable conviction in some quarters that the sector was simply not paying enough attention to the education of its students? There is some reason to believe this about some courses and institutions, and it is undeniable that the National Student Survey had been helpful for those of us arguing for more of a focus on the educational experience of students within institutions. One interesting aspect of this is the 'other people's children' phenomenon. Politicians and national newspaper columnists who rail against those awful Universities that devote too much of their effort to research and too little to education generally expend huge effort and resources into trying to get their own children into the most research-intense Universities.

These questions were never really resolved. The TEF trundled on with a life of its own, complete with various reviews and consultation on methodology. How much influence it has had on the complicated decision-making process of applicants remains unclear. The process—and the narrative statements in particular—absorbs vast amounts of effort both in creating the submissions and in the assessment panels. By 2021 an 'Independent review of TEF' led by Dame Shirley Pearce had brought great clarity to the questions surrounding

[9] https://www.gov.uk/government/consultations/higher-education-white-paper-students-at-the-heart-of-the-system.

the TEF, but no real agreement between the different stakeholder was ever achieved.[10]

So it remains. Every now and then the educational machinery in each institution goes through the difficult process of creating a narrative account to accompany its metrics, and then waits on edge for a crude outcome of, in its latest iteration, Gold, Silver, Bronze, or 'Requires Improvement'. Whose interests are served by the exercise remains unclear.

Autumn 2015

In the Autumn of 2015 I responded to an approach for a senior role at Warwick University, went through the application process and proceeded to interview early in November 2015. The attraction was partly the emotional bond I felt to my *alma mater*, though in fact a University evolves quickly and is barely recognisable as a community after several decades. This was as usual a lengthy and multi-stage process, and during the interview itself it became clear that both parties had some doubts—there was a specific institutional project they had committed to that raised questions for me, and my concerns perhaps made little sense to the panel. It was a positive experience nonetheless, and the right outcome was for them to appoint someone else. It had however been a distracting process, and the emotional journey of going right through the appointment process to a final interview had taken up a great deal of time.

This shaped my mood when I was contacted by the search consultants Perrett Laver on the 20th of November 2015 about the role of Deputy Vice-Chancellor (Student Education) at the University of Leeds. I had a few telephone conversations with them, and emailed on the 14th of December to say no—despite the significant attraction of working with the Leeds Vice-Chancellor Sir Alan Langlands. I had already come across him in two settings, once when he was Chief Executive of HEFCE, and once in September 2014 at a Universities UK (UUK) conference. I was there standing in for the Durham Vice-Chancellor, and at the opening plenary event Sir Alan welcomed the many delegates on behalf of the "five higher education institutions in Leeds". It struck me that this was a particular and deliberately inclusive form of words that would not be at all automatic—or even possible—in many Russell Group Universities.

[10] https://www.gov.uk/government/publications/independent-review-of-tef-report.

Fortunately—and I learned some of the background to this later—the search consultants rang again and said in effect "By no, what do you mean exactly?" After some to and fro I was persuaded to send in a rather sketchy application, and there were some further phone calls and exchanges. As I learned more about the role and about Leeds I became more enthusiastic, and by the time the real visits happened at the end of February 2016 I was keen to get the role. The formal interviews took place on the 1st of March, and I was offered the position that evening by phone.

Between the original approach and the interview some things had happened at Durham that also helped make the decision an easy one. My appointment when I moved to Durham in 2012 was described in the contract as 'for a period of 5 years in the first instance to 31st August 2017, renewable for subsequent periods as appropriate on the recommendation of the Vice-Chancellor.' This is a quite common formulation for roles like this, though practice varies widely. In one of several rather stormy episodes while in attendance at Council I discovered quite by chance that the Remunerations Committee had decided that all the PVC renewals coming up were to be externally advertised.

This is a delicate business without easy answers. A degree of security is comfortable for the individual, and may indeed attract someone to a role. Some PVCs (usually ones who had been appointed internally) may be expecting or even impatiently waiting to return happily to a regular academic role. Equally, it is easy to run out of energy and ideas in a demanding senior role, and important to keep fresh ideas coming into a University executive. The surrounding circumstances and the casual way this had been announced irritated me though, and I certainly had no intention of applying for the role I had been doing for some years if it were to be advertised externally. So I went home that evening with the unsettling news that we would in any case be leaving Durham by the Autumn of 2017.

In fact there were all sorts of issues in play behind this offhand announcement at Council, and within weeks I was told 'outside the room' that the terms of my original appointment would be held to, and my role would not be advertised in this way. Indeed it became clear that this was all about someone else on the executive, another of those moments that seemed to come out of the pages of a particularly Machiavellian C. P. Snow novel. However, by then I had rather lost confidence in anything I was told, so it all became immaterial.

We started planning in earnest for another move. There were many sad goodbyes to make at Durham, and we quickly decided that this would be a simpler move than the move from Norwich as we could not bear to say goodbye to our house Mishpocha. We made plans to rent it to some young friends who would use it as a base for their developing work on a project diverting food waste for (at the time) pop-up cafes. This worked out well, and a few years later they had started the REfUSE cafe in Chester-le-Street, with an inspiring vision.

> REfUSE's vision is show the value in things, places and people that are unjustly wasted or overlooked. Each month we intercept around 12 tonnes of food that would otherwise go to waste, from retailers around the North East. We then redistribute it to people, through our 'Pay As You Feel' community cafe in Chester-le-Street, our weekly themed 'Restaurant Nights', partnerships with other organisations and our Waste Not Box scheme.
> (From the REfUSE website https://refusedurham.org.uk/)

The Principal of St Cuthbert's Society arranged for a small dinner on the 4th of July 2016 to say goodbye properly, and this felt like the real moment of departure.

Tania and I made several visits to Leeds to look for somewhere to rent, and eventually settled on a rented house a few hundred metres from Headingley Stadium.

Confronting My Past

By 2016 the courteous pretence that my office in the Mathematics department at Durham had any meaning had long passed. As soon as we moved into Mishpocha in 2013 I brought my Mathematics books home, and the four steel filing cabinets were moved into a basement somewhere on the campus. In order to accommodate the books I lined half of the sitting room at home with bookshelves. These changes meant the office became available to the department, which was seriously pressed for space. It was almost the only kindness I could show to the department in return for the warmth of their welcome.

The filing cabinets contained Mathematics research papers accumulated over three decades. It was a vast collection, carefully arranged in alphabetical order, with many of the papers being extremely difficult to lay hands on. Indeed, some had never existed in the formal literature. There were some mimeographed translations of papers by the Russian ergodic theorists of the 1940s and 1950s, with hand-written annotations saying things like 'from the files of Roy Adler'. There were quite a large number of papers photocopied from the impressive holdings of the Ohio State University library which would not be found in the more constrained selection available in smaller universities.

A feature of Mathematics, pure Mathematics particularly, is that the research literature does not really go out of date, so there were also photocopied papers of journal articles going back to the seventeenth and eighteenth Centuries. Precious, evocative, irreplaceable—and unmanageable.

I had already moved the household's financial records into the cloud, and was used to scanning and careful use of metadata for later retrieval. So I decided to bite the bullet and get rid of the contents of the filing cabinets. I did this gradually, processing the papers of a few authors each evening after work. For each of many thousands of papers, I either found it electronically somewhere, scanned it or, on rare instances, decided it had no further use for me and could safely be discarded. The quantity of paper put into the recycling bin each night was mortifying, but it was some consolation that much of it came from an era before widespread digitization and the availability of high quality scanners. I became extremely adept at removing staples, patiently filing papers in electronic form, and finding suitable cloud services to deal with the resulting huge volume of electronic files.

This proved to be the first of two significant academic and physical lightenings of my life. As the end came in sight I felt a little like Guy Crouchback when he finally unburdens himself of 'Apthorpe's gear'.

> Guy folded the carbon paper in his field notebook and wrote: Received 7 November 1940 Apthorpe's gear.
>
> 'Sign here,' he said.
>
> Chatty took the book and studied it with his head first on one side and then on the other. Till the final moment Guy feared he would refuse. Then Chatty wrote, large and irregularly, J. P. Corner.
>
> Suddenly the wind dropped. It was a holy moment. Guy rose in silence and ritually received the book.
>
> 'Come back when you've time for a pow-wow,' said Chatty. 'I'd like to tell you about that village near Tambago.'
>
> Guy descended and let himself out. It was cold but the wind had lost all its hostility. The sky was clear. There was even a moon. He calmly made his way back to the hotel which was full of the Commando.
>
> (From 'Officers and Gentlemen' by Evelyn Waugh)

The process also amounted to a journey through many years of my own mathematical history. There were obscure papers that at some point had been stepping points on journeys into forests that ended in dead ends. There were others covered in annotations and stains from both wine and coffee that represented more productive steps. A great many brought no memory to mind at all, lost trains of thought that never really travelled beyond the photocopier.

15

Leeds

We moved to Leeds at the start of August 2016, though we spent the first week of my appointment on a pre-arranged holiday. This was a rather active sort of holiday, at a Christian event called Revive in Ashburnham Place, near Battle in East Sussex. It made for a real break between two intense roles. Leeds is one of the larger universities in the country, and everything about it is on a big scale. It also had at the time a long and highly successful balanced commitment to education and research. I was taking over from Viv Jones, who had been Pro Vice-Chancellor for Education for many years and had made Leeds a leading institution for the quality and coherence of its educational offer, its astute use of digital technology, and its exceptional professional services support for education. During the early stages of the appointment process I had raised various ideas and things I would wish to implement, and invariably they were either long since established there or were in train. I learned a great deal from my time at Leeds, and enjoyed it greatly.

One of many enjoyable things about Leeds was the exceptionally deep-rooted and genuine partnership with the Student's Union. This is an important part of the life of every University, but ran particularly deeply there. An event like a breakfast presentation from the Union about their plans would attract the entire University Executive and many of the Directors of Professional Services, for example.

The state of digital education at Leeds was impressive, well ahead of most of the sector. I inadvertently caused some disquiet among this community and for Neil Morris, the Director of the Digital Education Service, in particular by a casual choice. In advance of the move I was asked to send some pictures, and I

T. Ward, *People, Places, and Mathematics*, Springer Biographies,
https://doi.org/10.1007/978-3-031-39074-6_15

chose one taken at the Max Planck Institute in Bonn which showed me giving a seminar—standing in front of one of their magnificent blackboards, and manifestly using chalk. My reason was probably to simply avoid yet another standard corporate head shot, but this choice caused a small frisson of anxiety about whether I might not be supportive of digital education.

Tania and I started attending Leeds Vineyard Church straightaway, and discovered that the immediate neighbours of our rented house did so too. This quickly became a warm community, and we soon found ourselves hosting a small group in our rented house.

Philoxenos

The search for a more permanent home began in earnest, and by November 2016 we put in a somewhat half-hearted offer on a modern house, which happily did not succeed. At the time Tania was in London, and I did not want to email with an entirely negative story to tell, so did yet another quick search to send some new ideas as well. One of the houses in this search caught my attention particularly because it was a huge house, with a wide entrance hall. We wanted to live close to the University and the centre of the city, and as so often happens there were some 'correct' districts to accomplish this. The house whose details I sent in my email was in an 'incorrect' district, which the estate agents labelled 'urban' and 'vibrant'. The former word seemed to be a polite euphemism for multi-cultural, and the latter a coy reference to the fact that many years earlier the neighbourhood had been notorious for high crime levels.

It took some time to arrange, but we bought that house in the centre of Chapeltown at the end of March 2017. Tania had chosen the name Mishpocha for our house in Durham, and I chose the name Philoxenos for our house in Leeds. The word φιλοξενος appears in the Greek New Testament in the books 1 Timothy, Titus, and 1 Peter. Our use of it comes from 1 Timothy 3:2: "A bishop then must be blameless, the husband of one wife, vigilant, sober, of good behaviour, given to hospitality, apt to teach" (King James Version) where it is used to mean given to hospitality. As I have but one wife and taught Mathematics my score is then three out of seven, generally regarded as a passing mark in UK higher education—fortunately. The word also has the meaning of welcoming or loving the foreigner, and seemed particularly appropriate given the rise of febrile anti-immigrant forces unleashed in the UK by the Brexit vote of 2016.

It was a huge upright terraced house built in the 1890s, and the neighbourhood was indeed almost absurdly diverse. Chapeltown reflected different waves of new arrivals into Leeds both in its buildings and in its people, and had done so throughout its existence. The area had been studied from a sociological point of view, and we found an interesting publication in the Leeds University library by Max Farrar discussing this [124]. Our house was close to a former synagogue that had become the Northern School of Contemporary Dance, and our kitchen looked out on the Gurdwara Guru Kalgidhar Sahib. The house itself had an interesting history, as it had been a shebeen not many years earlier. Every now and then we would meet someone who knew it for that reason, or had played there in a band. As we slowly tidied it up we came across markers of various phases of its earlier history. After removing many layers of point from the hall door frames we found three tin *mezuzot*, still with intact *klaf* (a piece of parchment inscribed with the Jewish prayer *Shema Yisrael*, comprising the verses Deuteronomy 6:4–9 and 11:13–21). I managed to restore these and refit them. They were not possible to accurately date, as we found records of it being a Jewish household as far apart as 1890 and 1940. A less salubrious find of more recent vintage was the crudely daubed word 'Gents' under many layers of paint on the door out of the kitchen, one of many signs that there had been a great deal of chopping and changing inside the house over the years.

A significant number of the residents of Chapeltown were children and grandchildren of the so-called 'Windrush generation' of people who had been called to Britain to help rebuild the country after the war. There was a particularly strong link to Saint Kitts and Nevis, and one of the things people said was that there were more 'Saint Kittians' in Leeds than there were in Saint Kitts. The 50th Anniversary of the Leeds West Indian took place in 2017, and our house was in the centre of the festivities. A few months in advance someone knocked on our door and asked if we would host a few dancers, which we were more than happy to do. These turned out to be three courteous middle-aged gentlemen from Saint Kitts and Nevis who had managed to bring impressive headgear with them on the flight. The three dancers were great company, and it felt good to be able to help in that tiny way.

If you live in the centre of Carnival there are only two realistic options (Fig. 15.1). One is to simply be somewhere else for the whole weekend, the other is to embrace it completely, join in, and take pleasure in the unique history that it honours. We did the latter, and made friends with some of the stall-holders through letting them use our house for access to a toilet and shower when needed.

Fig. 15.1 Loudspeakers ready for Carnival a few metres from our front door in 2017

The Hostile Environment

We were still living there when what became known as the 'Windrush scandal' unfolded. It is embarrassing to look back and even think about this awful incident. During her long tenure as Home Secretary, Theresa May had been proud to lay claim to the creation of a 'hostile environment' policy. This policy—of making the UK a 'hostile' place for immigrants—was maintained by her successor Amber Rudd. Some of the victims of the Windrush Scandal were people who had come to the UK long ago, when they could do so without needing a passport. Legal changes in the 1960s and 1970s meant in some cases that as their country of origin became independent they acquired the right to citizenship of that country, but their situation in the UK was recognised and acknowledged: Anyone who had arrived in the UK before 1973 was granted the right to remain permanently in the UK automatically. Because it was automatic, many of the people involved did not acquire documents proving that: They simply continued to live and work in the UK as they had done for so many years already. The Immigration and Asylum Act 1999[1] gave explicit protection to long-standing residents from Commonwealth countries, including protection from deportation. When the new and more 'hostile' Immigration Act 2014[2] was passed during Theresa May's time as Home Secretary, all these protections were quietly dropped. This was challenged and the Home Office responded by declaring that there was no need to transfer those provisions because Commonwealth citizens who had lived in the UK since before the start of 1973 had 'sufficient' existing protection. Amber Rudd took over as Home Secretary in 2016, and was in post during the scandal.

During 2018 people in that type of situation started being arrested with an eye to deportation. It took some time for the country to become aware of what was happening: Significant numbers of people, all of them with long-distant origins in the Caribbean, were wrongly arrested and detained, denied the usual legal rights, and threatened with deportation. The numbers are difficult to get hold of definitively, but certainly more than 80 people were illegally deported by the Home Office through this process. Many of these were in law British citizens who had arrived in the UK prior to 1973, mostly from countries in the Caribbean. Much larger numbers were detained, denied access to the National Health Service, and had passports or identity documents confiscated. Some of the long-term residents who were wrongly deported were (also wrongly) denied re-entry to the UK. Of the cases I became aware of, all had stronger or longer links to the UK than I did.

Chapeltown was one of the areas targeted, and stickers started appearing on lampposts explaining what to do if you were picked up. Some individuals experienced absolute devastation of their lives, and in a few cases the local community had time to respond. You would hear of specific cases where, for example, a school was suddenly engaged in running a campaign to stop one of their pupils being whisked away and deported.

It took some time for the scale of the problem to become visible, but eventually some acknowledgement that this was happening started to emerge. By 2020 an independent review of the whole terrible business led by an Inspector of Constabulary called Wendy Williams had been conducted.[3] It pulled no punches, but the consequences on the people most responsible were minor: Amber Rudd stepped down in April 2018 in response to the scandal, but less than a year later was Secretary of State for Work and Pensions. Theresa

[1] https://www.legislation.gov.uk/ukpga/1999/33/contents.

[2] https://www.legislation.gov.uk/ukpga/2014/22/pdfs/ukpga_20140022_en.pdf.

[3] https://assets.publishing.service.gov.uk/government/uploads/system/uploads/attachment_data/file/876336/6.5577_HO_Windrush_Lessons_Learned_Review_LoResFinal.pdf.

May went on a few years later to become the second of five Conservative Prime Ministers between 2010 and 2022. For many of the victims, the consequences were life-changing destruction. For the country as a whole, we all had to confront a vivid and stark reminder of just how deeply ingrained racial prejudice really was, and that this resided not in little pockets but across an entire swathe of policy development, legislation, and practice on the ground.

> But over time those in power forgot about them and their circumstances, which meant that when successive governments wanted to demonstrate that they were being tough on immigration by tightening immigration control and passing laws creating, and then expanding the hostile environment, this was done with a complete disregard for the Windrush generation. [...] Ministers set the policy and the direction of travel and did not sufficiently question unintended consequences. Officials could and should have done more to examine, consider and explain the impacts of decisions. While I am unable to make a definitive finding of institutional racism within the department, I have serious concerns that these failings demonstrate an institutional ignorance and thoughtlessness towards the issue of race and the history of the Windrush generation within the department, which are consistent with some elements of the definition of institutional racism. (From the Executive Summary of the Independent Review led by Wendy Williams)

Mathematics Schools

At various times a political head of steam builds up over the idea that universities should play a bigger role in high schools. Most of the people I worked with in the higher education sector were in principle supportive but in practice somewhat cautious about this, largely for two reasons. One is the simple recognition that school teaching is a specialist business, and the expert practitioners are located in schools not in universities. Another is the question of distraction. A large university does have enormous capabilities, but a project that supports neither the education of our own students nor the research mission of the institution is rarely welcome—particularly if it absorbs a great deal of management time and effort.

One iteration of this came early on in my time at Leeds. Prime Minister Theresa May launched a campaign to encourage universities with high entry requirements to sponsor schools.

> So the government will reform university fair access requirements and say that universities should actively strengthen state school attainment—by sponsoring a

state school or setting up a new free school. And over time we will extend this to the sponsorship or establishment of more than one school, so that in the future we see our universities sponsoring thriving school chains in every town and city in the country. (From the speech 'The Great Meritocracy' by Theresa May on the 9th of September 2016)

At the start of 2017 some of these ideas became more focused on the STEM disciplines (Science, Technology, Engineering and Mathematics) in the launch of a new industrial strategy with a remarkably optimistic ambition: 'Prime Minister Theresa May will use her first regional Cabinet meeting to launch a modern Industrial Strategy for post-Brexit success.'

Plans to use the successful free school model to expand the provision of specialist maths education across the country [...] Working with local partners including top university maths departments to spread new specialist 'mathematics schools' building on high-performing Exeter and Kings College London Mathematics Schools. (From 'Technical education at heart of modern Industrial Strategy'[4])

This led to the University of Leeds being approached by several multi-academy trusts (MATs) with an eye to creating a sponsored specialist mathematics school. We were also approached directly by civil servants at the Department for Education, which led to one of the strangest engagements with government I have ever experienced. The issues in play for us were several. Above all, we had excellent relationships with many local schools and did not want to sponsor a school that might simply displace the strongest pupils and teachers from existing schools, resulting in no overall gain in attainment and damaging those relationships. So I asked the colleagues from the Department for Education to send us what evidence they had about modelling of the pipeline of younger pupils who might credibly be attracted into the school resulting in a growth in overall numbers, and to ask if they had any data relevant to the already severe shortage of mathematics teachers. I also wanted to know if anyone was paying attention to the progress of a much earlier wave of enthusiasm for sponsorship of schools by universities, which had led to the creation of a University Technical College sponsored by Leeds. My questions were clearly not welcome and the encounter ended with the extraordinary comment that "Liz Truss will be in touch".

[4]https://www.gov.uk/government/news/technical-education-at-heart-of-modern-industrial-strategy-for-post-brexit-success.

At the time the future prime minister Liz Truss was Chief Secretary to the Treasury and known for a certain robust style. I was most aware of her as one of the authors of a remarkable book called 'Britannia Unchained' [172], which famously contained a short passage that none of its authors ever seemed particularly eager to take responsibility for.

> Once they enter the workplace, the British are among the worst idlers in the world. We work among the lowest hours, we retire early and our productivity is poor. (From Britannia Unchained' [172, p. 61])

When the phone rang 2 days later it was indeed Liz Truss, who simply harangued me for half an hour. Wanting to see the evidence meant I had no ambition for the city of Leeds, did I not want to see pupils in Leeds succeed in mathematics, on and on it went. The idea of carefully understanding the numbers before acting seemed to irritate her most particularly.

In the end—and over some understandable opposition from the governing council of the University, who were particularly concerned about the distraction of management time—we did decide to sponsor a mathematics specialist school, having satisfied ourselves about the pipeline and displacement risks. It was an odd episode and a startling insight into the thinking and approach of a person who went on to lead the country for 49 turbulent days that caused lasting economic damage.

Other People's Children

In 2016 there was also a sudden reappearance in the press of the idea of allowing for new 'grammar' schools to be opened, ending a long period in which this was not allowed. These are state secondary schools that have academic selection on entry, and are controversial primarily because of the consequences for the life chances of children who do not pass the entry examinations. Prior to the creation of comprehensive (non-selective) state schools across much but not all of the country, grammar schools were for those who passed the 'eleven plus' examination, and secondary moderns for those who did not. When comprehensive schools were introduced this did not happen uniformly across the whole country, meaning that for many years the majority of secondary schools were non-selective but enough remained selective to allow robust statistical analysis of all aspects of the system. One of the major studies was carried out by Jonathan Cribb and Luke Sibieta from the Institute for Fiscal Studies and Anna Vignoles from the University of Cambridge. The resulting report, 'Entry into Grammar Schools in England' appeared in 2013 and helped populate some of the debates with solid data.[5]

Because the proposal to allow for new grammar schools had been floated first in the media, it surfaced in our formal political life not through a statement in Parliament but via a question followed by a perplexing (and robust) exchange on the 8th of September 2016.

> To ask the Secretary of State for Education to make a statement on the Government's plans to lift the statutory ban on opening new grammar schools in England. (Shadow Secretary of State for Education Angela Rayner)

> As the Prime Minister has said, this Government are committed to building a country that works for everyone, not just the privileged few. We believe that every person should have the opportunity to fulfil their potential, no matter what their background or where they are from [...] (Secretary of State for Education Justine Greening)

> Wow! Despite that waffle, the cat is finally out of the bag. The Government have revealed their plans for new grammar schools in England, but not in this House—we did not even hear the word "grammar" just then. Instead, they did it through leaks to the press and at a private meeting of Conservative Members. [...] Perhaps the Secretary of State can tell us the evidence base for this policy today. Has she read the Institute for Fiscal Studies report "Entry into Grammar Schools in England"? If so, perhaps she remembers the conclusion: "amongst high achievers, those who are eligible for" free school meals "or who live in poorer neighbourhoods are significantly less likely to go to a grammar school." (Shadow Secretary of State for Education Angela Rayner)

It is a remarkable feature of the political debates about this that they invariably feature the creation or expansion of grammar schools and scarcely mention the creation or growth of what would be the *de facto* secondary modern schools that the policy would entail. At times this felt like an instance of 'other people's children' thinking—opinion formers in government and in the national media tend to have children who did or would get into the grammar schools. Indeed this initial Parliamentary exchange on the proposal to lift the ban on the creation of new grammar schools seemed to include a belief that schools could be created for those who pass the 'eleven plus' examination

---

[5] https://www.suttontrust.com/wp-content/uploads/2020/01/grammarsifsvignoles.pdf.

without creating other schools for those who failed it, and initially managed to avoid using the loaded words 'grammar school' or 'secondary modern' entirely.

> There will be no return to the simplistic binary choice of the past, where schools separate children into winners and losers, successes or failures. This Government want to focus on the future, to build on our success since 2010 and to create a truly twenty-first-century school system. However, we want a system that can cater for the talent and the abilities of every single child. To achieve that, we need a truly diverse range of schools and specialisms. We need more good schools in more areas of the country responding to the needs of every child, regardless of their background. We are looking at a range of options, and I expect any new proposals to focus on what we can do to help everyone to go as far as their individual talents and capacity for hard work can take them. Education policy to that end will be set in due course. (Secretary of State for Education Justine Greening[6])

As the discussion widened it did start to grapple with the real issues, and the words 'secondary modern' even appeared, initially in an intervention by Ian Mearns, Labour Member of Parliament for Gateshead. In fairness it eventually became clear that some thought had been given to the entire cohort, and the aspiration was to somehow use the creation of more selective schools to raise standards everywhere. This optimistic approach made the careful work in the Institute for Fiscal Studies report all the more important and relevant.

Another interesting angle on the 'other people's children' phenomenon is the frequent appearance of national newspaper columnists arguing that higher education is not worth pursuing, or that many fewer people should be able to access it. A recent fine example of this tradition was an article in the *Daily Telegraph* on the 6th of June 2021 entitled 'Apart from dating one of my tutors, my university days were a complete waste of time'.[7] The author—a graduate of the London School of Economics—was Sophia Money-Coutts, daughter of the British peer Crispin James Alan Nevill Money-Coutts, the 9th Baron Latymer. I wish her well, and am always delighted when a graduate becomes a successful journalist on a national newspaper, but I want the same kind of opportunities, reserves of social capital that come with being the daughter of a Baron—and experience of higher education of the quality that the London

[6] https://hansard.parliament.uk/commons/2016-09-08/debates/70D5AB36-38ED-4DD3-8F7D-5B5DFAF8C4F5/NewGrammarSchools.

[7] https://www.telegraph.co.uk/family/life/apart-dating-one-tutors-university-days-complete-waste-time/.

Fig. 15.2 Sawian Jaidee and Klaus Schmidt in Philoxenos on the 22nd of July 2018 (With permission of Tania Barnett)

School of Economics provides—to be available for young people with less privileged backgrounds.

Sawian in Leeds

Sawian Jaidee, a former student, spent the Summer of 2018 visiting me in Leeds on a small research grant from Khon Kaen University in Thailand (Fig. 15.2). Sawian stayed with us, upstairs in our huge Chapeltown house, and in the evenings we worked away at something that Graham Everest and I had also worked on. Patrick Moss had shown in his 2003 thesis [224] that sampling any realizable sequence along the squares, or cubes, or fourth powers, and so on, produced another realizable sequence. Sawian and I decided to look at a strange object—the monoid $\mathscr{P}$ of all 'time changes' (maps on the natural numbers) that preserved realizability in this sense. Via email we managed to get hold of Patrick in retirement, and together were able to show that $\mathscr{P}$ is in one sense very small (specifically, that the only polynomials it contains are the monomials identified years earlier by Patrick), and in another sense very large (in that it contains uncountably many maps) [158]. Both Tania and I greatly enjoyed having Sawian visit, and she once again made us some splendid Thai food.

Gdańsk

During 2019 I had agreed to give the 'wandering seminar' in Poland. This brings together ergodic theory and dynamical systems research groups in Poland via short courses that move between the main research centres. The organisers on this occasion were Grzegorz Graff from Gdańsk and Jakub Byszewski from Kraków, and the plan was for me to give seven or eight seminars over the course of a weekend at the end of February 2020 at the Gdańsk University of Technology. Over the years I had made many visits to Poland, several to Kraków in particular. I was also one of the organisers of a special semester in Bonn planned for 2020, and had made the first of four planned visits to the Hausdorff Institute there already. The first half of 2020 was all mapped out with this series of research visits.

We were all just starting to understand the scale of impact that Covid might have, but decided to go ahead with the trip. Tania came along, and we had three wonderful days in Gdańsk. We were able to explore some of the legacies of the Jewish community there, and in particular its rapid growth during the 1920s when Gdańsk was the 'Free City' of Danzig under the protection of the League of Nations.

Grzegorz and Jakub had already been planning a survey on various aspects of realizable sequences and I had been preparing some lecture notes for the series of lectures. The three of us met up and decided to join forces on this small project. Over the next few months it was completed by email exchanges, and appeared as a survey in the *Bulletin of the London Mathematical Society* [42].

Covid

At the time we had no idea this would mark the end of easy travel for such a long period. We were already required to complete detailed contact tracing forms for the flight back from Poland, with records of where we sat in the aeroplane. We arrived back at the beginning of March 2020 and I returned to work at Leeds. Over the next week or two we rapidly went through plans to have half of the executive on campus on each day to minimise contact, and so on. I cancelled the planned trips to Bonn, both because it was clear travel was about to become difficult and because a great deal of work would need to be done on planning for the continuation of educational delivery.

The University was already in the midst of industrial action, and we needed to get the plans for what we would do in response to the pandemic through the next available meeting of Senate. A small but noisy student demonstration in support of the industrial action made it impossible to continue the business of Senate in its usual location, so the decision was taken to abandon it. It was frustrating that the one Senate item that could not be delayed was to approve a plan to ensure that several thousand students affected most by the pandemic could graduate, and a tiny vocal group of students were so intent on preventing this from being discussed and approved. After some deft reorganisation behind the scenes Senate reconvened in a rather crowded room in the Marjorie and Arnold Ziff Building with a low ceiling. Within days we were working remotely, and I do not know if that final Senate meeting led to my infection with Covid, but it was around that time.

Both Tania and I were unwell over Easter, and I became progressively less and less able to breathe easily. I tried to continue with online meetings, but was having to pause for breath after saying a few words, and something as simple as standing up made my pulse race wildly. Friends from Leeds Vineyard Church rallied round, with both moral and practical support. The latter ranged from delivering food to the door to dropping off a pulse oximeter. Being able to monitor my own oxygen levels at home made an enormous difference: In particular, it allowed me to show the oxygen levels live to various doctors doing video consultations. This meant I could stay at home, but it did take many weeks before I could work normally, and the shortness of breath and racing pulse came back several times over the next few months. We both lost much of our sense of taste and smell quickly, and I took to using huge quantities of raw garlic and ginger in the simplest of dishes just to try and experience the sensation of tasting something. We both ended up using essential oils to try and retrain our sense of smell, and the effect lingered for some years.

Through the early part of 2020 we had to make a large number of decisions about how to protect students from the impact of the pandemic. Our lead for digital education, Neil Morris, ended up writing a set of principles and a plan to move the education of forty thousand students online more or less overnight on a laptop in a hotel room in London. This extraordinary piece of determined and clear-sighted thinking became enormously valuable as a guiding template for how to deal with the many problems that came up. It was possible because Leeds already had a long track record of significant investment in digital education, so in some ways this was a ramping up in scale rather than something entirely new. Nonetheless it was difficult enough, and we all had to learn new ways of doing things at speed. I met with a small group of professional services leads and Faculty Pro-Deans frequently, and we all grappled with decision fatigue as events unfolded.[8] We were having to make multiple weighty decisions with limited information many times a day, and at the same time dealing with the anxieties and anger of thousands of students and staff. The resilience of the educational leadership throughout the University impressed me greatly, but we all suffered the consequences.

For students this was the beginning of a long period of immense difficulty and uncertainty. Specific words and phrases took on great symbolic meaning as they reached out for reassurance. When the University of Exeter used the words 'no detriment' (in relation to how assessments and degree classifications would be handled) in an announcement this became a phrase of almost mystical power, with petitions demanding the same elsewhere. The most painful aspect for many of us was how circumstances—and, no doubt, some poor or slow decisions—pushed this from a challenge to be faced together into a situation where some students felt they were in opposition to 'the University'. International students faced impossible decisions about when to try and get home before borders closed. In hindsight there are some things most of us would do differently, but in the heat of the moment with no knowledge of the future it is hard to see what one would change. We did quickly realise that this was a marathon not a sprint, but balancing the understandable anxieties of staff, students, the integrity of awards, and a responsible approach to the role of Universities in responding to the challenge was not at all straightforward. School pupils were equally affected but at least generally had the support of their own home life, and it is clear that it will be many years before the impact on younger people is fully understood.

October 2020

Alan Langlands retired at the end of the 2019–2020 academic year. I felt privileged to have worked with him for 4 years and had learned a great deal from the experience. He possessed extraordinary range, fully engaged with the largest national issues without ever losing authentic concern for individual students and colleagues. My mother had passed away within a few weeks of my starting work at Leeds, and I will never forget how kind he was around that time.

[8]This is a fairly well-understood process, in which multiple decisions even of little import diminish executive function [16].

Suddenly stepping down from the DVC role in November of 2020 at the behest of the newly appointed next Vice-Chancellor of Leeds came as a shock. It was not quite from one day to the next, because there were a few awkward meetings still to navigate, but after the dust had settled and the announcement had been made I suddenly had an empty diary for the next week—and possibly for many weeks beyond that. I felt like Charles Ryder on hearing where his army camp was located.

> He told me and, on the instant, it was as though someone had switched off the wireless, and a voice that had been bawling in my ears, incessantly, fatuously, for days beyond number, had been suddenly cut short; an immense silence followed, empty at first, but gradually, as my outraged sense regained authority, full of a multitude of sweet and natural and long forgotten sounds: for he had spoken a name that was so familiar to me, a conjuror's name of such ancient power, that, at its mere sound, the phantoms of those haunted late years began to take flight. (From 'Brideshead Revisited' by Evelyn Waugh)

There is no gentle way for a summary removal to happen. I was asked to stand down from the DVC role by the new Vice-Chancellor on Monday the 26th of October 2020 in a 10 minute Teams call. Tania was extremely shocked and quite frightened about what this would all mean in practice when I went upstairs to tell her the news.

We had been following a daily Bible reading so decided to look at the day's passage over a cup of tea to take our minds off the immediate sense of crisis, and it proved to be a remarkably appropriate passage from Ecclesiastes.

> What do people get for all the toil and anxious striving with which they labor under the sun? All their days their work is grief and pain; even at night their minds do not rest. This too is meaningless. A person can do nothing better than to eat and drink and find satisfaction in their own toil. This too, I see, is from the hand of God, for without him, who can eat or find enjoyment? To the person who pleases him, God gives wisdom, knowledge and happiness, but to the sinner he gives the task of gathering and storing up wealth to hand it over to the one who pleases God. This too is meaningless, a chasing after the wind. (Ecclesiastes 2, 22–26, NIV)

One of the other habits we had developed during the long periods of partial or complete lockdown due to Covid was to play table tennis. We were fortunate enough to have space for this in the house, which made it easy to maintain the habit. We were playing that same evening as usual when my rather slow thought process reached a conclusion. "I need to look for another job" I announced, abandoned the game for a moment, and rather randomly looked in my phone's Outlook client for the sub-folder called 'Headhunters'.

I had forgotten until I did so that the search consultants Perrett Laver had contacted me on the 15th of September to ask for suggestions of people to contact for the role of Pro Vice-Chancellor for Education at Newcastle. My habit is to always try and respond with some ideas, and I had done so the next day. So I wrote again, at 16.10 on the 26th of October, to say another suggestion—rather unexpectedly—was me. To my amazement they wrote back quickly to say "The longlist meeting is tomorrow morning. I am checking within the team about your potential late application. Meanwhile, would you like to send me a copy of your CV just in case we can still manage to present your profile?" Fortunately the relevant people seemed to work all hours, and by late that evening they had agreed to add my curriculum vitae as an addendum before the meeting the next morning. By the afternoon of 27 October I was in the process for what turned out to be my next role, a miraculous turn of events. It felt a little like the encounter between Guy Crouchback and the Electronic Personnel Selector.

> "Now here"—he picked a chit from his tray—"is a genuine enquiry. I've been asked to find an officer for special employment; under forty, with a university degree, who has lived in Italy and had Commando training–one, two, three, four, five"—Whirr, click, click, click, click, click. "Here we are. Now that *is* a remarkable coincidence."
>
> The card he held bore the name A/Ty. Captain Crouchback, G., R.C.H., att. H.O.O. HQ.
>
> Guy did not attempt to correct the machine on the point of his age, or of the extent of his Commando training.
>
> "I seem the only one."
>
> "Yes. I don't know what it's for, of course, but I will send your name in at once."
>
> (From 'The End of the Battle' by Evelyn Waugh)

I had an online conversation with the Director of Human Resources at Leeds (a wonderful colleague, who had suddenly been pitched into holding the University's corner against me) on the 27th of October, in which she wisely helped me to think through some of the implications but not commit to anything. I met with her again—online—on Friday the 30th, and finalised the arrangements. These were of course something of a pretence, driven in part by the fact that I did not approve of the profligate use of non-disclosure agreements which too many universities have used to brush the consequences of impetuous decisions—and worse—under the carpet. The agreement was that I would write a letter to the Vice-Chancellor on the 3rd of November with pre-agreed wording, asking to step down for a period of research leave, and there would be an immediate pre-agreed response agreeing to this, to take effect from the 7th of November. It felt farcical to bother to make sure that the response did not precede the request, but it seemed to matter to the University. The research leave was for 12 months, at the same DVC salary, with a small budget for research costs. It was a privileged and luxurious form of departure, and I would have been able to stay on as a professor in the Leeds Mathematics Department after the period of leave if I wanted to. In comparison to people really losing their employment I had multiple layers of protection, and I was acutely conscious that huge numbers of jobs were being lost as the economy shrank because of Covid in far less pampered circumstances.

Nonetheless, this all came as a shock. I was surprisingly untroubled by it all, in part because I was already uneasy about the direction in which the institution was moving under its new leadership. I was not the first, nor the last, senior figure to hastily depart over that period. Tania was extremely upset, as it meant a period of real uncertainty and almost certainly a move from Leeds, which had started to feel like home.

I had not worked in a regular academic position since 2008, and in fact had almost no connection to the Mathematics Department at Leeds. As soon as the announcement was made public the new and the previous head of department were in contact and made clear that I would have a warm welcome there nonetheless. This was a great kindness at a time of considerable turbulence in my life that I will not forget.

Realising that the new Vice-Chancellor, while gushingly fulsome and passionate in concern for people in the abstract had so little care for the individual was sobering and revealing. The chasm between people for whom the many is an aggregation of individuals each of infinite worth and people for whom the individual is an infinitesimal subset of the many is a large and at times unbridgeable one.

It is a distinction that has always interested me. Like the divide between people who are fundamentally honest and those who are not, the contrasting shades have grown in my mind from soft focus muddled ambiguity in childhood to something stark and vividly illuminated in adulthood. At the time it brought to mind a trip my friend Mark Eyeington and I made by coach from Coventry to the Barbican in London in 1983 to see 'Maydays', a monumental play by David Edgar. One—of several—beautifully depicted dividing lines in the multiple arcs of that extraordinary play concerned just this divide, played out on a large political stage. The theatre critic Mark Lawson, in response to the Royal Shakespeare Company's revival of the play, compared Maydays to one of the few novels I return to again and again [174]: He likened the weaving together of complex threads of personal histories woven into a specific and long period of history as something of an anti-establishment version of Anthony Powell's solidly establishment *roman-fleuve* 'Dance to the Music of Time'.

> It's a Golden Age. Utopia. It's nowhere. And, for the sake of that, I don't believe, I can't believe, I actually refuse to be required by anybody to believe, that anyone is human dust. (Spoken by the character Martin, in a scene set in May Day of 1975 in 'Maydays' by David Edgar [73])

I visited the Leeds campus again on Saturday the 31st of October to remove all my personal effects from the office on Floor 13 of the Marjorie and Arnold Ziff building, where I had spent four happy years. I took them away in a small wheeled suitcase: An absurd and embarrassing number of conference lanyards, ballpoint pens that did not work, plastic folders with sun-faded labels containing acronyms that no longer meant anything to me—and more significant things like a group photo of one of the Student Union executives I had worked with to be treasured and taken to a new home. I did not set foot on the Leeds campus again until almost two years had passed, when I returned for a meeting of the Russell Group PVC network.

I chaired my last Education Committee via Teams on the 4th of November, and arranged for the key people involved in education at Leeds to stay on for a later video conference, where I told them the official version of what was happening. An email was sent to Council that evening, and to the wider University the next day. My departure was sufficiently rapid that things like my key fob were never handed back.

Saying Goodbye

The networks and groups that formed my professional life and our home life reacted in various ways compatible with the conditions of lockdown that the country was under at the time. The Russell Group network of education leads were hugely supportive immediately—and somewhat taken aback. Alan Houston, who had taken over as PVC for Education at Durham after I moved on, arranged for a bottle of cask-strength Laphroaig to be delivered the very next day, an instinctive and kind gesture. The Leeds Student Union arranged a wonderful online send-off and thank you event for the 17th of

November 2020, during which a card and flowers for Tania and a rather splendid bottle of wine for me were delivered to the door. Working with the Student's Union at Leeds, both the elected sabbatical officers and the permanent staff, had been a great pleasure and a huge part of my life there. The education community of Pro-Deans, relevant Professional Services leads and so on also sent thoughtful gifts quickly, and later arranged a great online send-off on the 23rd of February 2021, during which a handsome gift of wine and a bottle of Tomatin were delivered to the house.

The usual niceties of a corporate send-off event for a senior departure of course did not happen. So, in contrast to these warm and moving events, in the case of the colleagues I had worked with most closely day by day for four years there was simply a sudden break from one day to the next and a handful of emailed notes.

One of many interesting University leaders whose writing I enjoy is Clark Kerr [165], and I was now in a position to speak—to close friends at any rate—my own version of one of his many pithy sayings.

> I left the presidency just as I had entered it—fired with enthusiasm. (Clark Kerr speaking afer his abrupt dismissal on the 20th of January 1967 by the Board of Regents of the University of California, shortly after Ronald Reagan became Governor)

At home the Church 'house group' we led organised another online send-off with a big group of church friends for the 7th of April 2021, which really marked the beginning of the final packing and clearing of Philoxenos, and the end of our life in Leeds. Just as the kind meal hosted by the Principal of St Cuthbert's Society in 2016 was emotionally the signifier of departure from Durham, this marked our departure from Leeds.

A strange detail was that a large library of Mathematics books remained in my office, about 30 shelf metres. These were boxed up and delivered to our home, where I stacked them all on the floor of the dining room. I had remained quite tranquil throughout this strange episode, though it had been hard for Tania. There were many sad goodbyes certainly, but almost immediately I was sensing that the Executive Team at Leeds was no longer a place where I would be able to work happily in any case. Instead of being upset, I felt a strong urge to do something practical, and the piles of books became a project. I systematically went through them—some I was able to obtain electronically, some had crucial sections that I scanned, some I decided I could live without. Then I got rid of them, to various book sellers and charities. This largely involved packaging them up and delivering them to local shops to be collected, a laborious and curiously satisfying task. Once started, the desire to complete the task was strong and when it was all over only a handful of books remained. Some of the remnants were accidental, but three were deliberate. I could not bear to part company with the copy of Klaus Schmidt's book [271] that he had given me, nor two of the books I had bought as a new graduate student: André Weil's number theory book [351] and Peter Walters' ergodic theory text from which he taught [303].

The wall of books behind me, a carefully arranged alphabetical sequence starting with one by Jon Aaronson on infinite ergodic theory and ending with Anton Zygmund's massive work on trigonometric series, had been a cumbersome physical manifestation of my professional identity for much of my life. The anticipation of shedding it was difficult, but it proved in the end to be a relatively painless exercise—a second lightening after the clearing out of my mathematical papers when we left Durham in 2016. The abrupt, involuntary, move away from Leeds needed a corresponding symbolic abrupt transition. Physical to digital, daunting bookshelves to empty office, perhaps external evidence of scholarship to greater internal confidence. This second shedding after confronting my accumulated papers when I left Durham was the more significant one, and it surprises me still that it was not only possible logistically but easy emotionally.

Patrick and Piotr

Once the turbulent few weeks of negotiating my departure and handing over had passed, my mind turned to Mathematics. The laboratory, primary document sources, raw materials, field study, semi-structured interview transcripts, samples, or data sets simply do not exist for a mathematician. There is 'the literature', one's own mind, and a blank sheet of paper. The first and last of these I had in abundance, but the middle one was unused to doing serious research.

As things turned out, I was only able to use 6 months of the research leave before taking up a new role at Newcastle University. My mental petrol tank was more or less empty after 11 years on university executives, and I definitely did not have the energy for anything requiring retaining a lot of Mathematics in my head all at once. Doing Mathematics involves many things, and one of them is simply holding all the ideas needed for the problem at hand in accessible memory. Early on in Maryland I had gained a small insight into this. I met Don Zagier in the lounge of the Mathematics department, with papers spread out all over one of the tables. He was still there the next day, and

I rather nervously asked him if he had sat there all night. 'Yes' he explained—'I simply need to get too many things clear in my head before I can think about this problem, and that takes time.' It was not the first time it was illuminated for me that mathematicians widely seen as exceptionally creative or brilliant are also possessed of formidable powers of sustained concentration and effort. For some of us this constrains sharply what can be done, and that certainly applied for me during that period. I had simply lost the habit of holding significant conceptual mathematical complexity in my head.

One project was ready and waiting however. Patrick Moss had already shown in his thesis [224] that the classical Fibonacci sequence, if sampled along the squares and multiplied by 5, became realizable, and with Sawian Jaidee we had explored a more abstract combinatorial set of questions motivated by this observation. With a little effort involving *ad hoc* arguments using special congruences satisfied by the Fibonacci numbers, Patrick and I were able to give a complete explanation of the phenomenon: Sampling along any even power produced a realizable sequence after multiplying by 5, sampling along any odd power produced a sequence that could never be realizable no matter what it was multiplied by. We felt it likely that this was a more general phenomenon for linear recurrence sequences, with the multiplier 5 replaced by the so-called 'discriminant', but did not have a feeling for what the 2 (from the squaring) might be a special case of, other than the trivial observation that the Fibonacci sequence is a linear recurrence of degree 2. The resulting publication appeared in the *Fibonacci Quarterly* [223], and I could not help in the strange circumstances enjoying one tiny, ridiculous, aspect of how the timing worked out. The journal is a small one, and survives on voluntary author 'page charges', which I paid using the agreed research fund from Leeds. On the other hand, by this stage I knew where I was going next, so the author affiliation address was Newcastle University.

I looked through an old list of problems I used to keep when Mathematics was my main concern. This was a plain text file called `ideas` which I had started in the far-off days when 'computer' meant a simple bash shell connected to a machine running Unix or VAX. The name meant that I could type the Unix command `more ideas` whenever I was bereft of them. Sadly on this occasion nothing looked tempting, and I have no clear memory of what made me start looking at the Stirling numbers. These are sequences that play a role in combinatorics, and I started to look at them from the point of view of 'realizability'—the arithmetic condition of counting periodic points for a map.

The Stirling numbers come in two infinite families. Some of the early sequences in the family are familiar in other settings, and it is easy to verify which are realizable. As I started to experiment with some of the later sequences, considerably helped by a computer algebra package called `GP-pari`, I convinced myself that they were certainly not realizable. This seemed easy to prove for any given sequence in the infinite family—but I had to develop a different argument for each one. With a little more effort, I was starting to be able to construct arguments that showed statements like 'every eighth sequence is not realizable'. However, all my attempts ended up with some—indeed, infinitely many—sequences missed out. Somewhat randomly, I searched on google using the keywords from my problem, something like `Stirling number divisibility Mobius p-adic`. This led to all sorts of papers and people, none of which seemed to contain quite the results I thought I needed. However, the name Piotr Miska cropped up several times. I sent an email to Piotr in Kraków, outlining what I thought to be true, and with my partial results. A few days later Piotr send a beautiful and lucid argument proving that all the Stirling sequences apart from the early simple cases were not realizable. I was delighted, but somewhere in my numerical experiments a trace of another, entirely unexpected, phenomenon had shown up. Roughly speaking, failing to be realizable meant that a certain operation on the sequence would throw up some rational numbers with a denominator in a certain infinite sequence, rather than (non-negative) integers. What seemed to be emerging was that these denominators did not behave in an uncontrolled way, but felt bounded. This was tricky territory, because it involved an inherently infinite calculation and I could only see the first few thousand terms in a computer calculation.

I did some more calculations and contacted Piotr again, this time asking if he thought it possible that the following was true. The Stirling numbers 'of the first kind' did not become realizable after multiplying by any number, but for each of the Stirling numbers 'of the second kind' there was a mysterious number that 'repaired' them: They become realizable once multiplied by this 'repair factor'. This indeed turned out to be true, resulting in a natural occurrence of this phenomenon of 'almost realizability' in a quite unexpected setting. The proof that eventually emerged, which we published in *Acta Arithmetica* [218], gives an upper bound on the mysterious repair factor: For the kth sequence it is a divisor of the factorial $(k-1)!$. In some cases the numerical calculation—computing the least common multiple of the first few thousand terms of an infinite sequence of denominators—gives exactly this answer, making the calculation for that case definitive. In general this is not the case, and it is a rather opaque business to really understand what this number is—beyond the certainty that it exists. With some editorial difficulty—because of the inherently infinite nature of the calculation—this sequence of repair factors was contributed to the Online Encyclopedia of Integer Sequences [232,

A341617]. There it remains, and if anyone can discern something about the arithmetic structure of this sequence do please let me know:

1, 1, 2, 6, 12, 60, 30, 210, 840, 2520, 1260, 13860, 13860, 180180, 90090, 30030, ...

Things progressed steadily at Newcastle, and I was appointed in the middle of December, with an agreement to start on the 1st of May 2021.

16

Newcastle

We moved back to Mishpocha in Durham at the start of May 2021. The move was made more difficult by a large number of plants in large pots and in the ground in Leeds. We had worked hard at bringing some vegetation to our house in Chapeltown, building raised wooden beds in the tarmac back yard, and digging through heavy clay and rubble at the front to make a varied hedge. In the end a separate truck was needed for the plants. At the other end, access to Mishpocha is impossible for a full-sized removal lorry, so the removal firm had to park up a little outside Durham and use a smaller truck to shuttle back and forth.

Once things settled down we both enjoyed being back in our old house, and connecting again with King's Church Durham where we still had many friends. Leaving Leeds was difficult, and we missed the friends we had made through Leeds Vineyard Church as well as the University particularly acutely. In fact we have been able to visit Leeds fairly often, some of those friends have visited here, and many links have been maintained.

Florian

Two of the strangest chance encounters that led to something mathematical were with the same person. In 2016 I had attended a conference to celebrate the 60th birthday of Igor Shparlinski, to be held at the beautiful Centre International de Rencontres Mathématiques (CIRM) in Luminy. I felt honoured to be invited, because this was really a conference for and about number

T. Ward, *People, Places, and Mathematics*, Springer Biographies,
https://doi.org/10.1007/978-3-031-39074-6_16

theorists in Igor's circles. The reason was the earlier collaboration on the book 'Recurrence Sequences' [110], which had become quite widely used. It was as always a pleasure to be in that beautiful place, but I knew very few people at the conference and in fact was enjoying the peace, until one evening I received an email from Igor a few floors down.

> Hi Tom - We are downstairs tasting some Romanian spirit. Please join us.
> Cheers IS

This turned out to be a lethal-looking viscous substance in a two litre plastic bottle, doubtless distilled from hooch by somebody's cousin in a shed in the mountains. It was, in fact, just my kind of thing, and the coy description of 'tasting' belied the reality. In this way I met Florian Luca for the first time; we were brought together deliberately by Igor who suggested we should discuss something. Together with Graham Everest I had done some work on two different natural classes of sequences, linear recurrence sequences that grow with a characteristic exponential rate, and bilinear or elliptic ones that grow with a characteristic quadratic-exponential rate. A question that had been in the air for some time was whether an elliptic sequence could ever simply be a linear one, but sampled along the squares. This would be most unexpected, and would contradict a natural conjecture about the arithmetic and Diophantine properties of some invariants called Néron–Tate heights attached to elliptic curves. Florian had indeed proved that no linear recurrence sequence sampled along the squares could eventually coincide with an elliptic sequence, but a journal referee had objected to the argument for several reasons. We talked about it briefly, and then over email found ways to present the ideas in a more palatable way—though a key step remains somewhat inscrutable. Rather than a note of thanks in an acknowledgement, Florian generously added me as a co-author [184]. Several questions were raised by this result. It should have a more quantitative formulation in several directions, with statements saying that they cannot coincide on a thin set, or possibly being able to specify that no single late term can coincide, with just how 'late' being defined in terms of the size of the coefficients of the recurrence and the initial terms of the sequences.

Six years later, some work with Patrick Moss done during the strange enforced period of research leave at the University of Leeds was in my mind [223]. This was based on Patrick's remarkable observation in his thesis [224] that the classical Fibonacci sequence, if sampled along the squares and multiplied by five, produced a sequence with the remarkable property of counting periodic points. We extended this to show that sampling along odd and even powers produced extremely different behaviours—it was satisfying to find something new to say about such an ancient sequence. We had some ideas for how to prove a similar property for other binary linear recurrences, with five replaced by a quantity called the 'discriminant' of the sequence. By chance Florian posted something on Facebook on the 14th of March 2022 which caught my eye. This was advertising a seminar entitled 'The p-adic order of a linear recurrence' to be given by Yuri Bilu at the Max Planck Institute in Bonn. Patrick and I had run out of steam partly because I had run out of time, but also because we needed more information about certain divisibility properties of binary linear recurrences. I took the opportunity to advertise our work, and the questions we had left open.

> Very nice questions! Can I advertise a related set of issues - when does a sequence (in particular a linear recurrence) satisfy the Dold congruence after you multiply it by some big number? Two examples, both not linear: https://arxiv.org/abs/2011.13068 and https://arxiv.org/abs/2102.07561 (My posting on Facebook, 14th March 2020)

Florian replied, and quickly saw a way to attack our problem—indeed, to do it in much greater generality—by using information from the Galois group associated to the characteristic polynomial of the linear recurrence sequence. In this more general context our 'squaring' for the Fibonacci sequence emerged as being 'raising to the power of the exponent of the associated Galois group', and the multiplier five—as expected—as the discriminant of the sequence. Within days, starting from this brief exchange on Facebook, this had become a short draft note, giving a satisfactory generalization to a much more general class of linear recurrence sequences [185]. Roughly speaking, this result says that any integer linear recurrence sequence is trying its best to be realizable, but needs some help in the form of sampling along a suitable power and clearing a finite denominator.

This is the result that I mentioned in the preface as 'one that Euler certainly could have asked in 1736'. The date is not random, as it marked the appearance of the first published proof of 'Fermat's little theorem', which says that for a prime p and integer $a \geqslant 1$ the expression $a^p - a$ is divisible by p [100]. Euler went on to prove the Euler–Fermat theorem for non-prime powers in 1763 [101]; it is a striking feature of Mathematics that both these papers are written in essentially modern mathematical notation, and could be read and understood by any Mathematics student possessed of a Latin dictionary. The result with Florian Luca involves congruences that are in principle of a similar sort.

Tomorrow and Tomorrow

I retired from the role at Newcastle University at the end of March 2023. The many wonderful colleagues I worked with there, and the issues we grappled with, are too close in time to be discussed. It is a remarkable University, with a strong thread of values and commitment to serving society both globally and regionally running through all it does.

The state of the sector as a whole is a less happy tale. English Universities (and, for slightly different reasons, Scottish and Welsh ones) are once again in a period of prolonged decline in the purchasing power of the unit of resource. For some institutions this will inevitably lead to redundancies, closures of programmes and departments, and a rapid deterioration of staff-student ratios. For others, the proportion of their activity that is exposed to regulated fees (that is, undergraduates from the UK) will decline, with lasting damage to the skills base of the country and to the role of the University as an anchor to local communities. Entire disciplines may come under threat outside a tiny elite of institutions with a global brand or resources of such power that they can ride out almost any storm. At some stage difficult decisions will need to be made about the size of the sector and how to find a financially viable approach to higher education. None of the likely options would be popular, and the temptation to kick the can a little further down the road must be overwhelming. Doubtless another major review will be set in train before the 17th of December 2024 (the absolute last date by which the current Parliament must be dissolved).

The logical paradoxes that are avoided while the can is kicked along are fundamental to any part of society. In pursuit of 'market dynamism' the sector grows ever more tightly regulated. In pursuit of 'competitive efficiencies' the most important unit of resource has been fixed in cash and allowed to decline rapidly in real terms. In pursuit of 'creative divergence' all programmes and institutions are measured through the same narrow set of metrics.

An economist might point out that there are lessons to be drawn from the different national approaches that led to 2,818,547 Trabant 601s being produced in one part of Europe and 21,529,464 Volkswagen 'Beetles' being produced globally, which might incline one to think about the wisdom of central regulation and rigidly controlled prices. On the other hand, anyone might register some concern about leaving entire institutions and their local areas or entire disciplines and their contribution to society to the mercy of unfettered market forces.

There are no easy answers. The closure of HEFCE, charged with protecting the health of the sector as a whole, lies behind the inadequate response to some of the problems the sector faces. However it is the flawed logic of replacing it with a market regulator tasked with ensuring the free operation of competition and at the same time exercising more and more control on what that competition produces that has more directly led to the current mess.

The inevitable consequences of the approach to regulated fees—underpaid and overworked staff, huge class sizes, impossible industrial relations, and financial instability—will certainly damage the experience of students, particularly as it comes on top of the after-effects of the pandemic response for many of them. I feel great sympathy for staff who year after year see inflation erode the value of their salaries, but the campaigns that pretend there is a simple solution are disingenuous. Higher Education Statistics Agency (HESA) data covering 144 higher education institutions for the 2021–2022 year[1] showed 100 in deficit, 37 in surplus (seven have not yet reported). Overall, the total income of the sector for that year was £46.9 billion and expenditure was £50.9 billion, 60% of which was expenditure on staff. Inflation has surged since the year 2021–2022, and the financial picture is much worse now. It does not take a sophisticated analysis to work out what would happen if, as everyone would like, pay settlements kept pace with inflation while income across much of the role that a University plays for the country does not.

> 'My other piece of advice, Copperfield,' said Mr. Micawber, 'you know. Annual income twenty pounds, annual expenditure nineteen nineteen and six, result happiness. Annual income twenty pounds, annual expenditure twenty pounds ought and six, result misery. The blossom is blighted, the leaf is withered, the god of day goes down upon the dreary scene, and—and in short you are for ever floored. (From 'David Copperfield' by Charles Dickens)

The understandable frustrations sometimes find expression in the idea that the underlying problem is that a University is run 'as a business'. Just as the efforts to frame the student 'as a consumer' has a multiplicity of possible meanings, some benign and some toxic, the phrase 'as a business' means different things to different people. Where 'consumer' entails well-informed, empowered decision maker with strong rights that sit alongside the responsibilities of being part of a University there is nothing wrong with it. Where 'consumer' entails the idea of purchase, the idea that I have paid you

[1] https://www.hesa.ac.uk/data-and-analysis/finances.

for a meal and it is now your job to cook it and serve it while I sit at the table, it has diminished the experience of everyone involved.

If 'as a business' means preposterous Chief Executive salaries and the wrong balance between shiny new buildings and the student experience then it is a legitimate challenge. If 'as a business' means generating profits to be divided among shareholders then the charge is being made by someone who simply does not understand what most Universities are as exempt educational charities. If 'as a business' means realistic financial modelling and planning, a sustainable balancing of surplus generation to invest in new facilities and maintain the old ones against recurrent expenditure, an understanding of the difference between capital and recurrent expenditure, and some grasp of the implications of the size of the sector then it is not just desirable, it is obligatory. The painful reality is that the Universities currently on the receiving end of the charge that they are doing the wicked thing of being run 'as a business' have manifestly not been doing so, and have fallen foul of some unrealistic financial modelling.

The unsustainable financial position for regulated fees will also damage the ability of Universities to continue the level of their contribution to the country's research and development base. Prior to the Higher Education and Research Act 2017, this might have been the topic of lively and thoughtful debate between HEFCE and Government. As it is, with education and research separated by HERA and education framed in policy and regulatory terms entirely around a consumer model of student purchasing choice, these fundamental questions are not only going unanswered—they are barely being asked.

17

Looking Back

Short of a deathbed it is always premature to attempt to understand the shape of influences on one's life, but retiring from Newcastle provides a reasonable moment. Whatever I do next it will not be a role on a University executive, but will I hope include contact with students in one way or another. Students and their optimism, energy, and passionate engagement with social justice are the best thing about working in higher education.

...or Perish

One defining feature of what it means to be 'an academic' is the practice of publishing research or scholarly papers and books. A. C. Coolidge's description of academic life as 'publish or perish' holds just as much weight now as it did in 1932 when he coined the phrase [58]. Indeed, 'win competitive research grants', 'build industry links', 'support high school attainment', 'embed employability in your teaching', 'ensure your students are active and informed supporters of the UN's Sustainable Development Goals', 'engage with continuing professional development in pedagogic practice', 'acquire expertise in distance learning', and 'become an effective substitute for the too thinly stretched national mental health care system' have been added to the list of things required of any academic.

Much of the second half of my career has involved encouraging colleagues to focus on collaborative educational activities instead of or alongside research publications, and doing my best to advance the careers of academics who

T. Ward, *People, Places, and Mathematics*, Springer Biographies,
https://doi.org/10.1007/978-3-031-39074-6_17

put more effort into student education and academic leadership than their own research. Needless to say I only ended up being appointed to roles where I could try to do this because I did not 'perish' professionally as an early career post-doctoral researcher, but published. In fact one of the memorable moments during the early days at Durham was a telling insight into the mechanics of this: It was understandably expected that the PVC for Research would be 'REFable' (that is, sufficiently research active to be part of the submission to the national Research Excellence Framework), but it was also clear that as PVC for Education I was expected to be 'REFable'. For completeness—really for the intended reader identified on Page xiii—and with some effort, I have collated more or less all my publications, broken down by the affiliation of the final published version. In several cases this is not where the work was done, because of the lengthy publication process in pure Mathematics. University of Warwick: [178, 181]. Ohio State University: [268, 269, 321–323]. University of East Anglia: [10, 23, 50, 52, 65, 74–79, 81–85, 103–117, 127, 159, 205–210, 220, 221, 237, 241, 249, 250, 324–337, 347, 348]. Durham University: [13, 19, 86, 204, 211, 238, 338, 339]. University of Leeds: [8, 43, 87, 88, 158, 184, 212, 286, 340–344]. Newcastle University: [42, 89, 218, 223, 345, 346]. Durham University (as a Visiting Professor, and including current projects): [80, 89–91, 185].

Students

Students and the role that education can play in their lives have been at the centre of my professional life from start to end. Being part of the journeys of many cohorts of students has been a joyous thing, and my admiration for their optimism and energy is boundless. It is painful to reflect that for someone contemplating full-time higher education from a less privileged background the obstacles in their path are higher, and the helping hand from the state weaker, than was the case when I started out as a lecturer.

Helping first year students through their first encounters with quantifiers and 'epsilon-delta' analysis is probably the most striking setting in which the 'threshold concept' and the excitement of helping students to reach it was manifest. Roughly speaking, most of the students need to get through a process that undermines things they thought they knew and then rebuilds them on firmer foundations. When the film 'Up in the Air' came out it had a certain resonance for anyone who has had the privilege of teaching anything like this first rigorous analysis module.

Fig. 17.1 Supervisory connections over the years: With my fellow student David Pask and our supervisor Klaus Schmidt (left); with my student Sawian Jaidee and my supervisor Klaus Schmidt (right), in Sydney on the 5th of February 2015

> We are here to make limbo tolerable, to ferry wounded souls across the river of dread until the point where hope is dimly visible. And then stop the boat, shove them in the water and make them swim. (The character Ryan Bingham in 'Up in the Air' [253])

I loved lecturing, and missed it very much when senior roles squeezed it to the corners of my life. I also loved being a Head of Department, serving on various University executives, doing Mathematics, and writing mathematics—but there was only one life to be lived. As my role shifted, the students I spent time with changed from being Mathematics undergraduates and postgraduates to Student Union sabbatical officers. The sabbatical officers I had the pleasure of working with are a remarkable group of people, and I have much enjoyed following the interesting things they have gone on to do.

The experience of postgraduate supervision is a complex mix of research collaboration, mentorship, counselling, friendship, shared learning, and apprenticeship (Fig. 17.1). I learned something from each of the research students I met, and in some cases continued collaborations long after they had completed. Almost all of them were completing doctorates, but Shaun Stevens and I had one 'MRes' (Masters by Research) student, Jonny Griffiths. He was a mathematics teacher in a local college, and did highly original and unconventional work on forms of recurrence phenomena from a geometric point of view (Fig. 17.2).

Jonny was possessed of more hinterland than most, as he had played the character 'Stringfellow' on children's television and was part of the 1980s jazz vocal group 'Harvey and the Wallbangers'. Most of our regular meetings involved Jonny updating us on his own latest ideas, and the whole experience was fascinating. He went on to write an account of his life as a teacher, which he described as 'A mixture of what happened, what I wish had happened, and what I'm glad did not happen in my classroom' [138].

Fig. 17.2 Celebrating Jonny's MRes with Shaun Stevens at Loch Fyne in Norwich on the 3rd of March 2012

A Reflection Principle

Motivations, attributes, opportunities, and pure chance all play a role in anyone's life. My motivation at each stage was not something I thought much about. At high school Mathematics either interested me because I was good at it or became something I could do well because I found it interesting. It was certainly an interest supported and encouraged by our parents. I encountered many dedicated teachers, lecturers, and a supervisor who both taught me and encouraged me. Undeniably I faced none of the barriers that many others do: Academia is too often a toxic environment for women and people whose ethnicity does not 'fit'. Taking on leadership roles was never an aspiration I had until the moments of decision arrived in 2002 and 2008, and as I have outlined a strong element of circumstance and chance played a role in my becoming a Head of Department and, later, a PVC. Moving away from the University of East Anglia and Norwich after 20 years was a huge disruption to the whole family in different ways, but none of us regret the consequences. Because of that move we have lived in some of the wonderful cities of parts of England new to us, encountered both Leeds Vineyard Church and King's Church Durham, and made lasting friendships in both places.

It is hardly an original observation that a curious psychological mirror stands at the beginning of an academic career, located more or less at the end of the doctorate and reflecting back and forward in time further and further.

The first reflected pair comes from the fact that many post-doctoral researchers call on skills and knowledge only recently acquired. The research done may often be a more or less direct continuation of ideas and questions thrown up in their doctorate. Some strong mathematicians have advocated moving field entirely as a new post-doc, and this may well be sound advice for sufficiently quick learners who have found a funded position that is not part of a project with pre-determined objectives. One of the consequences of rapid growth and declining trust in the sector is that more and more research funding streams require a high level of pre-determination of topic, and it is a rare privilege to have a free hand at such an early stage.

As a new lecturer the two reflected images move further apart, and the most relevant experience and knowledge is likely to come from your own experience of being taught as an undergraduate. Things have improved since I started as a lecturer, but there is still a striking contrast between the research and the education components of the role. We assume that a typical research background for a new lecturer might include 4 or 5 years of full-time study in a discipline area, followed by something like an apprenticeship in research as a postgraduate student lasting a minimum of 3 years, and then some years as a post-doctoral researcher building up a network of research links and deepening expertise. In many fields even after all this there would be mentoring and support for writing grant applications and reading of draft papers. The educational aspect of the role more or less assumes that somewhere in that journey an ability to be an effective educational practitioner was somehow picked up by osmosis.

Leaning on the last experience of undergraduate teaching can be a great help in finding one's feet and overcoming uncertainty about quite what to do. Unfortunately it can also be a source of real conservatism in the approach to education. In mathematics many current lectures and seminars would be immediately familiar to a student or lecturer from fifty or even a 100 years ago, certainly in format and possibly in content. That is both a strength and a weakness—there are good reasons beyond a simple habit of replication that lie behind this stability, but innovation and experimentation in pedagogy can be difficult in light of this context.

Taking on any sort of managerial role within a department pushes the two reflected images yet further apart. Most professional services and academic staff are wonderful, but it is a fortunate department that is entirely free of colleagues who are capable of bullying or of extraordinary resistance to change. Lessons learned in the school playground can be more useful in dealing with this than any of the laboriously acquired disciplinary knowledge or well-intentioned training for new managers.

More senior roles can push yet further back in time, as they call in part on reserves of emotional resilience, wellsprings of optimism and energy, and simple determination far more than they call on intellectual power or knowledge of any specific sort. Many of the attributes in play are probably formed in early childhood.

In trying to understand my own journey from the Yeatman hospital in Sherborne to our home in Durham, the influence of both our parents cannot be overstated. Each of the four reflected pairs in a mirror placed at the end of my doctorate bears witness to this.

Certainly our childhood household was a place where scholarly effort as well as practical effort was celebrated. Questions or ideas across the sciences met with a warm reception from both parents, and questions or discussion about more or less anything were welcomed by our mother. In some ways we learned more at home than at school, certainly in the diversity of topics. There were no train tracks taking me towards mathematics—indeed the four of us studied entirely different things once beyond high school. There were however fairly clear pathways leading me to further study of some sort certainly, but it could have been in another discipline. Our father at times expressed some disdain for the social sciences, but I imagine their belief in education would have overcome this had my leanings gone in that direction. I was supported and encouraged throughout my University studies, and our parents were a constant positive presence, albeit at a huge distance. Far more significant than the scientific and mathematical knowledge they passed on, our parents had a special gift for encouragement. Any achievement or positive feedback was not only celebrated but remembered. At Christmas or birthdays a card might recall something positive from the past, carefully recorded months or even years earlier. The thin blue airmail letters, laboriously typed through three carbon papers, arrived regularly and kept the family connected with each other's achievements.

Serving on University executives involved some quite deep-rooted attributes from the early years—beyond my own memories. Whatever happened in that languid, sun-drenched, outdoor childhood I have generally found it easy to access reserves of optimism, calm, and resilience.

For all but a small handful of the years in which I had responsibility for education at institutional level there has been either industrial action or a pandemic, and the last decade I worked saw the worst industrial relations higher education has ever experienced. It is painful to reflect on the multiple ways in which staff and management were pitted against each other with generations of students bearing the consequences—while the actual drivers of the fundamental issues were remote from the fray, in actuarial and pension firms, at the pension regulator, and above all in government.

To some extent I have found it relatively easy to persevere at finding positive aspects of colleagues who at times hide those aspects particularly well. On the other hand I have never become particularly good at being the driver of events, nor have I found it easy to act with the decisive brutality that senior roles occasionally entail. I tend to see too many sides to most questions, and have never found a way to access the confident belief—or delusion—that my way forward is always the wisest one.

Antlions

The hours spent gently nudging grains of sand into the conical traps laid by antlions in the sunlit garden of Leopard's Head prepared me for something. It was not to become the voracious ruthless creature who had carefully prepared the sand trap, imagining that as a result they were in control of events—while really remaining entirely reliant on the behaviour of passing insects. Nor was it to be the hapless wandering ant who might innocently fall into its trap. It was instead to remain the suntanned child, nudging the grains of sand with a twig, innocently enjoying the languid pleasure of appreciating one of many wonders of the world.

Universities Must...

A research university in the UK is simultaneously many different things to many different people. An autonomous institution equipped to enable students to become independent individuals and citizens by developing their own critical thinking powers in an environment of academic freedom, in the great European tradition of Wilhelm von Humboldt's *Bildungsideal*. A place where people of diverse interests come together to pursue knowledge for its own sake and avoid narrow disciplinary boundaries, in the spirit of John Henry Newman—though in a modernised form, built not on religious principles but on something more inclusive. As a great engine of both research power for the nation and driver of social mobility for society in the tradition of Clark Kerr. The source of a steady flow of nurses, doctors, teachers, engineers, writers, thinkers, researchers that the country needs. Diplomatically, it is one of the most significant parts of society in which soft political power is projected around the world. For many people it is the place where they meet their life partners, as it was for me. At times it is genuinely a *universitas magistrorum et scholarium*, a community of teachers and scholars. And of course it is a part of a society and a political economy, an engine of economic stability in many towns and regions: Custodian and, one hopes, responsible user of vast amounts of public resource. Some attach almost mystical weight to the meaning of the word.

> [the University] owns no property, pays no salaries and receives no material dues. The real University is a state of mind. It is that great heritage of rational thought that has been brought down to us through the centuries and which does not exist at any specific location. It's a state of mind which is regenerated throughout the centuries by a body of people who traditionally carry the title of professor, but even that title is not part of the real University. The real University is nothing less than the continuing body of reason itself.
>
> In addition to this state of mind, 'reason', there's a legal entity which is unfortunately called by the same name but which is quite another thing. This is a nonprofit corporation, a branch of the state with a specific address. It owns property, is capable of paying salaries, of receiving money and of responding to legislative pressures in the process.
>
> But this second university, the legal corporation, cannot teach, does not generate new knowledge or evaluate ideas. It is not the real University at all. It is just a church building, the setting, the location at which conditions have been made favorable for the real church to exist.
>
> (Robert M. Pirsig, writing as Phaedrus in 'Zen and the art of Motorcycle Maintenance')

It would be an abdication to express no opinion here. For me a University is a distinct thing, not a blank slate onto which we can project all our desires for research capacity, skills development, school improvement, culture wars, regional economic anchoring, environmental sustainability, social progress, and industrial development. It is close to von Humboldt's *Bildungsideal*, but with a dynamic evolution in disciplines and in how the University and its graduates can serve modern society. The compass by which its direction is charted should come from the distributed creativity and passion of its diverse academic interests, and resources and capabilities of all sorts from government backed funding to the internet to the physical infrastructure of the campus to the latest interactive video technology should be placed at the service of that direction. The transformation from stable public funding to fierce competition for shrinking resources has certainly made the life of those in leadership different and more difficult, but it is a delusion to imagine that it has shifted the location of the beating heart of the institution. That heart is geometrically impossible but culturally powerful: It is in the seminar room, the laboratory, the lecture theatre, the personal tutorial, the online teaching session, the student support office—not in the executive offices. Leadership has a role in shaping strategic directions and seeing and responding to major changes in the external environment from a distance, but leaders should always have the humility to understand that a University is ultimately neither more nor less than the aggregate of the abilities, attitudes, and efforts of its many thousand students and staff. The reality is that 'L'État, c'est moi' from on high, be it implicitly or explicitly stated, has never made sense to me in a University context.

It is perhaps encouraging that despite fundamental shifts in how they are funded and how government exercises control over them in the UK, Universities are still looked to for the solutions to a great variety of problems. One of the ways in which this is manifest is that the word 'Universities' in a headline or political pronouncement is frequently followed by the injunction 'must'. Universities must…solve some of the problems we face. Universities must…enable what I want to see in the world. Universities must…do better. We are collectively eager to do so, but sometimes the mutually incompatible demands at the end of the sentence are a problem, as is the great diversity of these imperatives. In 2017 this prompted me to compile a short prose poem of headlines from a few months of this din, with which I end.

Universities Must…

Universities must be taught a hard lesson (*The Times* 12/6/17)
Universities 'must reconnect with society' in a sceptical world (*Times Higher Educational Supplement* 15/6/17)
SPLC president to Congress: Universities must protect free speech but also speak out for other democratic values (*Southern Poverty Law Centre* 19/6/17)
Universities must not shy away from protecting free speech (*The Times* 26/6/17)
Higher education institutions must learn to play politics or their role in civil society will be determined by public opinion and financial pressures (*Royal Society of Arts Journal* 30/6/17)
Universities must unequivocally protect free speech on their campuses (*The Denver Post* 1/7/17)
Universities Must Do More To Protect 'At Risk' Students From Suicide, Experts Say (*HuffPost* 13/7/17)
As a deaf student, I'm used to being excluded. Universities must do better (*Guardian* 14/7/17)

Universities must value their students more than their reputation (*Times Higher Educational Supplement* 15/7/17)
Why Public Universities Must Lead in Solving the Global Food Crisis (*Harvest Plus* 17/7/17)
More universities must confront sexual harassment (*Nature* 26/7/17)
Courts Hold Colleges & Universities Must Balance Due Process in Sex Assault Cases With Need to Protect Campus (*HuffPost* 27/7/17)
Universities must focus on prevention of sexual assault not response (*The Australian* 1/8/17)
Why universities must keep pace with knowledge revolution (*TEDx* 8/8/17)
'It's a tough transition': why universities must plan for generation alpha (*Guardian* 10/8/17)
Universities must not be exporter of talents (*The Hindu* 13/8/17)
To be truly inclusive, universities must help prisoners feel they belong (*Guardian* 16/8/17)
Not good enough: universities 'must pay more to clear up after unruly students' (*Daily Echo* 16/8/17)
Universities must move from teaching to transforming (*University World News* 18/8/17)
Employers and universities must join forces to prepare graduates for work (*Yorkshire Post* 22/8/17)
Colleges Must Change To Educate Generation iPhone (*Forbes* 24/8/17)
Universities must produce postgraduates the new world needs (*Financial Review* 30/8/17)
Why Universities Must Continue to Fight for DACA and Their Undocumented Students (*BuzzFlash* 31/8/17)
Bath universities must take more responsibility says Bath councillor (*Bath Chronicle* 1/9/17)
Universities must act now on sustainability goals (*The Conversation* 4/9/17)
Universities must react to falling enrolment (*The Daily Campus* 6/9/17)
Universities must embrace accountability or lose public confidence (*Daily Telegraph* 7/9/17)
Universities must be more digitally savvy to stay competitive (*Education Technology* 10/9/17)
Universities must be government accredited (*Tribune*, Pakistan 15/9/17)
Scottish universities must investigate race divide (*The Herald* 25/9/17)
Universities must welcome conservative speakers (*The Cavalier Daily* 20/9/17)
Universities must do more to stop the graduate brain drain (*Guardian* 25/9/17)
Universities must do better to promote and protect free speech (*Varsity* 2/10/17)
Universities must adapt to the future (*Mail & Guardian*, South Africa 6/10/17)
Diversity crisis in higher education. Universities must make change a priority (*Times Higher Educational Supplement* 7/10/17)

Avila University is right: Colleges must cut tuition costs or face extinction (*Kansas City Star* 8/10/17)
Universities must prioritise students over their research interests, new regulator says (*Telegraph* 9/10/17)
Universities must rediscover the passion for knowledge (*spiked-online.com* 9/10/17)
Universities must do more to support students' mental health (*Jewish News Online* 10/10/17)
Why universities must adapt to always-on students and support BYOD policies (*Tech Republic* 11/10/17)
How Australian universities must change in a digital world (*Acuity Magazine* 11/10/17)
Universities must prioritise students over their research interests, new regulator says (*Qualifi* 11/10/17)
Universities must adopt best practices in hiring academic staff (*Standard Digital*, Kenya 14/10/17)
India's Best Universities Must Discard the Practice of Academic Inbreeding (*The Wire* 15/10/17)
The Asian century is gaining momentum: universities must prepare (*Guardian* 18/10/17)
Universities must do better to promote and protect free speech (*PoliticsHome* 19/10/17)
Universities must allow anti-transgender speakers, minister demands (*Pink News* 19/10/17)
Universities must protect free speech and debate or face fines (*Breitbart* 19/10/17)
Why universities must innovate in global engagement (*World University News* 21/10/17)
Universities Must Challenge Richard Spencer's 'Right' to Incite a Race War in America (*Haaretz*, Israel 22/10/17)

We thank with brief thanksgiving
Whatever gods may be
That no life lives for ever;
That dead men rise up never;
That even the weariest river
Winds somewhere safe to sea.
(from 'The Garden of Proserpine' by Algernon Charles Swinburne)

References

1. 'The Fourah Bay College, Sierra Leone', *Durham University Journal* **I** (1876), no. 2, 6–7. https://palimpsest.dur.ac.uk/slp/dujournal.html.
2. 'Review: Calculus made easy', *Nature* **86** (1911), no. 2158, 41. https://doi.org/10.1038/086041c0.
3. *The Universal Encyclopedia of Mathematics* (Pan Books Ltd, London, 1976). With a Foreword by James A. Newman.
4. 'Anatole Katok', *Mosc. Math. J.* **18** (2018), no. 3, 599–600. https://doi.org/10.17323/1609-4514-2017-18-3-599-600.
5. E. A. Abbott, *Flatland: A Romance of Many Dimensions*, (Seeley & Company, London, 1884).
6. L. M. Abramov and V. A. Rohlin, 'Entropy of a skew product of mappings with invariant measure', *Vestnik Leningrad. Univ.* **17** (1962), no. 7, 5–13.
7. A. Ahmed, 'The Asquith Tradition, the Ashby Reform, and the Development of Higher Education in Nigeria', *Minerva* **27** (1989), no. 1, 1–20. https://www.jstor.org/stable/41820754.
8. M. Aka, M. Einsiedler, and T. Ward, *A Journey Through the Realm of Numbers—from Quadratic Equations to Quadratic Reciprocity*, in *Springer Undergraduate Mathematics Series* (Springer, Cham, 2020). https://doi.org/10.1007/978-3-030-55233-6.
9. R. Anderson, 'University fees in historical perspective', *History & Policy* (8 February 2016).
10. M. Baake and T. Ward, 'Planar dynamical systems with pure Lebesgue diffraction spectrum', *J. Stat. Phys.* **140** (2010), no. 1, 90–102. https://doi.org/10.1007/s10955-010-9984-x.

T. Ward, *People, Places, and Mathematics*, Springer Biographies,
https://doi.org/10.1007/978-3-031-39074-6

11. M. Baake and N. Neumärker, 'A note on the relation between fixed point and orbit count sequences', *J. Integer Seq.* **12** (2009), no. 4, Article 09.4.4, 12. https://doi.org/10.1007/bf03024025.
12. S. Baggott and M. Dick, *Matthew Boulton: Enterprising Industrialist of the Enlightenment (Science, Technology and Culture, 1700–1945)* (Routledge, 2013).
13. S. Baier, S. Jaidee, S. Stevens, and T. Ward, 'Automorphisms with exotic orbit growth', *Acta Arith.* **158** (2013), no. 2, 173–197. https://doi.org/10.4064/aa158-2-5.
14. R. Barnett, *Person of no nationality: A story of childhood loss and recovery* (David Paul, London, 2010).
15. Y. Bauman and E. Rose, 'Why are Economics Students More Selfish than the Rest?', *IZA Discussion Paper No. 4625*. https://doi.org/10.2139/ssrn.1522693.
16. R. F. Baumeister, 'Ego Depletion and Self-Control Failure: An Energy Model of the Self's Executive Function', *Self and Identity* **1** (2002), no. 2, 129–136. doi.org/10.1080/152988602317319302.
17. P. R. Baxandall and H. Liebeck, *Differential Vector Calculus*, in *Longman Mathematical Texts* (Longman, London-New York, 1981).
18. E. T. Bell, *Men of Mathematics* (Victor Gollancz, London, 1939).
19. J. Bell, R. Miles, and T. Ward, 'Towards a Pólya-Carlson dichotomy for algebraic dynamics', *Indag. Math. (N.S.)* **25** (2014), no. 4, 652–668. https://doi.org/10.1016/j.indag.2014.04.005.
20. M. Beloff, *The Plateglass Universities* (Secker & Warburg, 1968).
21. V. Bergelson, 'Weakly mixing PET', *Ergodic Theory Dynam. Systems* **7** (1987), no. 3, 337–349. https://doi.org/10.1017/S0143385700004090.
22. I. Berlin, *The Hedgehog and the Fox: An Essay on Tolstoy's View of History* (Wiedenfeld & Nicolson, London, 1953).
23. S. Bhattacharya and T. Ward, 'Finite entropy characterizes topological rigidity on connected groups', *Ergodic Theory Dynam. Systems* **25** (2005), no. 2, 365–373. https://doi.org/10.1017/S0143385704000501.
24. S. Bhattacharya, 'Isomorphism rigidity of commuting automorphisms', *Trans. Amer. Math. Soc.* **360** (2008), no. 12, 6319–6329. https://doi.org/10.1090/S0002-9947-08-04597-2.
25. H. Bhorat, N. Kachingwe, M. Oosthuizen, and D. Yu, 'Understanding growth–income inequality interactions in Zambia', in *Inequality in Zambia*, Ed. C. Cheelo, M. Hinfelaar, and M. Ndulo, pp. 120–178 (Routledge Contemporary Africa Series, Routledge, London, 2022). https://doi.org/10.4324/9781003241027-9.
26. T. Blair, *A Journey* (Hutchinson, London, 2010).
27. J. Blanden, P. Gregg, and S. Machin, *Intergenerational Mobility in Europe and North America: A Report Supported by the Sutton Trust* (Centre for Economic Performance, 2005). https://www.suttontrust.com.
28. M. Bond, *From Northern Rhodesia to Zambia: Recollections of a DO/DC 1962–73* (Gadsden Publishers, Lusaka, 2014).

29. A. Boring, K. Ottoboni and P. B. Stark, 'Student evaluations of teaching (mostly) do not measure teaching effectiveness' *ScienceOpen Research* (2016), 12pp. https://doi.org/10.14293/S2199-1006.1.SOR-EDU.AETBZC.v1
30. R. Bowen and O. E. Lanford, III, 'Zeta functions of restrictions of the shift transformation', in *Global Analysis (Proc. Sympos. Pure Math., Vol. XIV, Berkeley, Calif., 1968)* (Amer. Math. Soc., Providence, R.I., 1970), 43–49.
31. D. W. Boyd, 'Kronecker's theorem and Lehmer's problem for polynomials in several variables', *J. Number Theory* **13** (1981), no. 1, 116–121. https://doi.org/10.1016/0022-314X(81)90033-0.
32. D. W. Boyd, 'Mahler's measure and special values of L-functions', *Experiment. Math.* **7** (1998), no. 1, 37–82. http://projecteuclid.org/euclid.em/1047674271.
33. M. Boyle and D. Lind, 'Expansive subdynamics', *Trans. Amer. Math. Soc.* **349** (1997), no. 1, 55–102. https://doi.org/10.1090/S0002-9947-97-01634-6.
34. M. Boyle and B. Weiss, 'Remembering Dan Rudolph', *Ergodic Theory Dynam. Systems* **32** (2012), no. 2, 319–322. https://doi.org/10.1017/S014338571100112X.
35. K. Bradley, *Once a District Officer* (Macmillan, London, 1966).
36. K. Bradley, *Lusaka: The New Capital of Northern Rhodesia*, in *Studies in International Planning History* (Routledge, London, 2013). With an introduction by Robert Home.
37. E. Brink, G. Malungane, S. Lebelo, D. Ntshangase, and S. Krige, *Soweto, 16 June 1976: Personal Accounts of the Uprising* (Kwela Books, Cape Town, 2001).
38. W. J. Broad, 'Rubber Bible Turns 60', *Science* **204** (1979), 1181. https://www.science.org/doi/10.1126/science.204.4398.1181.
39. N. D. Burke and I. F. Putnam, 'Markov partitions and homology for n/m-solenoids', *Ergodic Theory Dynam. Systems* **37** (2017), no. 3, 716–738. https://doi.org/10.1017/etds.2015.71.
40. R. A. Butler, J. Stuart, and D. Eccles, *Technical Education* (Her Majesty's Stationery Office, London, 1956). file-store.nationalarchives.gov.uk/pdfs/small/cab-129-79-cp-56-40-40.pdf.
41. K. Buzzard, 'Proving theorems with computers', *Notices Amer. Math. Soc.* **67** (2020), no. 11, 1791–1799. doi.org/10.1090/noti.
42. J. Byszewski, G. Graff, and T. Ward, 'Dold sequences, periodic points, and dynamics', *Bull. Lond. Math. Soc.* **53** (2021), no. 5, 1263–1298. https://doi.org/10.1112/blms.12531.
43. J. Byszewski and G. Cornelissen, 'Dynamics on abelian varieties in positive characteristic', *Algebra Number Theory* **12** (2018), no. 9, 2185–2235. https://doi.org/10.2140/ant.2018.12.2185. With an appendix by Robert Royals and Thomas Ward.
44. É. Cammaerts, *The Laughing Prophet: The Seven Virtues and G.K. Chesterton* (Methuen & Company, London, 1937).
45. A. Carnevale and C. Voll, 'Orbit Dirichlet series and multiset permutations', *Monatsh. Math.* **186** (2018), no. 2, 215–233. https://doi.org/10.1007/s00605-017-1128-9.

46. J. Chaplin, 'Africa Christian Press', *The Journal of Modern African Studies* **6** (1968), no. 1, 108–109. https://doi.org/10.1017/S0022278X00016748.
47. E. Chargaff, review of *The Double Helix* by James D. Watson (Atheneum, New York, 1968) in *Science,* **159** (1968), issue 3822, 1448–1449. https://www.science.org/doi/10.1126/science.159.3822.1448.
48. C. Chevalley, 'La théorie du corps de classes', *Ann. of Math. (2)* **41** (1940), 394–418. https://doi.org/10.2307/1969013.
49. D. Chillingworth (ed.), *Proceedings of the Symposium on Differential Equations and Dynamical Systems,* in *Lecture Notes in Mathematics, Vol. 206* (Springer-Verlag, Berlin-New York, 1971). Held at the University of Warwick, Coventry, September 1968-August 1969. Summer School, July 15–25, 1969.
50. V. Chothi, G. Everest, and T. Ward, 'S-integer dynamical systems: periodic points', *J. Reine Angew. Math.* **489** (1997), 99–132. https://doi.org/10.1515/crll.1997.489.99.
51. V. Chothi, *Periodic points in S-integer dynamical systems* (Ph.D. thesis, University of East Anglia, 1996).
52. V. Chothi, G. Everest, and T. Ward, 'Oriented local entropies for expansive actions by commuting automorphisms', *Israel J. Math.* **93** (1996), 281–301. https://doi.org/10.1007/BF02761107.
53. D. V. Chudnovsky and G. V. Chudnovsky, 'Sequences of numbers generated by addition in formal groups and new primality and factorization tests', *Adv. in Appl. Math.* 7 (1986), no. 4, 385–434. https://doi.org/10.1016/0196-8858(86)90023-0.
54. A. Clark, *Diaries: In Power 1983–1992* (Weidenfeld & Nicolson, London, 1993).
55. A. C. Clarke, *Rendezvous with Rama* (Gollancz, London, 1973).
56. E. Coen and J. Coen, *Burn After Reading,* (Working Title Films, 2008).
57. C. Connolly, *Enemies of Promise* (George Routledge & Sons, London, 1938).
58. H. J. Coolidge and R. H. Lord, *Archibald Cary Coolidge: Life and Letters* (Kessinger Publishing, Montana, 2010). Facsimile reprint of the 1932 original.
59. L. Cottrell, *The Bull of Minos* (Pan Books Ltd., London, 1955).
60. P. Cowley, D. Denver, J. Fisher, and C. Pattie (eds.), *The General Election of 1997,* in *British Elections and Parties Review* **8** (Frank Cass & Co. Ltd., London, 1998).
61. A. Crosland, *A plan for Polytechnics and other Colleges* (Her Majesty's Stationery Office, London, 1966). filestore.nationalarchives.gov.uk/pdfs/small/cab-129-125-c-70.pdf.
62. R. F. Curtain and D. Hinrichsen, 'Obituary to A. J. Pritchard', *Internat. J. Control* **81** (2008), no. 4, 537–545. https://doi.org/10.1080/00207170801927791.
63. R. F. Curtain (ed.), *Stability of stochastic dynamical systems,* in *Lecture Notes in Mathematics, Vol. 294* (Springer-Verlag, Berlin-New York, 1972).
64. D. Cutts and A. Russell, 'From coalition to catastrophe: the electoral meltdown of the Liberal Democrats', *Parliamentary Affairs* **68** (2015), no. 1, 70–87. https://doi.org/10.1093/pa/gsv028.

65. P. D'Ambros, G. Everest, R. Miles, and T. Ward, 'Dynamical systems arising from elliptic curves', *Colloq. Math.* **84/85** (2000), no. part 1, 95–107. https://doi.org/10.4064/cm-84/85-1-95-107. Dedicated to the memory of Anzelm Iwanik.
66. A. de Saint-Exupéry, *Letter to a Hostage* (Pushkin Press, London, 2013).
67. S. R. Dearing, *National Committee of Inquiry into Higher Education* (Her Majesty's Stationery Office, London, 1997). www.educationengland.org.uk/documents/dearing1997/dearing1997.html.
68. C. Deninger, 'Mahler measures and Fuglede–Kadison determinants', *Münster J. Math.* **2** (2009), 45–63.
69. F. Dyson, 'Birds and frogs', *Notices Amer. Math. Soc.* **56** (2009), no. 2, 212–223. https://www.ams.org/notices/200902/rtx090200212p.pdf.
70. G. Eddington, J. Judd, R. H. Mole, and A. Ward, 'The acute lethal effects in monkeys of radiostrontium', *J. Pathol. Bacteriol.* **71** (1956), 277–293. https://doi.org/10.1002/path.1700710203.
71. G. Eddington, J. Judd, and A. Ward, 'Toxicity of Radiostrontium in monkeys', *Nature* **172** (1953), 122–123. https://www.nature.com/articles/172122b0.
72. G. Eddington, J. Judd, and A. Ward, 'Delayed Toxicity of Radiostrontium in monkeys', *Nature* **175** (1955), 33. https://www.nature.com/articles/175033a0.
73. D. Edgar, *Maydays* (Methuen, London, 1983).
74. M. Einsiedler, G. Everest, and T. Ward, 'Primes in sequences associated to polynomials (after Lehmer)', *LMS J. Comput. Math.* **3** (2000), 125–139. https://doi.org/10.1112/S1461157000000255.
75. M. Einsiedler, G. Everest, and T. Ward, 'Entropy and the canonical height', *J. Number Theory* **91** (2001), no. 2, 256–273. https://doi.org/10.1006/jnth.2001.2682.
76. M. Einsiedler, G. Everest, and T. Ward, 'Primes in elliptic divisibility sequences', *LMS J. Comput. Math.* **4** (2001), 1–13. https://doi.org/10.1112/S1461157000000772.
77. M. Einsiedler, G. Everest, and T. Ward, 'Morphic heights and periodic points', in *Number theory (New York, 2003)*, pp. 167–177 (Springer, New York, 2004).
78. M. Einsiedler, G. Everest, and T. Ward, 'Periodic points for good reduction maps on curves', *Geom. Dedicata* **106** (2004), 29–41. https://doi.org/10.1023/B:GEOM.0000033838.15992.72.
79. M. Einsiedler, D. Lind, R. Miles, and T. Ward, 'Expansive subdynamics for algebraic $\mathbb{Z}^d$-actions', *Ergodic Theory Dynam. Systems* **21** (2001), no. 6, 1695–1729. https://doi.org/10.1017/S014338570100181X.
80. M. Einsiedler, E. Lindenstrauss, and T. Ward, *Entropy in ergodic theory and topological dynamics* (to appear). https://tbward0.wixsite.com/books.
81. M. Einsiedler and T. Ward, 'Fitting ideals for finitely presented algebraic dynamical systems', *Aequationes Math.* **60** (2000), no. 1–2, 57–71. https://doi.org/10.1007/s000100050135.

82. M. Einsiedler and T. Ward, 'Asymptotic geometry of non-mixing sequences', *Ergodic Theory Dynam. Systems* **23** (2003), no. 1, 75–85. https://doi.org/10.1017/S0143385702000950.
83. M. Einsiedler and T. Ward, 'Entropy geometry and disjointness for zero-dimensional algebraic actions', *J. Reine Angew. Math.* **584** (2005), 195–214. https://doi.org/10.1515/crll.2005.2005.584.195.
84. M. Einsiedler and T. Ward, 'Isomorphism rigidity in entropy rank two', *Israel J. Math.* **147** (2005), 269–284. https://doi.org/10.1007/BF02785368.
85. M. Einsiedler and T. Ward, *Ergodic theory with a view towards number theory*, in *Graduate Texts in Mathematics* **259** (Springer-Verlag London, Ltd., London, 2011). https://doi.org/10.1007/978-0-85729-021-2.
86. M. Einsiedler and T. Ward, 'Homogeneous dynamics: a study guide', in *Introduction to modern mathematics*, in *Adv. Lect. Math. (ALM)* **33**, pp. 171–201 (Int. Press, Somerville, MA, 2015). https://academic.hep.com.cn/btu/EN/chapter/978-7-04-042141-5/chapter06.
87. M. Einsiedler and T. Ward, 'Diophantine problems and homogeneous dynamics', in *Dynamics and analytic number theory*, in *London Math. Soc. Lecture Note Ser.* **437**, pp. 258–288 (Cambridge Univ. Press, Cambridge, 2016). https://doi.org/10.1017/9781316402696.006.
88. M. Einsiedler and T. Ward, *Functional analysis, spectral theory, and applications*, in *Graduate Texts in Mathematics* **276** (Springer, Cham, 2017). https://doi.org/10.1007/978-3-319-58540-6.
89. M. Einsiedler and T. Ward, 'Homogeneous dynamics and its connection to Diophantine approximation', (2022). Springer Lecture Notes, to appear
90. M. Einsiedler and T. Ward, *Homogeneous dynamics and applications* (to appear). https://tbward0.wixsite.com/books.
91. M. Einsiedler and T. Ward, *Unitary representations and unitary duals* (to appear). https://tbward0.wixsite.com/books.
92. M. Einsiedler, 'A generalisation of Mahler measure and its application in algebraic dynamical systems', *Acta Arith.* **88** (1999), no. 1, 15–29. https://doi.org/10.4064/aa-88-1-15-29.
93. M. Einsiedler, *Problems in higher dimensional dynamics* (Ph.D. thesis, Universität Wien, 1999).
94. M. Einsiedler, M. Kapranov, and D. Lind, 'Non-Archimedean amoebas and tropical varieties', *J. Reine Angew. Math.* **601** (2006), 139–157. https://doi.org/10.1515/CRELLE.2006.097.
95. M. Einsiedler, A. Katok, and E. Lindenstrauss, 'Invariant measures and the set of exceptions to Littlewood's conjecture', *Ann. of Math. (2)* **164** (2006), no. 2, 513–560. https://doi.org/10.4007/annals.2006.164.513.
96. A. Einstein, 'Die Ursache der Mäanderbildung der Flußläufe und des sogenannten Baerschen Gesetzes', *Die Naturwissenschaften* **14** (1926), no. 11, 223–224. https://doi.org/10.1007/BF01510300.

97. C. P. Emudong, 'The Gold Coast Nationalist Reaction to the Controversy over Higher Education in Anglophone West Africa and its Impact on Decision Making in the Colonial Office, 1945–47', *The Journal of Negro Education* **66** (1997), no. 2, 137–146. https://doi.org/10.2307/2967223.
98. J. W. England and R. L. Smith, 'The zeta function of automorphisms of solenoid groups', *J. Math. Anal. Appl.* **39** (1972), 112–121. https://doi.org/10.1016/0022-247X(72)90228-4.
99. D. Epstein and M. Pollicott, 'Bill Parry obituary', *The Guardian* (13 October 2006). www.theguardian.com/news/2006/oct/13/guardianobituaries.obituaries.
100. L. Euler, 'Theorematum quorundam ad numeros primos spectantium demonstratio', *Commentarii academiae scientiarum Petropolitanae* **8** (1741), 141–146. scholarlycommons.pacific.edu/euler-works/54. (presented on August 2, 1736).
101. L. Euler, 'Theoremata arithmetica nova methodo demonstrata', *Novi Commentarii academiae scientiarum Petropolitanae* **8** (1763), 74–104. scholarlycommons.pacific.edu/euler-works/271. (presented on June 8, 1758).
102. G. Everest and B. N. Fhlathúin, 'The elliptic Mahler measure', *Math. Proc. Cambridge Philos. Soc.* **120** (1996), no. 1, 13–25. https://doi.org/10.1017/S0305004100074624.
103. G. Everest, G. Mclaren, and T. Ward, 'Primitive divisors of elliptic divisibility sequences', *J. Number Theory* **118** (2006), no. 1, 71–89. https://doi.org/10.1016/j.jnt.2005.08.002.
104. G. Everest, R. Miles, S. Stevens, and T. Ward, 'Orbit-counting in non-hyperbolic dynamical systems', *J. Reine Angew. Math.* **608** (2007), 155–182. https://doi.org/10.1515/CRELLE.2007.056.
105. G. Everest, R. Miles, S. Stevens, and T. Ward, 'Dirichlet series for finite combinatorial rank dynamics', *Trans. Amer. Math. Soc.* **362** (2010), no. 1, 199–227. https://doi.org/10.1090/S0002-9947-09-04962-9.
106. G. Everest, P. Rogers, and T. Ward, 'A higher-rank Mersenne problem', in *Algorithmic number theory (Sydney, 2002)*, in *Lecture Notes in Comput. Sci.* **2369**, pp. 95–107 (Springer, Berlin, 2002). https://doi.org/10.1007/3-540-45455-1_8.
107. G. Everest, C. Röttger, and T. Ward, 'The continuing story of zeta', *Math. Intelligencer* **31** (2009), no. 3, 13–17. https://doi.org/10.1007/s00283-009-9053-y.
108. G. Everest, V. Stangoe, and T. Ward, 'Orbit counting with an isometric direction', in *Algebraic and topological dynamics*, in *Contemp. Math.* **385**, pp. 293–302 (Amer. Math. Soc., Providence, RI, 2005). https://doi.org/10.1090/conm/385/07202.
109. G. Everest, S. Stevens, D. Tamsett, and T. Ward, 'Primes generated by recurrence sequences', *Amer. Math. Monthly* **114** (2007), no. 5, 417–431. https://doi.org/10.1080/00029890.2007.11920430.
110. G. Everest, A. van der Poorten, I. Shparlinski, and T. Ward, *Recurrence Sequences*, in *Mathematical Surveys and Monographs* **104** (American Mathematical Society, Providence, RI, 2003). https://doi.org/10.1090/surv/104.

111. G. Everest, A. J. van der Poorten, Y. Puri, and T. Ward, 'Integer sequences and periodic points', *J. Integer Seq.* **5** (2002), no. 2, Article 02.2.3, 10.
112. G. Everest and T. Ward, 'A dynamical interpretation of the global canonical height on an elliptic curve', *Experiment. Math.* **7** (1998), no. 4, 305–316. http://projecteuclid.org/euclid.em/1047674148.
113. G. Everest and T. Ward, *Heights of Polynomials and Entropy in Algebraic Dynamics*, in *Universitext* (Springer-Verlag London, Ltd., London, 1999). https://doi.org/10.1007/978-1-4471-3898-3.
114. G. Everest and T. Ward, 'The canonical height of an algebraic point on an elliptic curve', *New York J. Math.* **6** (2000), 331–342. http://nyjm.albany.edu:8000/j/2000/6_331.html.
115. G. Everest and T. Ward, 'Primes in divisibility sequences', *Cubo Mat. Educ.* **3** (2001), no. 2, 245–259.
116. G. Everest and T. Ward, *An Introduction to Number Theory*, in *Graduate Texts in Mathematics* **232** (Springer-Verlag London, Ltd., London, 2005). https://doi.org/10.1007/b137432.
117. G. Everest and T. Ward, 'A repulsion motif in Diophantine equations', *Amer. Math. Monthly* **118** (2011), no. 7, 584–598. https://doi.org/10.4169/amer.math.monthly.118.07.584.
118. G. Everest, P. Ingram, V. Mahé, and S. Stevens, 'The uniform primality conjecture for elliptic curves', *Acta Arith.* **134** (2008), no. 2, 157–181. https://doi.org/10.4064/aa134-2-7.
119. G. Everest and H. King, 'Prime powers in elliptic divisibility sequences', *Math. Comp.* **74** (2005), no. 252, 2061–2071. https://doi.org/10.1090/S0025-5718-05-01737-0.
120. G. Everest, V. Miller, and N. Stephens, 'Primes generated by elliptic curves', *Proc. Amer. Math. Soc.* **132** (2004), no. 4, 955–963. https://doi.org/10.1090/S0002-9939-03-07311-8.
121. G. Everest, J. Reynolds, and S. Stevens, 'On the denominators of rational points on elliptic curves', *Bull. Lond. Math. Soc.* **39** (2007), no. 5, 762–770. https://doi.org/10.1112/blms/bdm061.
122. J.-H. Evertse, H. P. Schlickewei, and W. M. Schmidt, 'Linear equations in variables which lie in a multiplicative group', *Ann. of Math. (2)* **155** (2002), no. 3, 807–836. https://doi.org/10.2307/3062133.
123. A. Eyles, L. Elliot Major, and S. Machin, *Social Mobility – Past, Present and Future: The state of play in social mobility, on the 25th anniversary of the Sutton Trust* (Centre for Economic Performance, 2022). https://www.suttontrust.com.
124. M. Farrar, *The Struggle for Community in a British Multi-ethnic Inner-city Area: Paradise in the Making* (Edwin Mellen Press Ltd., London, 2002).
125. T. Feder, 'Joanne Cohn and the email list that led to arXiv', *Physics Today* (2021), no. 4. https://doi.org/10.1063/2FPT.6.4.20211108a.
126. N. Ferguson, *Empire: How Britain Made the Modern World* (Penguin Books, London, 2018).

127. A. Flatters and T. Ward, 'A polynomial Zsigmondy theorem', *J. Algebra* **343** (2011), 138–142. https://doi.org/10.1016/j.jalgebra.2011.07.010.
128. A. Flatters, 'Primitive divisors of some Lehmer–Pierce sequences', *J. Number Theory* **129** (2009), no. 1, 209–219. https://doi.org/10.1016/j.jnt.2008.05.008.
129. A. Flatters, *Arithmetic properties of recurrence sequences* (Ph.D. thesis, University of East Anglia, 2010).
130. H. W. French, *Born in Blackness: Africa, Africans, and the Making of the Modern World, 1471 to the Second World War* (Liveright Publishing Corporation, 2021).
131. J. Gaither, G. Louchard, S. Wagner, and M. D. Ward, 'Resolution of T. Ward's question and the Israel-Finch conjecture: precise analysis of an integer sequence arising in dynamics', *Combin. Probab. Comput.* **24** (2015), no. 1, 195–215. https://doi.org/10.1017/S0963548314000455.
132. J. S. Galbraith, *Crown and Charter: The early Years of the British South Africa Company* (University of California Press, Berkeley, 1974).
133. G. Gamow, *One Two Three...Infinity: Facts and Speculations of Science* (Viking Press, 1947).
134. G. Gamow, *Mr Tompkins in Paperback* (Cambridge University Press, 1953).
135. P. Gerlach, 'The games economists play: Why economics students behave more selfishly than other students', *PLoS One* **12(9)** (2017). https://doi.org/10.1371/journal.pone.0183814.
136. A. Ghosh and R. Royals, 'An extension of the Khinchin–Groshev theorem', *Acta Arith.* **167** (2015), no. 1, 1–17. https://doi.org/10.4064/aa167-1-1.
137. S. Gorard, 'A Re-Consideration of Rates of 'Social Mobility' in Britain: Or Why Research Impact Is Not Always a Good Thing', *British Journal of Sociology of Education* **29** (2008), no. 3, 317–324.
138. J. Griffiths, *Denominator* (Lulu Press, 2018).
139. R. Grinter, 'Andrew Thomson obituary', *The Guardian* (4 May 2021). www.theguardian.com/science/2021/may/04/andrew-thomson-obituary.
140. J. Grove, 'Thatcher had 'immense impact' on higher education', *Times Higher Education* (8th April 2013).
141. R. M. Hain and S. Zucker, 'Unipotent variations of mixed Hodge structure', *Invent. Math.* **88** (1987), no. 1, 83–124. https://doi.org/10.1007/BF01405093.
142. U. Haley and A. Jack, *Measuring societal impact in business and management research: From challenges to change* (2023). http://dx.doi.org/10.2139/ssrn.4306589.
143. G. H. Hardy, *A Mathematician's Apology* (Cambridge University Press, 1940).
144. E. M. Hartley, *Cartesian Geometry of the Plane* (Cambridge University Press, 1960).
145. E. M. Hartley, *A finite collineation group in five dimensions, its maximal and other subgroups and its simplest invariant primal* (Ph.D. thesis, University of Cambridge, 1951).
146. B. Hasselblatt, 'Anatole Katok', *Ergodic Theory Dynam. Systems* **42** (2022), no. 2, 321–388. https://doi.org/10.1017/etds.2021.72.

147. T. Hatton, *Phoenix Rising: A Memoir of Waterford Kamhlaba's Early Years* (Kamhlaba Publishing, Mbabane, 2013).
148. T. Heffernan and P. Harpur, 'Discrimination against academics and career implications of student evaluations: university policy versus legal compliance', *Assessment & Evaluation in Higher Education* (2023), 12pp. https://www.tandfonline.com/doi/full/10.1080/02602938.2023.2225806.
149. C. F. Higgins, 'ABC transporters: physiology, structure and mechanism–an overview', *Research in Microbiology* **152** (2001), no. 3–4, 205–10. doi.org/10.1016/s0923-2508(01)01193-7.
150. N. Hillman, *How higher education changed during the Queen's reign – and why we should now consider establishing a new university in her name* (Higher Education Policy Institute, 2022). https://www.hepi.ac.uk/2022/09/14.
151. N. Hoad, 'Cosmetic Surgeons of the Social: Darwin, Freud, and Wells and the Limits of Sympathy on The Island of Dr. Moreau', in *Compassion: The Culture and Politics of an Emotion*, Ed. L. Berlant, pp. 187–218 (Routledge, New York, 2004).
152. L. Hogben, *Mathematics for the Million* (W. W. Norton & Company, New York, 1937).
153. B. Host and B. Kra, 'Nonconventional ergodic averages and nilmanifolds', *Ann. Math. (2)* **161** (2005), no. 1, 397–488.
154. D. Hunt, 'Obituary: Alfred Jacobus (Alf) van der Poorten', *Gazette of the Australian Mathematical Society* **38** (2011), no. 1, 33–36.
155. Z. N. Hurston, *Barracoon: The Story of the Last "Black Cargo"* (Amistad, Harper Collins Publishers, 2018).
156. A. Jackson, 'A labor of love: the Mathematics Genealogy Project', *Notices Amer. Math. Soc.* **54** (2007), no. 8, 1002–1003. https://www.ams.org/notices/200708/tx070801002p.pdf.
157. F. Jacob, *The Statue Within* (Basic Books, New York, 1988).
158. S. Jaidee, P. Moss, and T. Ward, 'Time-changes preserving zeta functions', *Proc. Amer. Math. Soc.* **147** (2019), no. 10, 4425–4438. https://doi.org/10.1090/proc/14574.
159. S. Jaidee, S. Stevens, and T. Ward, 'Mertens' theorem for toral automorphisms', *Proc. Amer. Math. Soc.* **139** (2011), no. 5, 1819–1824. https://doi.org/10.1090/S0002-9939-2010-10632-9.
160. S. Jaidee, *Mertens' theorem for arithmetical dynamical systems* (Ph.D. thesis, University of East Anglia, 2010).
161. J. P. Jones, D. Sato, H. Wada, and D. Wiens, 'Diophantine representation of the set of prime numbers', *Amer. Math. Monthly* **83** (1976), no. 6, 449–464. https://doi.org/10.2307/2318339.
162. A. E. Juola, *Grade Inflation in Higher Education: What Can or Should We Do?* (1976). https://eric.ed.gov/?id=ED129917. ERIC Institute of Education Sciences.

163. M. Kennedy and A. Magennis, *Ireland, the United Nations and the Congo: A Military and Diplomatic History, 1960–1* (Four Courts Press, Dublin, 2014).
164. D. Kernohan, 'The Melting of the MOOCs', in *Digital Futures: Expert Briefings on Digital Technologies for Education and Research*, Ed. M. Hall, M. Harrow, and L. Estelle (Chandos Publishing, 2015).
165. C. Kerr, *The Uses of the University* (Harvard University Press, Cambridge MA, 1963).
166. H. King, *Prime appearance in elliptic divisibility sequences* (Ph.D. thesis, University of East Anglia, 2005).
167. B. Kitchens and K. Schmidt, 'Automorphisms of compact groups', *Ergodic Theory Dynam. Systems* **9** (1989), no. 4, 691–735. https://doi.org/10.1017/S0143385700005290.
168. S. Kolyada, M. Möller, P. Moree, and T. Ward (eds.), *Dynamics and numbers*, in *Contemporary Mathematics* **669** (American Mathematical Society, Providence, RI, 2016). https://doi.org/10.1090/conm/669.
169. S. Kolyada, Y. Manin, M. Möller, P. Moree, and T. Ward (eds.), *Dynamical numbers—interplay between dynamical systems and number theory*, in *Contemporary Mathematics* **532** (American Mathematical Society, Providence, RI, 2010). https://doi.org/10.1090/conm/532.
170. S. Kolyada, Y. Manin, and T. Ward (eds.), *Algebraic and topological dynamics*, in *Contemporary Mathematics* **385** (American Mathematical Society, Providence, RI, 2005). https://doi.org/10.1090/conm/385.
171. M. Kontsevich and D. Zagier, 'Periods', in *Mathematics unlimited—2001 and beyond*, pp. 771–808 (Springer, Berlin, 2001).
172. K. Kwarteng, P. Patel, D. Raab, C. Skidmore, and E. Truss, *Britannia unchained: Global lessons for growth and prosperity* (Palgrave Macmillan, London, 2012).
173. Y. Lacroix, 'Natural extensions and mixing for semi-group actions', in *Séminaires de Probabilités de Rennes (1995)*, in *Publ. Inst. Rech. Math. Rennes* **1995**, p. 10 (Univ. Rennes I, Rennes, 1995).
174. M. Lawson, 'The vital return of David Edgar's Maydays – and the best plays about the left', *The Guardian* (8 October 2018).
175. F. Ledrappier, 'Un champ markovien peut être d'entropie nulle et mélangeant', *C. R. Acad. Sci. Paris Sér. A-B* **287** (1978), no. 7, A561–A563.
176. D. H. Lehmer, 'Factorization of certain cyclotomic functions', *Ann. of Math. (2)* **34** (1933), no. 3, 461–479. https://doi.org/10.2307/1968172.
177. W. S. Leslie, *Touchdown to Adventure* (Kingfisher Books, London, 1958).
178. D. Lind, K. Schmidt, and T. Ward, 'Mahler measure and entropy for commuting automorphisms of compact groups', *Invent. Math.* **101** (1990), no. 3, 593–629. https://doi.org/10.1007/BF01231517.
179. D. A. Lind, 'Ergodic group automorphisms are exponentially recurrent', *Israel J. Math.* **41** (1982), no. 4, 313–320. https://doi.org/10.1007/BF02760537.
180. D. A. Lind, 'A zeta function for $\mathbf{Z}^d$-actions', in *Ergodic theory of* $\mathbf{Z}^d$ *actions (Warwick, 1993–1994)*, in *London Math. Soc. Lecture Note Ser.* **228**, pp. 433–

450 (Cambridge Univ. Press, Cambridge, 1996). https://doi.org/10.1017/CBO9780511662812.019.

181. D. A. Lind and T. Ward, 'Automorphisms of solenoids and p-adic entropy', *Ergodic Theory Dynam. Systems* **8** (1988), no. 3, 411–419. https://doi.org/10.1017/S0143385700004545.
182. D. Lind and K. Schmidt, 'Bernoullicity of solenoidal automorphisms and global fields', *Israel J. Math.* **87** (1994), no. 1–3, 33–35. https://doi.org/10.1007/BF02772981.
183. J. E. Littlewood, *A mathematician's miscellany* (Methuen & Co. Ltd., London, 1953).
184. F. Luca and T. Ward, 'An elliptic sequence is not a sampled linear recurrence sequence', *New York J. Math.* **22** (2016), 1319–1338. http://nyjm.albany.edu:8000/j/2016/22_1319.html.
185. F. Luca and T. Ward, 'On (almost) realizable subsequences of linearly recurrent sequences', *J. Integer Seq.* **26** (2023), no. 4, Article 23.4.6. https://cs.uwaterloo.ca/journals/JIS/VOL26/Ward/ward9.html.
186. G. F. Lungu, 'Educational Policy-Making in Colonial Zambia: The Case of Higher Education for Africans from 1924 to 1964', *The Journal of Negro History* **78** (1993), no. 4, 207–232. www.jstor.org/stable/2717416.
187. A. L. Mackay (ed.), *The Harvest of a Quiet Eye: a Selection of Scientific Quotations* (Institute of Physics, 1977).
188. K. Mahler, 'An application of Jensen's formula to polynomials', *Mathematika* **7** (1960), 98–100. https://doi.org/10.1112/S0025579300001637.
189. K. Mahler, 'On some inequalities for polynomials in several variables', *J. London Math. Soc.* **37** (1962), 341–344. doi.org/10.1112/jlms/s1-37.1.341.
190. A. Manning (ed.), *Dynamical systems—Warwick 1974*, in *Lecture Notes in Mathematics, Vol. 468* (Springer-Verlag, Berlin-New York, 1975). Presented to Professor E. C. Zeeman on his fiftieth birthday, 4th February 1975.
191. J. H. Marsh, *Skeleton Coast* (Hodder and Stoughton Ltd., London, 1944).
192. B. Mathiszik, *Zetafunktionen und periodische Orbits in Bittermodellen* (Ph.D. thesis, Martin–Luther–Universität Halle–Wittenberg, 1988).
193. B. Mathiszik, 'Zetafunktionen fur $\mathbf{Z}^d$-Subshifts endlichen Typs', in *Prof. Dr. sc. Horst Michel zum Gedenken (Halle, 1989)*, in *Wissensch. Beitr.* **89**, pp. 68–72 (Martin-Luther-Univ. Halle-Wittenberg, Halle, 1989).
194. J. V. Matijasevič, 'A Diophantine representation of the set of prime numbers', *Dokl. Akad. Nauk SSSR* **196** (1971), 770–773.
195. T. McLeish, *Faith and Wisdom in Science* (Oxford University Press, 2014).
196. T. McLeish, *The Poetry and Music of Science: Comparing Creativity in Science and Art* (Oxford University Press, 2019).
197. A. Mee, *The Children's Encyclopædia* (The Educational Book Company, London, 1910).
198. A. Mee, *The Children's Encyclopedia* (Waverley Book Company, 28 ed., 1960).

199. N. D. Mermin, *Boojums all the way through: communicating science in a prosaic age* (Cambridge University Press, 1990).
200. F. Mertens, 'Über eine zahlentheoretische funktion', *Sitzungsberichte der Kaiserlichen Akademie der Wissenschaften* **106** (1897), 761–830.
201. J. Meyer and R. Land, 'Threshold concepts and troublesome knowledge: Linkages to ways of thinking and practising within the disciplines', in *ISL10 Improving Student Learning* (Oxford Brookes University, United Kingdom, Jan. 2003), 412–424.
202. T. Meyerovitch, 'Finite entropy for multidimensional cellular automata', *Ergodic Theory Dynam. Systems* **28** (2008), no. 4, 1243–1260. https://doi.org/10.1017/S0143385707000855.
203. P. Mihăilescu, 'Primary cyclotomic units and a proof of Catalan's conjecture', *J. Reine Angew. Math.* **572** (2004), 167–195. https://doi.org/10.1515/crll.2004.048.
204. R. Miles, M. Staines, and T. Ward, 'Dynamical invariants for group automorphisms', in *Recent trends in ergodic theory and dynamical systems*, in *Contemp. Math.* **631**, pp. 231–258 (Amer. Math. Soc., Providence, RI, 2015). https://doi.org/10.1090/conm/631/12606.
205. R. Miles and T. Ward, 'Mixing actions of the rationals', *Ergodic Theory Dynam. Systems* **26** (2006), no. 6, 1905–1911. https://doi.org/10.1017/S0143385706000356.
206. R. Miles and T. Ward, 'Periodic point data detects subdynamics in entropy rank one', *Ergodic Theory Dynam. Systems* **26** (2006), no. 6, 1913–1930. https://doi.org/10.1017/S014338570600054X.
207. R. Miles and T. Ward, 'Uniform periodic point growth in entropy rank one', *Proc. Amer. Math. Soc.* **136** (2008), no. 1, 359–365. https://doi.org/10.1090/S0002-9939-07-09018-1.
208. R. Miles and T. Ward, 'Orbit-counting for nilpotent group shifts', *Proc. Amer. Math. Soc.* **137** (2009), no. 4, 1499–1507. https://doi.org/10.1090/S0002-9939-08-09649-4.
209. R. Miles and T. Ward, 'A dichotomy in orbit growth for commuting automorphisms', *J. Lond. Math. Soc. (2)* **81** (2010), no. 3, 715–726. https://doi.org/10.1112/jlms/jdq010.
210. R. Miles and T. Ward, 'A directional uniformity of periodic point distribution and mixing', *Discrete Contin. Dyn. Syst.* **30** (2011), no. 4, 1181–1189. https://doi.org/10.3934/dcds.2011.30.1181.
211. R. Miles and T. Ward, 'Directional uniformities, periodic points, and entropy', *Discrete Contin. Dyn. Syst. Ser. B* **20** (2015), no. 10, 3525–3545. https://doi.org/10.3934/dcdsb.2015.20.3525.
212. R. Miles and T. Ward, 'The dynamical zeta function for commuting automorphisms of zero-dimensional groups', *Ergodic Theory Dynam. Systems* **38** (2018), no. 4, 1564–1587. https://doi.org/10.1017/etds.2016.77.

213. R. Miles, *Arithmetic dynamical systems* (Ph.D. thesis, University of East Anglia, 2000).
214. R. Miles, 'Dynamical systems arising from units in Krull rings', *Aequationes Math.* **61** (2001), no. 1–2, 113–127. https://doi.org/10.1007/s000100050164.
215. R. Miles, 'Periodic points of endomorphisms on solenoids and related groups', *Bull. Lond. Math. Soc.* **40** (2008), no. 4, 696–704. https://doi.org/10.1112/blms/bdn052.
216. J. B. Milner-Thornton, *The Long Shadow of the British Empire: The Ongoing Legacies of Race and Class in Zambia* (Palgrave Macmillan, New York, 2012).
217. G. T. Minton, 'Linear recurrence sequences satisfying congruence conditions', *Proc. Amer. Math. Soc.* **142** (2014), no. 7, 2337–2352. https://doi.org/10.1090/S0002-9939-2014-12168-X.
218. P. Miska and T. Ward, 'Stirling numbers and periodic points', *Acta Arith.* **201** (2021), no. 4, 421–435. https://doi.org/10.4064/aa210309-9-8.
219. P. Moree, A. Pohl, L. Snoha, and T. Ward (eds.), *Dynamics: topology and numbers*, in *Contemporary Mathematics* **744** (American Mathematical Society, Providence, RI, 2020). https://doi.org/10.1090/conm/744.
220. G. Morris and T. Ward, 'A note on mixing properties of invertible extensions', *Acta Math. Univ. Comenian. (N.S.)* **66** (1997), no. 2, 307–311.
221. G. Morris and T. Ward, 'Entropy bounds for endomorphisms commuting with K actions', *Israel J. Math.* **106** (1998), 1–11. https://doi.org/10.1007/BF02773458.
222. G. Morris, *Dynamical constraints on group actions* (Ph.D. thesis, University of East Anglia, 1998).
223. P. Moss and T. Ward, 'Fibonacci along even powers is (almost) realizable', *Fibonacci Quart.* **60** (2022), no. 1, 40–47.
224. P. Moss, *The arithmetic of realizable sequences* (Ph.D. thesis, University of East Anglia, 2003).
225. M. J. Mossinghoff, G. Rhin, and Q. Wu, 'Minimal Mahler measures', *Experiment. Math.* **17** (2008), no. 4, 451–458. http://projecteuclid.org/euclid.em/1243429958.
226. H. K. Mutumba, *The History of the Barotse National School, 1907–1934* (Master's thesis, The University of Zambia, 1984).
227. N. Neumärker, 'Realizability of integer sequences as differences of fixed point count sequences', *J. Integer Seq.* **12** (2009), no. 4, Article 09.4.5, 8.
228. H. Newby, *Strength in variety*, *Times Higher Education* (April 18th 2013).
229. M. Nsibandze and C. Green, 'In-service teacher training in Swaziland', *Prospects: Quarterly Review of Education (UNESCO)* **III** (1978), no. 1, 110–116.
230. C. C. O'Brien, *To Katanga and Back: a UN case history* (Simon & Schuster, 1962).
231. A. M. Odlyzko and H. J. J. te Riele, 'Disproof of the Mertens conjecture', *J. Reine Angew. Math.* **357** (1985), 138–160. https://doi.org/10.1515/crll.1985.357.138.

232. OEIS Foundation Inc., *The On-Line Encyclopedia of Integer Sequences* (2022). Published electronically at http://oeis.org.
233. D. S. Ornstein and B. Weiss, 'Entropy and isomorphism theorems for actions of amenable groups', *J. Analyse Math.* **48** (1987), 1–141. https://doi.org/10.1007/BF02790325.
234. G. Orwell, *Books v. Cigarettes* (Tribune, 1946). February 8.
235. G. Osae-Addo and H. Ward, *Newtown Families* (Africa Christian Press, Kumasi, 1967).
236. A. D. Osseo-Asare, *Atomic Junction: Nuclear Power in Africa after Independence* (Cambridge University Press, 2019).
237. A. Pakapongpun and T. Ward, 'Functorial orbit counting', *J. Integer Seq.* **12** (2009), no. 2, Article 09.2.4, 20. https://cs.uwaterloo.ca/journals/JIS/VOL12/Ward/ward17.pdf.
238. A. Pakapongpun and T. Ward, 'Orbits for products of maps', *Thai J. Math.* **12** (2014), no. 1, 33–44. http://thaijmath.in.cmu.ac.th/index.php/thaijmath/article/view/565.
239. A. Pakapongpun, *Functorial orbit counting* (Ph.D. thesis, University of East Anglia, 2010).
240. C. Parikh, *The unreal life of Oscar Zariski* (Academic Press, Inc., Boston, MA, 1991). With a foreword by David Mumford.
241. J. Phillips, T. Ward, and P. Q. Montgomery, 'Variations in post-tonsillectomy haemorrhage rates are scale-invariant', *The Laryngoscope* **118** (2008), no. 6, 1096–1098.
242. J. A. Piliavin and H.-W. Charng, 'Altruism: A review of recent theory and research', *Annual Review of Sociology* **16** (1990), 27–65. http://www.jstor.org/stable/2083262.
243. C. Pincher, *Dangerous To Know: The Autobiography of Harry Chapman Pincher* (Biteback Publishing, 2014).
244. J. Pintz, *An effective disproof of the Mertens conjecture* (Journées arithmétiques, Besançon/France 1985, Astérisque 147/148, 325–333, 1987).
245. C. Pirtle, *Engineering the World: Stories from the First 75 Years of Texas Instruments* (Southern Methodist University Press, 2005).
246. M. Pollicott, R. Sharp, S. Tuncel, and P. Walters, 'The mathematical research of William Parry FRS', *Ergodic Theory Dynam. Systems* **28** (2008), no. 2, 321–337. https://doi.org/10.1017/S0143385708000102.
247. A. Powell, *A Question of Upbringing* (Heinemann, London, first ed., 1951).
248. J. Pratt, *The Polytechnic Experiment 1965–1992* (The Society for Research into Higher Education & Open University Press, 1997). https://eric.ed.gov/?id=ED415724.
249. Y. Puri and T. Ward, 'Arithmetic and growth of periodic orbits', *J. Integer Seq.* **4** (2001), no. 2, Article 01.2.1, 18.
250. Y. Puri and T. Ward, 'A dynamical property unique to the Lucas sequence', *Fibonacci Quart.* **39** (2001), no. 5, 398–402.

251. Y. Puri, *Arithmetic properties of periodic orbits* (Ph.D. thesis, University of East Anglia, 2000).
252. D. A. Rand and L. S. Young (eds.), *Dynamical systems and turbulence, Warwick 1980*, in *Lecture Notes in Mathematics* **898** (Springer-Verlag, Berlin-New York, 1981).
253. J. Reitmen, *Up in the Air* (DreamWorks Pictures, 2009).
254. W. Reyburn, *Gilbert Harding: A Candid Portrayal* (Angus & Robertson, London, 1978).
255. N. Richert, 'Mathematical Reviews celebrates 75 years', *Notices Amer. Math. Soc.* **61** (2014), no. 11, 1355–1356. https://doi.org/10.1090/noti1200.
256. L. C. Robbins, *Higher education: report of the Committee appointed by the Prime Minister under the Chairmanship of Lord Robbins 1961–63* (Her Majesty's Stationery Office, London, 1963). www.educationengland.org.uk/documents/robbins/robbins1963.html.
257. B. Roberts, 'The in-service teacher-education project in Swaziland (1973–77)', in *The Mathematical Education of Primary-School Teachers*, Ed. R. Morris, in *Studies in Mathematics Education* **3**, pp. 205–212 (UNESCO, Paris, 1984).
258. W. A. Rodney, *How Europe Underdeveloped Africa* (Bogle–L'Ouverture Publications, London, 1972).
259. C. G. J. Roettger, 'Periodic points classify a family of Markov shifts', *J. Number Theory* **113** (2005), no. 1, 69–83. https://doi.org/10.1016/j.jnt.2004.11.012.
260. A. Rosenthal, 'Finite uniform generators for ergodic, finite entropy, free actions of amenable groups', *Probab. Theory Related Fields* **77** (1988), no. 2, 147–166. https://doi.org/10.1007/BF00334034.
261. R. Royals, *Arithmetic and dynamical systems* (Ph.D. thesis, University of East Anglia, 2015).
262. D. J. Rudolph and K. Schmidt, 'Almost block independence and Bernoullicity of $\mathbf{Z}^d$-actions by automorphisms of compact abelian groups', *Invent. Math.* **120** (1995), no. 3, 455–488. https://doi.org/10.1007/BF01241139.
263. D. J. Rudolph and B. Weiss, 'Entropy and mixing for amenable group actions', *Ann. of Math. (2)* **151** (2000), no. 3, 1119–1150. https://doi.org/10.2307/121130.
264. A. M. Samoĭlenko, O. M. Sharkovsʹ kiĭ, and M. e. a. Mīsyurevich, 'Sergīĭ Fedorovich Kolyada [obituary]', *Nelīnīĭnī Koliv.* **21** (2018), no. 3, 431–432.
265. O. Scharmer, *Theory U: Leading from the Future as it Emerges* (Society for Organizational Learning, Cambridge Massachusetts, 2007).
266. H. P. Schlickewei, 'An explicit upper bound for the number of solutions of the S-unit equation', *J. Reine Angew. Math.* **406** (1990), 109–120. https://doi.org/10.1515/crll.1990.406.109.
267. H. P. Schlickewei, 'S-unit equations over number fields', *Invent. Math.* **102** (1990), no. 1, 95–107. https://doi.org/10.1007/BF01233421.
268. K. Schmidt and T. Ward, 'Mixing automorphisms of compact groups and a theorem of Schlickewei', *Invent. Math.* **111** (1993), no. 1, 69–76. https://doi.org/10.1007/BF01231280.

269. K. Schmidt, 'Mixing automorphisms of compact groups and a theorem by Kurt Mahler', *Pacific J. Math.* **137** (1989), no. 2, 371–385. projecteuclid.org/euclid.pjm/1102650389.
270. K. Schmidt, *Algebraic ideas in ergodic theory*, in *CBMS Regional Conference Series in Mathematics* **76** (Published for the Conference Board of the Mathematical Sciences, Washington, DC; by the American Mathematical Society, Providence, RI, 1990). https://doi.org/10.1090/cbms/076.
271. K. Schmidt, *Dynamical Systems of Algebraic Origin*, in *Progress in Mathematics* **128** (Birkhäuser Verlag, Basel, 1995).
272. S. L. Segal, *Nine introductions in complex analysis*, in *Notas de Matemática [Mathematical Notes]* **80** (North-Holland Publishing Co., Amsterdam-New York, 1981).
273. M. A. Shereshevsky, 'Expansiveness, entropy and polynomial growth for groups acting on subshifts by automorphisms', *Indag. Math. (N.S.)* **4** (1993), no. 2, 203–210. https://doi.org/10.1016/0019-3577(93)90040-6.
274. J. H. Silverman, *The arithmetic of elliptic curves*, in *Graduate Texts in Mathematics* **106** (Springer-Verlag, New York, 1986). https://doi.org/10.1007/978-1-4757-1920-8.
275. J. H. Silverman and N. Stephens, 'The sign of an elliptic divisibility sequence', *J. Ramanujan Math. Soc.* **21** (2006), no. 1, 1–17.
276. R. A. Slaughter, 'Review of Education 2,000: A Consultative Document on Hypotheses for Education in AD 2,000', *British Journal of Educational Studies* **32** (1984), no. 3, 271–274.
277. C. J. Smyth, 'A Kronecker-type theorem for complex polynomials in several variables', *Canad. Math. Bull.* **24** (1981), no. 4, 447–452. https://doi.org/10.4153/CMB-1981-068-8.
278. C. J. Smyth, 'On measures of polynomials in several variables', *Bull. Austral. Math. Soc.* **23** (1981), no. 1, 49–63. https://doi.org/10.1017/S0004972700006894.
279. C. P. Snow, *The Two Cultures and the Scientific Revolution* (Cambridge University Press, 1959).
280. C. P. Snow, *The Physicists* (House of Stratus, 2000).
281. M. Spivak, *Calculus* (W. A. Benjamin Inc., New York, 1967).
282. M. Spivak, *The Joy of TEX: A Gourmet Guide to Typesetting with the AMS-TEX Macro Package* (American Mathematical Society, USA, 1st ed., 1997).
283. M. Staines, *On the inverse problem for Mahler measure* (Ph.D. thesis, University of East Anglia, 2013).
284. V. Stangoe, *Orbit counting far from hyperbolicity* (Ph.D. thesis, University of East Anglia, 2005).
285. G. Stephenson, *An Introduction to Partial Differential Equations for Science Students* (Longman, 1971).
286. S. Stevens, T. Ward, and S. Zegowitz, 'Halving dynamical systems', in *Dynamics and numbers*, in *Contemp. Math.* **669**, pp. 285–298 (Amer. Math. Soc., Providence, RI, 2016). https://doi.org/10.1090/conm/669/13433.

287. I. Stewart and D. Tall, *The foundations of mathematics* (Oxford, 1977).
288. I. Stewart, *Concepts of modern mathematics* (Penguin Books, 1981).
289. P. T. Strait, *A First Course in Probability and Statistics with Applications* (W. B. Saunders, 2 ed., 1988).
290. D. Sullivan, 'Rufus Bowen (1947–1978)', *Inst. Hautes Études Sci. Publ. Math.* (1979), no. 50, 7–9 (1 plate). http://www.numdam.org/item?id=PMIHES_1979__50__7_0.
291. G. Supran, S. Rahmstorf, and N. Oreskes, 'Assessing ExxonMobil's global warming projections', *Science* **379** (2023). https://www.science.org/doi/10.1126/science.abk0063.
292. J. T. Tate, 'Fourier analysis in number fields, and Hecke's zeta-functions', in *Algebraic Number Theory (Proc. Instructional Conf., Brighton, 1965)*, Ed. J. W. S. Cassels and A. Fröhlich, pp. 305–347 (Thompson, Washington, D.C., 1967).
293. E. P. Thompson, *Warwick University Ltd: Industry, Management & the Universities* (Penguin, 1970).
294. S. P. Thompson, *Calculus Made Easy* (Macmillan & Co., London, 2nd ed., 1914).
295. J. Thomson, 'On the grand currents of atmospheric circulation', *Proceedings of the Royal Society of London* **51** (1892), 42–46. https://www.jstor.org/stable/115088.
296. N. Thrift, *The Pursuit of Possibility: Redesigning Research Universities* (Bristol University Press, 2022).
297. B. Thwaites and C. Wysock-Wright (eds.), *Education 2,000: A consultative document on hypotheses for education in AD 2,000* (Cambridge University Press, 1983).
298. N. Tredell, 'C. P. Snow: The Dynamics of Hope', (2012).
299. S. B. Trend, *Report of the Committee of Enquiry into the Organisation of Civil Science* (Her Majesty's Stationery Office, London, 1965). https://discovery.nationalarchives.gov.uk/details/r/C1976289.
300. J. T. Trowbridge, 'Reminiscences of Walt Whitman', *The Atlantic Monthly* (1902). https://www.theatlantic.com/past/docs/unbound/poetry/whitman/walt.htm.
301. A. J. Tuominen, 'Pygmies, Bushmen, and savage numbers – a case study in a sequence of bad citations', *Historia Mathematica* **62** (2023), 51–72. https://www.sciencedirect.com/science/article/pii/S0315086022000672.
302. J. Uglow, *The Lunar Men: the Friends who made the Future, 1730–1810* (Faber, 2002).
303. P. Walters, *An introduction to ergodic theory*, in *Graduate Texts in Mathematics* **79** (Springer-Verlag, New York-Berlin, 1982).
304. A. H. Ward, 'A crystal-photomultiplier gamma counter for hospital use', *J. Scientific Instruments* **29** (1952), 181–183.
305. A. Ward, 'Retention and excretion of radiostrontium in monkeys', *J. Nuclear Energy* **5** (1957), 192–202. https://doi.org/10.1016/0891-3919(57)90245-0.

306. A. Ward, 'Measurement of Radioactivity in human breath', *Nature* **167** (1959), 600.
307. A. Ward and J. Marr, 'Radioactive fallout in Ghana', *Nature* **187** (1960), 299–300.
308. E. H. Ward, *Senior Physics 1* (Nelson, Sunbury-on-Thames, 1965).
309. E. H. Ward, *Senior Physics 2* (Nelson, Sunbury-on-Thames, 1965).
310. E. H. Ward and A. H. Ward, *Energy, Heat and the Structure of Matter* (Nelson, Sunbury-on-Thames, 1977).
311. E. H. Ward and A. H. Ward, *Essential senior physics in SI units* (Nelson, Sunbury-on-Thames, 1977).
312. E. H. Ward and A. H. Ward, *Light and Sound* (Nelson, Sunbury-on-Thames, 1977).
313. E. H. Ward and A. H. Ward, *Magnetism, Electricity and Atomic Physics* (Nelson, Sunbury-on-Thames, 1977).
314. E. H. Ward and A. H. Ward, *Mechanics* (Nelson, Sunbury-on-Thames, 1977).
315. E. H. Ward, *The Girl's Own Paper Index*. www.lutterworth.com/gop/index.php.
316. E. H. Ward, *Girl's Own Guide: An Index of All the Fiction Stories Ever to Appear in the Girl's Own Paper from 1880–1941* (A.&B. Whitworth, Manchester, 1992).
317. H. Ward, *The Story of Creation* (Denholm House Press, Nutfield, Redhill, Surrey, 1972).
318. K. Ward and H. Ward, *AIDS, sex and family planning—a Christian view* (Africa Christian Press, 1989).
319. T. Ward, *The entropy of automorphisms of solenoidal groups* (Master's thesis, University of Warwick, 1986).
320. T. Ward, *Topological entropy and periodic points for* $\mathbb{Z}^d$ *actions on compact abelian groups with the descending chain condition* (Ph.D. thesis, University of Warwick, 1990).
321. T. Ward, 'Almost block independence for the three dot $\mathbf{Z}^2$ dynamical system', *Israel J. Math.* **76** (1991), no. 1–2, 237–256. https://doi.org/10.1007/BF02782855.
322. T. Ward, 'The Bernoulli property for expansive $\mathbf{Z}^2$ actions on compact groups', *Israel J. Math.* **79** (1992), no. 2–3, 225–249. https://doi.org/10.1007/BF02808217.
323. T. Ward, 'Periodic points for expansive actions of $\mathbf{Z}^d$ on compact abelian groups', *Bull. London Math. Soc.* **24** (1992), no. 4, 317–324. https://doi.org/10.1112/blms/24.4.317.
324. T. Ward, 'An algebraic obstruction to isomorphism of Markov shifts with group alphabets', *Bull. London Math. Soc.* **25** (1993), no. 3, 240–246. https://doi.org/10.1112/blms/25.3.240.
325. T. Ward, 'Automorphisms of $\mathbf{Z}^d$-subshifts of finite type', *Indag. Math. (N.S.)* **5** (1994), no. 4, 495–504. https://doi.org/10.1016/0019-3577(94)90020-5.
326. T. Ward, 'Additive relations in fields: an entropy approach', *J. Number Theory* **53** (1995), no. 1, 137–143. https://doi.org/10.1006/jnth.1995.1082.

327. T. Ward, 'Rescaling of Markov shifts', *Acta Math. Univ. Comenian. (N.S.)* **65** (1996), no. 1, 149–157.
328. T. Ward, 'An uncountable family of group automorphisms, and a typical member', *Bull. London Math. Soc.* **29** (1997), no. 5, 577–584. https://doi.org/10.1112/S0024609397003330.
329. T. Ward, 'Three results on mixing shapes', *New York J. Math.* **3A** (1997/98), no. Proceedings of the New York Journal of Mathematics Conference, June 9–13, 1997, 1–10. http://nyjm.albany.edu:8000/j/1997/3A_1.html.
330. T. Ward, 'Almost all S-integer dynamical systems have many periodic points', *Ergodic Theory Dynam. Systems* **18** (1998), no. 2, 471–486. https://doi.org/10.1017/S0143385798113378.
331. T. Ward, 'A family of Markov shifts (almost) classified by periodic points', *J. Number Theory* **71** (1998), no. 1, 1–11. https://doi.org/10.1006/jnth.1998.2242.
332. T. Ward, 'Dynamical zeta functions for typical extensions of full shifts', *Finite Fields Appl.* **5** (1999), no. 3, 232–239. https://doi.org/10.1006/ffta.1999.0250.
333. T. Ward, 'Additive cellular automata and volume growth', *Entropy* **2** (2000), no. 3, 142–167. https://doi.org/10.3390/e2030142.
334. T. Ward, *Ergodentheorie: von Planetenbahnen zur Zahlentheorie* (Max-Planck-Institut für Mathematik, 2004).
335. T. Ward, 'Group automorphisms with few and with many periodic points', *Proc. Amer. Math. Soc.* **133** (2005), no. 1, 91–96. https://doi.org/10.1090/S0002-9939-04-07626-9.
336. T. Ward, 'Mixing and tight polyhedra', in *Dynamics & stochastics*, in *IMS Lecture Notes Monogr. Ser.* **48**, pp. 169–175 (Inst. Math. Statist., Beachwood, OH, 2006). https://doi.org/10.1214/074921706000000194.
337. T. Ward, 'Ergodic theory: interactions with combinatorics and number theory', in *Encyclopedia of Complexity and Systems Science*, pp. 3040–3053 (Springer Verlag, New York, 2009).
338. T. Ward, 'Graham Everest 1957–2010', *Bull. Lond. Math. Soc.* **45** (2013), no. 5, 1110–1118. https://doi.org/10.1112/blms/bdt053.
339. T. Ward, 'Mathematics in the UK', *Mathematics Today* **49** (2013), 105–107.
340. T. Ward, 'Mixing, wine, and serendipity', *Nieuw Arch. Wiskd. (5)* **18** (2017), no. 4, 235–236. www.nieuwarchief.nl/serie5/pdf/naw5-2017-18-4-235.pdf.
341. T. Ward, 'Integer sequences and dynamics', in *TCDM 2018—2nd IMA Conference on Theoretical and Computational Discrete Mathematics*, in *Electron. Notes Discrete Math.* **70**, pp. 83–88 (Elsevier Sci. B. V., Amsterdam, 2018 ©2019). https://doi.org/10.1016/j.indag.2014.04.005.
342. T. Ward, 'Brilliant teachers and changing lives', *The Scholar* **11** (2019). thebrilliantclub.org/evaluation/the-scholar/. Guest editorial.
343. T. Ward, 'An Augar too late?', *Research Professional News* (12 May 2019). https://www.researchprofessionalnews.com/rr-he-views-2019-5-an-augar-too-late/.
344. T. Ward, 'Verba volant, scripta manent', *Journal of Humanistic Mathematics* **10** (2020). doi.org/10.5642/jhummath.202001.36.

345. T. Ward, *Defining what we mean by 'Quality' education* (The QAA at 25, 2022). www.qaa.ac.uk/news-events/blog/defining-what-we-mean-by-quality-education.
346. T. Ward, 'Alan Ward obituary', *The Guardian* (28 December 2021). www.theguardian.com/science/2021/dec/28/alan-ward-obituary.
347. T. Ward and Y. Yayama, 'Markov partitions reflecting the geometry of $\times 2$, $\times 3$', *Discrete Contin. Dyn. Syst.* **24** (2009), no. 2, 613–624. https://doi.org/10.3934/dcds.2009.24.613.
348. T. Ward and Q. Zhang, 'The Abramov-Rokhlin entropy addition formula for amenable group actions', *Monatsh. Math.* **114** (1992), no. 3–4, 317–329. https://doi.org/10.1007/BF01299386.
349. R. C. Weast (ed.), *CRC Handbook of Chemistry and Physics* (CRC Press, Cleveland, 56 ed., 1975).
350. R. L. Weber (ed.), *A Random Walk in Science* (Institute of Physics, 1973).
351. A. Weil, *Basic number theory*, in *Die Grundlehren der mathematischen Wissenschaften, Band 144* (Springer-Verlag, New York-Berlin, third ed., 1974).
352. A. Wieferich, 'Zum letzten Fermatschen Theorem', *J. Reine Angew. Math.* **136** (1909), 293–302. https://doi.org/10.1515/crll.1909.136.293.
353. Wikipedia contributors, *Ruth Barnett — Wikipedia, The Free Encyclopedia* (2021). https://en.wikipedia.org/wiki/Ruth_Barnett_(Holocaust_survivor). [Online; accessed 11-December-2022].
354. Wikipedia contributors, *1994 group — Wikipedia, the free encyclopedia* (https://en.wikipedia.org/w/index.php?title=1994_Group, 2022). [Online; accessed 29-September-2022].
355. Wikipedia contributors, *Alan Howard Ward — Wikipedia, The Free Encyclopedia* (2022). https://en.wikipedia.org/w/index.php?title=Alan_Howard_Ward. [Online; accessed 1-July-2022].
356. Wikipedia contributors, *Alfred van der Poorten — Wikipedia, The Free Encyclopedia* (https://en.wikipedia.org/w/index.php?title=Alfred_van_der_Poorten&oldid=1087261558, 2022). [Online; accessed 7-October-2022].
357. Wikipedia contributors, *Erdős number — Wikipedia, the free encyclopedia* (https://en.wikipedia.org/w/index.php?title=Erd%C5%91s_number&oldid=1107735023, 2022). [Online; accessed 9-October-2022].
358. Wikipedia contributors, *Scholarship level — Wikipedia, The Free Encyclopedia* (2022). https://en.wikipedia.org/w/index.php?title=Scholarship_level. [Online; accessed 28-June-2022].
359. Wikipedia contributors, *School Mathematics Project — Wikipedia, The Free Encyclopedia* (2022). https://en.wikipedia.org/w/index.php?title=School_Mathematics_Project [Online; accessed 28-June-2022].
360. Wikipedia contributors, *Tom McLeish — Wikipedia, The Free Encyclopedia* (2022). https://en.wikipedia.org/w/index.php?title=Tom_McLeish&oldid=1074108508. [Online; accessed 27-September-2022].
361. D. Willetts, *A University Education* (Oxford University Press, Oxford, 2017).

362. P. Williams, 'Britain's Full-Cost Policy for Overseas Students', *Comparative Education Review* **28** (1984), no. 2, 258–278. http://www.jstor.org/stable/1187351.
363. A. M. Wilson, 'On endomorphisms of a solenoid', *Proc. Amer. Math. Soc.* **55** (1976), no. 1, 69–74. https://doi.org/10.2307/2041844.
364. A. J. Windsor, 'Smoothness is not an obstruction to realizability', *Ergodic Theory Dynam. Systems* **28** (2008), no. 3, 1037–1041. https://doi.org/10.1017/S0143385707000715.
365. K. Wright, 'See No Bias, Hear No Bias, Speak for No Change', *Physics* **16** (2023), no. 33. https://physics.aps.org/articles/v16/33.
366. C. Young, *Politics in Congo: Decolonization and Independence* (Princeton University Press, 2015).
367. C. Zeeman, J. Harrison, and J. Smith, *The Histories of Mathematics & Statistics at Warwick* (2004). warwick.ac.uk/fac/sci/maths/general/institute/histories-small.pdf.
368. E. C. Zeeman, *Catastrophe Theory: Selected Papers 1972–1977* (Addison–Wesley, Reading (Massachusetts), 1977).
369. S. Zegowitz, *Closed orbits in quotient systems* (Ph.D. thesis, University of East Anglia, 2015).
370. S. Zegowitz, 'Closed orbits in quotient systems', *Ergodic Theory Dynam. Systems* **37** (2017), no. 7, 2337–2352. https://doi.org/10.1017/etds.2016.3.
371. S. Zeki, J. P. Romaya, D. M. T. Benincasa, and M. F. Atiyah, 'The experience of mathematical beauty and its neural correlates', *Frontiers in Human Neuroscience* **8** (2014). https://doi.org/10.3389/fnhum.2014.00068.
372. R. J. Zimmer, *Ergodic theory and semisimple groups*, in *Monographs in Mathematics* **81** (Birkhäuser Verlag, Basel, 1984). https://doi.org/10.1007/978-1-4684-9488-4.

Index

T. Ward, *People, Places, and Mathematics*, Springer Biographies,
https://doi.org/10.1007/978-3-031-39074-6

C

V

W